CULTURES 1840–1940

Beyond the great nd world fairs in London, Paris or Cl us exhibitions took place in provincial cities and towns across the world. Focusing on the period between 1840 and 1940, this volume takes a novel look at the exhibitionary cultures of this period and examines the motivations, scope, and impact of lesser-known exhibitions in, for example, Australia, Japan, Brazil, as well as a number of European countries.

The individual case studies included explore the role of these exhibitions in the global exhibitionary network and consider their 'marginality' related to their location and omission by academic research so far. The chapters also highlight a number of important issues from regional or national identities, the role of modernization and tradition, to the relationship between capital cities and provincial towns present in these exhibitions. They also address the key topic of colonial exhibitions as well as the displays of arts and design in the context of the so-called marginal fairs. *Cultures of International Exhibitions 1840–1940: Great Exhibitions in the Margins* therefore opens up new angles in the way the global phenomenon of a great exhibition can be examined through the prism of the regional, and will make a vital contribution to those interested in exhibition studies and related fields.

Marta Filipová is Honorary Research Fellow,
University of Birmingham, UK

Cultures of International Exhibitions 1840–1940

Great Exhibitions in the Margins

Edited by
Marta Filipová

Routledge
Taylor & Francis Group
LONDON AND NEW YORK

First published 2015 by Ashgate Publishing

2 Park Square, Milton Park, Abingdon, Oxfordshire OX14 4RN
52 Vanderbilt Avenue, New York, NY 10017

Routledge is an imprint of the Taylor & Francis Group, an informa business

First issued in paperback 2020

British Library Cataloguing in Publication Data
A catalogue record for this book is available from the British Library

The Library of Congress has cataloged the printed edition as follows:
Cultures of international exhibitions 1840-1940 : great exhibitions in the margins / edited by Marta Filipová.
pages cm
Includes bibliographical references and index.
ISBN 978-1-4724-3281-0 (hardcover : alk. paper)
1. Exhibitions--History--19th century. 2. Exhibitions--History--20th century.
I. Filipová, Marta.

T395.C85 2015
607'.34--dc23

2014045778

ISBN 978-1-4724-3281-0 (hbk)
ISBN 978-1-138-57508-0 (pbk)

Contents

Figures and Tables

11 International Exhibitions and Urban Aspirations: Launceston, Tasmania, in the Nineteenth Century

11.1 The Albert Hall, Launceston, Tasmania, erected for the Tasmanian International Exhibition of 1891–92. Spurling's Studios, Roger Holdsworth (compiler), n.d. (1890s?). State Library of Victoria, H2012.90/10

11.2 Fountain, Prince's Square, Launceston, Tasmania, erected in 1859. Anon., Roger Holdsworth (compiler), n.d. (1890s?). State Library of Victoria, H2012.90/59

11.3 Fountain, City Park, Launceston, Tasmania, erected in 1861. Anon., H. Grattan-Guinness (compiler), n.d. (1880s?). National Library of Australia, PIC/8510/12 LOC Album1040

12 'Urbi et Orbi': Decentralization and Design in Nancy's International Exposition of Eastern France, 1909

12.1 Official poster for the 1909 International Exposition of Eastern France, designed by Pierre Roger Claudin, printed by Gédovius. Musée de l'Ecole de Nancy

12.2 Plan of the 1909 International Exposition of Eastern France, Nancy. Musée de l'Ecole de Nancy

12.3 Monumental gate to the 1909 Exposition, designed by Paul Charbonnier and manufactured by Pompey firm, photograph. Musée de l'Ecole de Nancy

12.4 Interior displays in the transplanted Zutzendorf farmhouse in the Alsatian Village, 1909, postcard. Author's collection

13 Merging Peripheries and Centres: The Transnational Interconnectedness of the Helsinki National Exhibition of 1876

13.1 Helsinki exhibition pavilion, from *Yleinen Suomen Näyttelö Helsingissä Vuonna 1876. Wirallinen luettelo*, front cover, Helsinki, 1876

13.2 The Stockholm Pavilion, from *Expositions-Tidningen*, 1866

13.3 Charles Riis, interior of the Helsinki exhibition with a fountain and a display of ceramics in the centre of the rotunda, photograph. City Museum of Helsinki

14 The Last Exhibition of the Italian Colonial Empire: Naples 1938–40

14.1 Poster of the *I Mostra triennale delle terre italiane d'Oltremare* (1940), designed by Corrado Mancioli and Giammusso Ugo, Naples, 1940. Author's collection

14.2 Planimetry with drawings of *I Mostra triennale delle terre italiane d'Oltremare* (1940), from *The Guide to Individual Exhibitions*, Editions Triennale d'Oltremare, 1940. Author's collection

14.3 Naples, *I Mostra triennale delle terre italiane d'Oltremare* (1940), restaurant with an outdoor pool, the Arena Flegrea, next to the Pavilion of Sanità, razza e cultura (Health, Race and Culture). Archives Plinio Marconi, Rome

TABLES

Contributors

Samuel D. Albert teaches art history at the Fashion Institute of Technology in New York City. He has written on architecture and urbanism in the Austro-Hungarian Empire and its successor states, focusing on representations of national identity. He is currently working on two projects: a monograph on architecture and urbanism in the British Mandate of Palestine and a study on Transylvanian modernity in the inter-war period.

Giovanni Arena is an art historian and curator, member of the Mission of the Second University of Naples for the preservation of archaeological, historical and artistic heritage of Libya. He received his doctorate in 'Cultural Heritage Conservation and Valorization' at the SUN University, Italy, in 2008. He taught social history of art there (2013) and since 2011 he has worked for the National Research Center, National Institute of Optics, Italy, in the field of non-destructive artwork diagnostics by advanced optical methods. He has authored two books and a number of papers concerning the assessments of ancient and modern art.

John R. Davis is Professor of History and International Relations at Kingston University, London. His research has focused on Anglo-German relations, British and German history, and exhibitions. His books include *Britain and the German Zollverein, 1848–1866* (Basingstoke, 1997); *The Great Exhibition* (Stroud, 1999); *The Great Exhibition of 1851* (co-edited with Franz Bosbach) (Munich, 2002); and *The Victorians and Germany* (Frankfurt am Main, 2007). He is currently working on a general history of the *Zollverein*.

Jeffer Daykin teaches Japanese and Chinese history at Portland Community College in Portland, Oregon. His thesis, 'Losing "The Trail" on the Road to Modernity: Japanese Participation in the Lewis and Clark Centennial and American Pacific Exposition and Oriental Fair, Portland, Oregon, 1905' (Portland State University, 2008), examined the relationship of the 1903 Fifth National Industrial Exposition held in Osaka, Japan and Portland's 1905 international exposition. He is currently researching the transfer of the exposition form from Europe and the US to Japan and China in the long nineteenth century.

Davy Depelchin is Researcher for the Belgian Science Policy Office at the Royal Museums of Fine Arts of Belgium in Brussels and Associate Lecturer at Ghent University. He co-curated the exhibitions *From Delacroix to Kandinsky, Orientalism in Europe* (Brussels, Munich and Marseille, 2010–11); *Unveiling India, The Early Lensmen (1850–1910)* (Brussels and New Delhi, 2013–14); and *Jean Portaels and the Call of the Orient (1841–1847)* (Brussels, 2015); and co-edited/wrote for the accompanying catalogues. He is also the author of journal articles and book chapters on orientalism in nineteenth-century architecture and fine arts.

Marta Filipová is Honorary Research Fellow at the University of Birmingham, UK. She received her PhD from the University of Glasgow in 2009 with a thesis on Czech art historiography. Her current research focuses on the history and theory of exhibitions and on the construction of regional and national identities in art and art history, especially in Central Europe. She has published on these topics in, for example, *The Journal of Design History, The Riha Journal, Austrian History Yearbook* and *The Journal of Art Historiography* and in a number of edited volumes.

Marina Muñoz Torreblanca is Lecturer in Art History in the Hispanic and European Studies Program (HESP) at the Pompeu Fabra University where she teaches eighteenth- and nineteenth-century Catalan art. She defended her doctoral thesis in 2010 with the title 'The "Primitive" Reception in Exhibitions Held in Spain before 1929'. This project is currently being turned into a publication. In her teaching and research, Muñoz has focused on art and cultural studies, especially as an influence in world's exhibitions and primitive art collections, and has published several articles on this topic.

Anne Neale is a cultural historian and was formerly Coordinator of History and Theory in the School of Architecture and Design at the University of Tasmania, where she remains an honorary associate. Her research in Tasmanian architecture and public parks resulted in numerous entries in *The Encyclopaedia of Australian Architecture*, edited by Philip Goad and Julie Willis (Port Melbourne, Victoria, 2012), and *The Oxford Companion to Australian Gardens*, edited by Richard Aitken and Michael Looker (Melbourne, 2002). Her ongoing research into the history of early Pre-Raphaelite art has been published in the *Burlington Magazine*, the *British Art Journal* and the *Journal of Pre-Raphaelite Studies*.

Tomáš Okurka is Lecturer in Modern History at the University of Ústí nad Labem, Czech Republic, and Historian of the Municipal Museum of Ústí nad Labem. His field is the economic and social history of the nineteenth and twentieth centuries and the history of Germans in Bohemia. He published a monograph about the German General Exhibition in Ústí nad Labem 1903. Currently, the manuscript of his doctoral thesis on industrial and general exhibitions in Bohemia 1891–1914 is being prepared for publication.

Claire O'Mahony is a Course Director of the MSt in the History of Design and Director of Graduate Studies for the Department for Continuing Education at the University of Oxford and she is a Fellow of Kellogg College. Recent publications include articles and chapters: 'War within the walls: Conflict and Citizenship in the murals of the Hôtel de Ville, Paris', in the *Journal of War and Culture Studies*; 'The colony within: Cultural regeneration and the troubadours of Toulouse's Capitole', in K. Griffiths and D. Evans, eds, *Institutions and Power*; 'Cubist Chameleons: André Mare, the *camoufleurs*, and the canons of art history', *Journal of War and Culture Studies*; and she was editor of *Symbolist Objects: Subjectivity and Materiality at the Fin de Siècle* (High Wycombe, 2009).

Livia Rezende teaches at the History of Design postgraduate programme jointly run by the Victoria and Albert Museum (V&A) and the Royal College of Art in London, where she also received her PhD in 2010. She undertook a postdoctoral fellowship and a visiting professorship at the Escola Superior de Desenho Industrial (College of Industrial Design – ESDI) of the State University of Rio de Janeiro. She has published widely on the history of international exhibitions with a special focus on designed displays of nature and the status of raw materials at industrial exhibitions. Her latest research project explores the history and historiography of the institutionalization of design education and profession in Latin America in the post-war period. She is a Trustee and an Honorary Member of the Design History Society.

Rebecca Rice is Curator of Historical New Zealand Art at the Museum of New Zealand Te Papa Tongarewa. She researches and publishes in the field of colonial New Zealand art with a particular focus on the histories of exhibition and display. She has contributed entries on colonial New Zealand artists to *Art at Te Papa* (Wellington, 2009); a chapter exploring the colonial context of 'Nineteenth-Century Auckland' for *Art Toi: New Zealand Art at Auckland Art Gallery Toi o Tamaki* (2011); and has written an essay, 'Lindauer's Maori at Home and Abroad', for the catalogue accompanying the exhibition *Gottfried Lindauer's Maori Portraits* (Berlin, 2014). She curated *Van der Velden: An Art of Two Halves* at Te Papa (2013) and convened a symposium considering his impact on New Zealand art. She is currently researching towards an exhibition of the visual legacy of the nineteenth-century New Zealand Wars at Te Papa for 2016.

Wilson Smith is a PhD candidate in Economic and Social History at the University of Edinburgh, working on the thesis 'Displaying Edinburgh in

1886: The International Exhibition of Industry, Science and Art'. He previously pursued a career in academic librarianship in the course of which he was Principal Librarian at Edinburgh College of Art between 1999 and 2010.

TAINA SYRJÄMAA is Professor of European History at the University of Turku, Finland. She has published widely on the history of exhibition medium, historical spatiality and urban history. Her publications include *Edistyksen luvattu maailma. Edistysusko maailmannäyttelyissä 1851–1915* (The Promised Land of Progress. The Belief in Progress in the World Exhibitions 1851–1915) (Helsinki, 2007) and *Constructing Unity. Living in Diversity. A Roman Decade* (Helsinki, 2006). She is co-editor of *Nordic Perspectives on Encountering Foreignness* (Turku, 2009) and *Power and Culture: New Perspectives on Spatiality in European History* (Pisa, 2008).

MIKLÓS SZÉKELY is Research Fellow at the Institute for Art History at the Hungarian Academy of Sciences and Lecturer in nineteenth-century Hungarian art and museum studies at the Pázmány Péter Catholic University. Fields of his scholarly interests include nineteenth-century Hungarian art, the history of universal exhibitions, museum studies, contemporary museum architecture and cultural politics. He is the author of 'Rebuilding History. The Political Meaning of the Hungarian Historical Pavilion at the 1900 Paris Universal Exhibition', in *Cultural Diplomacy and Cultural Imperialism. European Perspective(s)*, edited by Martina Topic and Sinisa Rodin (Frankfurt am Main, 2012) and of *Contemporary Art Museums in Central Europe: Between International Discourse and Nation (Re)building Strategies*, Etudes du CEFRES 17 (Prague, 2014).

Acknowledgements

The essays in this volume originated in a symposium on 'Great Exhibitions in the Margins' held at the University of Wolverhampton in 2012. The event was made possible by the Early Researcher Award Scheme from the University of Wolverhampton. The publication of this book was supported, in part, with funds provided by the Faculty of Arts, University of Wolverhampton.

This book would not have been possible without the intellectual input of many people and the support of a number of institutions. I am particularly grateful to Chris Long, Rina Arya and Richard Woodfield, as well as to all the contributors for their cooperation. Margaret Michniewicz, Jacqui Cornish and others at Ashgate Publishing Ltd must be thanked for their professionalism and guidance in the production of the volume. Above all, though, I would like to thank Matthew, whose insight helped this project in many ways.

Introduction: The Margins of Exhibitions and Exhibitions Studies

Marta Filipová

Summarizing his visit to the New York Exhibition in 1853, the English art educator and curator George Wallis (1811–1891) noted the basic differences between American and European products, technologies and work attitudes he found at the Exhibition and in American factories he saw.[1] Describing the section on decorative furniture and upholstery, for instance, he observed that 'elaborate carving is the rule rather than the exception, and this is fully borne out by the examples shown in the Exhibition, the best of which are evidently the work of Europeans, and the designs partake more of French and German than of English characteristics.'[2]

Furthering such claims about national differences, he went on to note that some of the 'best and most intelligent artizans ... are the sons of Englishmen .... German workmen ..., being generally an intelligent and orderly class of men, are highly esteemed' here.[3] His observations, contained in a report for the Committee on the Machinery of the United States and presented to the House of Commons included a number of these comparisons of national characteristics alongside the detailed descriptions of the exhibited objects. The primary aim of his report was to map out the competition that the US increasingly represented and first displayed to the British public at the Great Exhibition of 1851.

Directly inspired by its namesake in London, the Crystal Palace in New York was one of the earliest evidences of the influence the Great Exhibition had on subsequent fairs of various sizes. The presence of Wallis and his colleague, the entrepreneur and inventor Joseph Whitworth (1803–1887), in the role of British commissioners, sent out by the British government, also documents how quickly international exhibitions became places of contest and comparison between nations.

Born in Wolverhampton in 1811, George Wallis was an ardent promoter of art education, decorative and industrial art and became the first keeper of the Art Collections at the South Kensington Museum.[4] He helped to organize several small exhibitions of art and industries, first in Wolverhampton in 1839 and 1840 and subsequently in Manchester in 1845, which were inspired by similar events in Paris.[5] For this initiative, he was appointed superintendent of the British textile division and a deputy commissioner of juries of the Great Exhibition in 1851. He also promoted the Exhibition in the northern regions of the UK and, apart from the 1853 New York Exhibition, he was involved in the Parisian *expositions universelles* of 1855 and 1867 and in the London International Exhibition of 1862.[6]

Wallis was a great believer in national and international exhibitions and their role in stimulating competition, promoting the latest inventions and informing the public about good design, in a way similar to his more famous colleagues, Christopher Dresser or William Morris. His involvement in exhibitions of various sizes, ranging from regional to international, indicates the growing importance of this medium around the mid-1800s. In the second half of the nineteenth and the first half of the twentieth century it was hard to find a country in the Western world that did not host a great exhibition or a large-scale international fair in its capital or metropolis. World fairs, *expositions universelles* and international or universal exhibitions predominantly took place in the capital cities of Europe and the large metropolises of the US.[7] The Great Exhibition in London of 1851, the Parisian *expositions universelles* and a number of world's fairs in the US, including those in Chicago (1893) and St. Louis (1904), set precedents for other, often smaller events, that replicated and developed the structural and organizational schemes of the largest events.[8]

Almost every metropolis had the ambition to host a carefully orchestrated exhibition that would showcase the best achievements and inventions to date and secure its status of a world city. International exhibitions soon established themselves as important yet 'ephemeral' events which played a major role in creating political ideas about nations, states and peoples, both exhibiting and exhibited, which were proclaimed to the rest of the world by means of the tangible exhibits on display, events associated with the exhibitions and intangible ideological messages they were conveying.

In their links with industrialization, progress and the growth of consumer culture, these events served as crucial vehicles of modernity.[9] The evolution of exhibitions tells us the history of progress, to paraphrase Robert W. Rydell, and in many respects demonstrates the rapid changes in culture, society, politics and economy that took place in the second half of the nineteenth and the first half of the twentieth century.[10] Crucial inventions and industrial advancements, the increasing importance of international trade and competition, or the rise and decline of colonial empires were projected into exhibitions and in many ways summarized the modern experience in the confined space of the exhibition ground.

Examinations of exhibitions have therefore often been limited to metropolitan modernity and the major sites of capitalist culture, which most obviously demonstrate such changes. Yet while the best-known examples of these events, which have attracted most research, were staged in the largest cities of Europe and North America, exhibitions outside these locations spread quickly. The better-known and still relatively large shows include exhibitions in, for example, Glasgow (1888) or the Universal and International Exposition in Brussels (1910), the scope of which, including their political agenda, and the amount of capital invested in them, turned them into mega events similar to the established fairs in Western Europe and the US.[11]

The 'exhibition network' was, however, even larger and wider than is often acknowledged and the attempt to achieve recognition through exhibitionary practices was not limited to the wealthiest and largest cities and nations.[12] Fairs and exhibitions frequently took place in what can often be seen as the political margins and also be understood as evidence that this medium and its agendas found universal popularity. At the same time, exhibitions became part of the growing awareness of modernity in the regions and provinces that often underwent important transformations in the second half of the nineteenth and first half of the twentieth century.

Emphasizing some of the elementary features of the modern period, such as rapid urban growth, improving communication and increasing interconnectedness of the world, this volume takes a novel approach and turns attention to what may be called regional modernity. Addressing academic neglect of this topic is what lies at heart of this volume, which seeks to draw attention to the importance of these smaller and lesser-known exhibitions. And although they were, in most cases, not the biggest participants in the universal network of exhibitions, such events made a significant contribution to the creation of municipal or regional identities and cultures, which they helped to consolidate.

Aims and Goals

The role of international exhibitions in promoting ideas of national identity has often been discussed as an important element of these events.[13] Exhibits were often divided into sections and, later, entire pavilions were dedicated to individual nations to allow clear division and comparison between them. First fully implemented at the *Exposition Universelle* of 1867 in Paris, such organization supported competition between the different participants and emphasized the creation of difference between the participating nations. It is no coincidence that such organizational and structural development of the exhibition grounds appeared alongside the creation of new nation-states, such as Italy or Germany, and an increase in national awareness on the part of the peoples of multinational states and empires, including the UK and Austria-Hungary. What can be seen as the golden age of great exhibitions in terms of their number and size thus coincided with the 'frenzied flowering of nationhood' of the second half of the nineteenth and the first half of the twentieth century.[14]

At the same time, this period experienced a growth in the awareness of regional distinctiveness based on economic, social or ethnic factors that were also often projected onto exhibition grounds. Like the idea of the nations, the concept of region is based on a shared sense of belonging together derived from a shared dialect, culture, religion and history of a specific geographical area that was not necessarily identical with an administrative or political unit.[15] It may also be claimed that throughout the nineteenth century regional identities were in many ways challenged by the political attempts to introduce an all-embracing national identity. Local exhibitions therefore often put emphasis on sustaining the individual character of the region by emphasizing its history and traditions, while also demonstrating its ability to be national and international by hosting a universally recognized event.

Given these starting points, this book on lesser-known regional, national and international exhibitions has several goals. First, it seeks to point to the fact that exhibitions were also hosted in 'peripheral' locations with specific political and cultural agendas. And although these exhibitions were not the major players in establishing the exhibitions canon, they were all important contributors to it that demonstrate the rising importance of provinces and regions in the global and increasingly globalized world.[16] Many of the case studies in this book thus emphasize the originality of the regional exhibitions that related to the specific – regional – circumstances under which they were organized, while they acknowledge the efforts of the exhibitors and organizers to establish an internationally recognizable culture of trading, manufacture and arts. The book therefore aims to expand the understanding and knowledge of the exhibition network beyond the conventional and limiting geographical and political focus of exhibition studies on centres and capitals.

In this context, the 'marginality' of the exhibitions becomes a rather broad concept – many of the case studies examine events that took place on the borders of monarchies and empires, where cultural and ethnic tensions were strong and where new centres were created.[17] The contributors therefore challenge the traditional view of what constitutes the hinterland and what is considered a capital in their detailed examination of the individual exhibitions. Consequently, the selection of the case studies in this volume reflects their alleged marginality not only in a strictly political or geographical sense but also in terms of the research they have attracted. Their marginality often stems from the fact that they stand outside of the established canon of the literature on exhibitions.

Second, this book brings case studies that seek to analyse the similarities and differences between exhibitions in the less prominent locations and their greater models. For organizers of the former, comparison with an established exhibition in for example London or Paris, was often necessary in order to legitimize what they were trying to achieve. At the same time, this volume aims to demonstrate that these 'marginal' exhibitions made an important contribution in their impact on local economies, societies and cultures. The provincial character of these events can therefore be seen in the fact they were embedded in the local communities much more than the largest universal exhibitions.

Third, the studies in this volume also extend the period of examination beyond the dates often used in literature on exhibitions to describe the 'century of progress'. This period is often assumed to have started with the Great Exhibition in London of 1851 and to have closed with the New York World Fair of 1939.[18] At the same time, authors such as Rydell, Greenhalgh and Geppert describe the period until about 1900 as the golden age of modern international exhibitions but also focus on a period that extended beyond this date and ended up in 'exhibition fatigue' before WWII.[19] While this volume still takes the Second World War as its end point, the contributors also look before 1851 and beyond 1939, to examine the important prehistory of the great exhibitions as well as events staged during the Second World War. The individual texts thus trace the development of the exhibition medium from the early exhibitions of the 1840s, which focused primarily on trade, through the codification of the basic features of exhibitions in the second half of the nineteenth century, up to the increasingly more implicit political motivation behind exhibitions of the first half of the twentieth century.

Each contribution to this volume therefore examines the microhistories of one or a series of exhibitions on the 'margins' and their relationship to the macrohistory of the local and global circumstances in which they were organized. Authors of the individual studies present these events as important ideological tools, the ambitions of which may be compared with the more canonical events, and in this sense become part of the wider exhibitionary culture. Such culture is understood here in its broadest sense as referring to the political, social and economic circumstances that shaped the event and its aftermath, the visual, popular and material culture that was linked with exhibitions through display or representation, as well as the local and global principles of exhibiting that were followed by the organizers.

Exhibition as a Concept

It is widely acknowledged that due to its scale, its use of architectural or technological innovations and the sheer amount of money invested in the event, the Great Exhibition of 1851 acted as an influential model for subsequent exhibitions that were organized on a similarly large, if not larger, scale.[20] As was mentioned above, the organizers of the New York Exhibition of 1853, for example, copied many elements of the Great Exhibition or gave them a local character. The analyses in the first section of this volume, 'Exhibition as a Concept', focus on the historical development of the national and international exhibitions in different geographical locations and examine how the exhibition medium is transferred across political and cultural contexts.

The case of the exhibition plans in Hungary and its capital of Budapest at the end of the nineteenth and the beginning of the twentieth century opens the discussion about the importance of national and international exhibitions outside the main sites of the largest events. It also provides a pertinent example of the mechanics of exhibitions preparation and shows how individuals and local organizations attempted to advance regional economy by attempting to organize a universal exhibition and by participating at world exhibitions. Miklós Székely, focusing on Hungary's complex political history and ethnic composition in this period, explores how and why the attempts to present a unified Hungarian identity at home and abroad became crucial for the local government as well as for various individuals and societies. The Millennium Exhibition of 1896 in Budapest was planned as an important commemoration of the foundation of the Hungarian state. It was strongly motivated by the determination to put the city on the map of world exhibitions and to reinforce Hungarian identity in the context of the multinational Habsburg Empire. Székely, however, points out that this was only one of many exhibitions planned both in Hungary and abroad for such purposes. He therefore examines the background of the political negotiations that instigated organization of these showcases of Hungarian identity and pays special attentions to other exhibition proposals that were never implemented.

This analysis invites a consideration of the intentions and motivations of organizers behind the complicated endeavour of putting together an international exhibition. It has been claimed that the greatest of exhibitions always strove to become more impressive and extensive than the previous one.[21] Such an evolutionary approach, however, does not always apply to exhibitions that were organized outside of the main centres, where the ambitions to recreate a universal exhibition clashed with a lack of finances. The links between exhibitions in capital cities and exhibitions in regions, nevertheless, cannot be reduced to a simple centre–periphery or original–copy relationship as many exhibitions in the margins developed an individual, regional version with a localized agenda.

The Universal Exhibition that took place in Barcelona in 1888, discussed by Marina Muñoz Torreblanca, provides an opportunity to examine such dynamics. Muñoz explains that Barcelona was in good position to hold a universal exhibition which would be comparable to those in London or Paris. It made an effort to replicate some of the basic exhibition elements, including the emphasis on machinery displays, inclusion of funfairs and other entertainment features, and the design of the pavilions. It also had the ambition to become the first city in Spain to host an international exhibition of this kind. Yet the exhibition in the Catalonian regional capital failed to achieve the significance and status enjoyed by others elsewhere. Muñoz overviews the possible reasons for such failure and ponders the possibility of transferability of the 'typical features' of a great exhibition across geographical and cultural locations. The 'greatness' of the most recognized exhibitions, according to the author, lay in their contribution to narratives of universal progress, which was usually ascribed to the number of exhibitors they attracted, the scientific dimension of the event, and the display of overseas possession. None of these was fully achieved in Barcelona and the exhibition therefore made its greatest mark on the urban transformation of Barcelona rather than on the history of universal exhibitions.

Such history often starts with the London exhibition of 1851, yet as we are reminded by John Davis in his analysis of the All German Exhibition that took place in 1844 in Berlin, it was by no means the first such event.[22] Very few researchers have looked at the important predecessors to the Great Exhibition that can be regarded as the prehistory of the exhibition boom and that include single trade fairs, national exhibitions and shows organized by the mechanics institutes in, for example, France or the UK.[23] The Berlin 1844 Exhibition exercised a considerable influence on the organizers of the Great Exhibition in London and other cities with its successes as well as failings. Already in Berlin, one of the main reasons behind the decision to put so much effort into carrying out such a complex project was the recognition of the increasing importance of trading and industrialization of cities. Promotion of local and national economies but also regional cultures thus stood behind a number of exhibitions across Europe, regardless of their size.

Although it remained a national affair in its scope and impact, the Berlin exhibition represented an important step in the development of the international exhibition medium by the way it handled the logistical and financial complexities of the preparation alongside the diplomatic negotiations of what constituted the politically controversial concept of 'all Germans'. This section therefore also opens up the discussion about the importance of the terms used in the titles of these smaller exhibitions. References to regional, national, international and universal character of the exhibition suggest geographical as well as political ambitions of the organizers, while mentioning a specific location (New Zealand, South Staffordshire) where the event took place was understood as creating a direct link with the particular place. Similarly, allusions to specific ethnic groups (All-Germans, Italian) drew on affiliations with larger national units, whereas commemorative exhibitions (Millennium, Centennial) used references to an historical moment that bridged the past and the present-day. This suggests that the political reasons and motivations for the exhibitions in the margins were made obvious already in the events' titles.

Constructing Identities

In many cases the theoretical frameworks customarily used for the study of the large events can be applied to the study of regional and local exhibitions with only a limited number of adjustments. Tony Bennett's idea of the exhibitionary complex, a theoretical approach grounded in Foucault's visualization of discipline, analyses exhibitions as sites which install order among people and exhibits.[24] This idea, that an exhibition is a place where control is linked to the dominant ideology of the state power, was clearly applicable to colonial and imperial exhibitions and world fairs of developed nation-states. A number of nations in the making, however, found other uses for exhibitions and these often included turning them into sites of resistance against the dominant power in the specific state. Minority or regional interest groups often challenged state imperatives in order to promote new or reinvented identities, demonstrating that rival agencies tried to claim the exhibition for their own purposes. Many exhibitions discussed in this volume may be seen as tools for reinforcing an imagined community with a carefully crafted narrative for outsiders of its history, present and future but which contradicted officially sanctioned accounts promoted by central political authorities.

The contributions gathered together in the section 'Constructing Identities' focus on the attempts by different interest groups and individuals involved in exhibitions to construct and display specific ideas of local, regional and national identity. Exhibitions in Central and Eastern European clearly demonstrate this phenomenon, for the second half of the nineteenth century was a period of increasing national awareness amongst a number of ethnic groups of the region which took place on the background of the industrial transformation. Various interest groups, including the Hungarians, Romanians and Bohemian Germans, started using the exhibition medium to achieve their goals of national recognition and showcase their technical and industrial advancements.

Exhibitions organized in the regions of the Habsburg monarchy where national revival was taking place can often be seen as examples of a site of resistance to the dominant ideology which was often promoted by world's fairs. They did not comply with the political dimension that the official Habsburg exhibitions, such as the World Exhibition of 1873 and the series of Jubilee exhibitions in 1898 and 1908, celebrating 50 and 60 years respectively of the reign of Emperor Franz Joseph, adopted.[25] Organized to promote industrial progress and free trade, the secondary function of especially the 1873 World Exhibition was to showcase the achievements of the Habsburg empire, confirm a sense of common state identity and emphasize Vienna as a modern capital of the monarchy. In the same geopolitical environment of the Habsburg monarchy, trade fairs and regional and national exhibitions that took place outside Vienna often promoted political visions that were independent of the official ideology. A number of relatively large exhibitions were held in many places within Austria-Hungary (Prague, 1891 and 1895, Cracow, 1887 and Lviv, 1894), in order to promote a sense of belonging to a nation or ethnic group, whose history, culture or economy could be easily and comprehensibly summarized in an exhibition and which, crucially, differed from imperial visions of Austro-Hungarian identity.

The northern and north-western areas of Austrian Bohemia, inhabited primarily by a German-speaking population, were among the main industrial regions of the monarchy and therefore sites of many trade fairs between the 1870s and the beginning of the First World War. Although advancement of local trade and culture were the official and primary aims, these regional exhibitions became closely tied to the promotion of the German population of Bohemia, where clashes between Czechs and Germans took many forms. Examining a series of exhibitions organized by ethnic Germans in Bohemia, Tomáš Okurka concentrates on the economic and political motivations behind these exhibitions and focuses on the nationalistic and ethnic conflicts between the Bohemian Germans and the Bohemian Czechs. The author shows that these events consciously created political narratives that supported the status of the German minorities in Bohemia and that sought to encourage popular audiences to identify with nationalist political and social visions.

In order to further point out the tensions between national and ethnic groups, Samuel D. Albert analyses the Millennium Exhibition in Budapest (1896) alongside the Bucharest National General Exposition (1906), which was also held to commemorate several events in the history of the Kingdom of Romania (25 years since King Carol I's coronation, 40 years since his accession to the Princeship of the United Principalities of Romania, and 1,800 years since Trajan's Dacian campaign). The exhibitions, however, were turned into a battleground for territorial and cultural possession, as the same ethnographic material, mined from border regions with disputed identities, was used in both exhibitions to reflect an ethnic ideal that the respective organizers held about their historic and contemporary existence. As Albert points out, it was these local circumstances, in this case the disputed histories and on-going ethnic tensions, that shaped the politics of these events and helped to turn what were intended to be international exhibitions into peripheral affairs. The marginal status of the Bucharest and Budapest exhibitions was a function of the limited range of objects selected for displays, which were purely targeted at national audiences, and this has been reinforced due to the historical marginalization of these events by subsequent commentators.

Part of the rationale behind the organization of industrial fairs and exhibitions in places with mobilized minorities was encouragement of a sense of identity and belonging to the region where they took place. Moreover, by staging an event that was increasingly recognized internationally as an important sign of progress and prosperity, the host city was claiming its place in the international exhibition network. With various degrees of success, smaller exhibitions were driven by the ambition to step out beyond the geographical, cultural or political limitations of their location in order to put themselves on the international map.

Two exhibitions held in England's Black Country in 1869 and 1902 illustrate this phenomenon clearly. The organizers of these events in the town of Wolverhampton, small in size and diminishing in industrial importance in the second half of the nineteenth century, used the exhibition to change its image, which was primarily based on coal mining and steel production. Many of the common features of exhibitions established in London or Paris, including a display of fine art, musical performances and funfairs as well as the involvement of the local aristocracy, were employed here on a smaller scale in an attempt to present Wolverhampton as a city on a par with other cultural and manufacturing centres of England. The impact of the two exhibitions was, however, ultimately limited to the local area. Despite this failure to become recognized international events, their success in mobilizing local audiences in order to increase their awareness of region's material and visual culture demonstrates the importance that exhibitions in the geographical margins had on local populations.

Historicity and Modernity

Many contributors to this volume deploy Mitchell's idea that universal exhibitions were organized to 'represent, to recall … some larger meaning'.[26] In a number of cases such understanding may even be extended to consider exhibitions as hyperreal images of the world outside the exhibition grounds, or as an attempted construction of the world as it should be. The various visual representations of the exhibition grounds, including bird's-eye views or exhibition interiors, depict these places as ideal or idealized microcosms of the outside world where objects were put into order, categories and classifications.[27]

In this way, exhibitions tended to impose meaning on the world and become an effective tool for communicating a specific ideological vision to their audiences. The main message that exhibitions tried to convey has often been linked to the utopic vision of progress.[28] Progress – whether political, social, cultural or economic – was put on display so that nations, regions or cities could be compared and contrasted.[29] The section 'Historicity and Modernity' addresses these issues and focuses in particular on exhibitions as sites of modernity which embraced a narrative of history that was incorporated into the modern exhibition project.

The attitude to history and memory was an important feature of international exhibitions which took diverse forms in specific political contexts. The case of the Rio de Janeiro Centennial Exposition of 1922, explored by Livia Rezende, brings attention to another aspect in the relationship between the past and the future-oriented projection intended for a modern metropolis. In Rio de Janeiro, an important historic site on the Castello Hill linked to the Portuguese colonial past was physically erased to provide room for the grounds of an exhibition which was consciously used by the Brazilian government to demonstrate the nation's modernity. The exhibition, as a universal medium, illustrated the anxiety of the Brazilian government to portray the nation as modern in a competitive environment of an international fair. The organizers put the desired modernity of the city and of the nation on display for the benefit of domestic as well as international audiences.

Importantly, the competing displays of contemporary industries and technological advancements at exhibitions were designed as educational and instructional for the varied audiences. Such need was perhaps even more prominent in regional exhibitions, which served as places where inventions and innovations were first introduced to the professional and general public and where the impact of nationally or regionally manufactured products on local artisans or students could be expected to be greater. It was also desirable for exhibitors to elevate 'the general taste of the working classes of the district' and inform them of the historical context of their locality which had necessary consequences in the present and the future.[30] Instruction about progress therefore took various forms and one of them was to juxtapose present and future on the one hand with the past on the other. The urban setting of an exhibition included 'frozen times and distant places' which took the visitors on a journey through time and space.[31] References to a glorious history could take the form of re-creations of historic buildings, or whole streets often played the role that emphasized the contrasting modernity of the rest of the exhibition site.[32]

As Davy Depelchin explains, the city of Ghent tried to reconcile the city's past and present in the Universal and International Exhibition of 1913. This was based on an attempt at modernizing its historical centre, inserting new buildings that supported a modern but also picturesque image of the city. The desire to emphasize local history by the choice of exhibits and the architecture of the pavilions in the exhibition thus led the organizers to suppress another popular aspect that was a frequent inclusion at the exhibitions, namely the colonial displays. Whereas, for example, at The Brussels Universal and International Exhibition of 1910, great emphasis was placed on the display of the 'exotic' against the technological advancements, Ghent found the modernity of the city in local heritage which was given visual prominence at the exhibition.

The acknowledgement of a city's historic character, which could be turned into a commercialized spectacle and therefore utilized for the general success of the exhibition, took many forms. Wilson Smith, in his study of The International Health Exhibition at South Kensington in 1884 and of the 1886 International Exhibition of Industry, Science and Art in Edinburgh, examines the attempts to recreate a historic town within the confines of the exhibitions, carried out in order to highlight the modern character of the contemporary city. The two exhibitions included the construction of Old London and Old Edinburgh respectively, as they were imagined by the exhibitions' organizers and designers. Functioning to a great extent as popular tourist attractions, which included costumed performances and sale of souvenirs, these historical illusions were designed to provide a sharp contrast with the displays of industrial, sanitary or cultural advancements of the present.

Art and Design Exhibited

As Victor Hugo pointed out in 1867, the world came to exhibitions to confront products, compare ideals as well as utopian visions.[33] Such visions were constructed by means of a complex set of exhibits that drew on objects ranging from industrial production to works of art. As a number of researchers have agreed, France played a leading role in including the visual arts into international exhibitions.[34] Displays of art at exhibitions in Paris were used as a symbol of French civilization and supremacy over other European and non-European countries.[35] Starting in 1855 and the Palais des Beaux Arts, Paris set out a precedent in providing a place for national displays of fine art. Here, 'the glory of the French School' was manifested in a show that included contemporary works of art ranging from the Romantics to the Barbizon painters, alongside retrospective exhibitions of, for instance, Ingres or Delacroix.[36]

Displays of fine art quickly gained in importance because their inclusion next to industrial and scientific exhibits addressed different but complementary goals pursued by the organizers. Their presence meant that an exhibition was no longer merely a trade fair, but was also a cultural event that showcased the best of the country's artistic past and present.[37] The art to be displayed here was put together in a way that presented a desired narrative about the historical and contemporary artistic heritage of the host city or country and which was aimed at international audiences as well as a local public. These representations were indeed a kind of invented tradition in the way they reinforced political and state ideologies through the inclusion or exclusion of art works.[38]

Fine art quickly found place at exhibitions on the geographical periphery which have been outside of scholarly focus, when judged against the global exhibitions in Paris or London. The studies in the section 'Art and Design Exhibited' thus address the question of the relationship between the visual culture of the exhibition and the global exhibition medium. Rebecca Rice has focused on two exhibitions in Dunedin, New Zealand, which took place in 1865 and 1889. Both were remarkably large-scale international exhibitions and the earlier one even was the first of its kind in the country. The geographical location created a special environment for the medium of a great exhibition where many sections, but especially the fine art one, highlighted the ambiguities of a colonial culture on the margins. This included the desire to participate in and emulate metropolitan examples, as well as an attempt to assert the colony's uniqueness within the model of exhibition. Rice explores in particular the search for the canon of 'New Zealand art' and of New Zealand artists at the exhibitions in the context of the local art's position on the international (or colonial) art scene.

When considering the world map of international exhibitions, one can hardly imagine a more geographically peripheral place than Launceston in Tasmania. Yet its inhabitants did not share such Eurocentric bias, and their regional pride translated into their effort to hold an international exhibition in 1891/92, the focus of Anne Neale's contribution. The fifth exhibition in Australia and the first outside the capital city, Launceston's remarkable connection with the international exhibition network was manifested in a number of architectural features, ordered from Europe following the display of their prototypes at international industrial exhibitions. Neale analyses the legacies of exhibitions that have survived in art and design – the objects of her focus bear witness not only to the extraordinary urban aspirations of a small and remote antipodean community, but also to the power and influence of the international industrial exhibition movement.

Aspiration can be seen as another key element of exhibitions 'in the margins'. The displays at the major exposition in Paris, mentioned above, projected a centralized image of French art as well as of the French nation and often failed to incorporate the cultural and regional diversity of the French Republic. Claire O'Mahony uses the example of the 1909 International Exposition of Eastern France held in Nancy in Lorraine to demonstrate how the idea of a common national identity was challenged through art and design artefacts displayed at this exhibition. Lorraine, annexed by the German Empire in 1871, was a disputed territory and the tensions between the French and German allegiances informed the decision of the exhibition organizers to propagate local production and culture as regionally distinctive. O'Mahony focuses particularly on the decorative and industrial arts and the crucial part they played in decentralizing the national framework of great exhibitions in favour of the exposition's regionalism.

International Ambitions

The ambition to gain national and international recognition as well as establish international communication was common to many exhibitions explored in this book. Exhibitions were not built in isolation but developed a model the organizers found most suitable for their purposes. At the same time, exhibitions served as positive or negative examples for future events, contributing to active international exchanges. The section 'International Ambitions' brings studies of exhibitions that put emphasis on establishing links with the exhibition network through exchanges of ideas that ranged from the structural and organizational design of the site to the adoption of political motivations behind the exhibition.

The Finnish General Exhibition of 1876, discussed by Taina Syrjämaa, illustrates the complex background of the inter- and transnational exhibition experiences. The way the Helsinki exhibition was designed, organized and implemented suggests that the main influences came from various sources. Its organizers were familiar with exhibitions in the neighbouring countries, such as Sweden, as well as more remote models in Vienna and Paris, and these all served as important points of reference. Syrjämaa approaches exhibitions as a transnational medium in which single exhibitions had numerous interconnections with each other that shaped their final form.

The examinations of how exhibitionary cultures spread around the globe bring attention to another topic for a long time neglected in exhibition studies

– international exhibitions organized in the colonies, both in those existing at the time, for instance in the British colonies of Australia and New Zealand, and in those that had already gained freedom, which was the case of, for example, the former Portuguese colony of Brazil. The focus turned traditionally on colonial displays at world's fairs has only recently been accompanied by more attention paid to the adoption of the exhibition medium in these remote locations.[39] These fairs provide a wealth of subjects to be explored, as they show the dynamics between the colonizer and the colonized from a different perspective. The fairs in, for example, Australia (Melbourne 1854, 1880–81 and 1888–89, Sydney 1879–80, Adelaide 1887, Hobart 1894–95), New Zealand (Christchurch 1906–07), Canada (Montreal 1860), India (Calcutta 1883–84, Lahore, 1881–82) and other colonies such as Jamaica or South Africa were organized to 'promote settlement, investment, and development' in these countries and often served as displays of the British dominance and imperial authority.[40]

Still, the main body of literature that has examined colonial and ethnographic displays at international exhibitions has put emphasis on a few of the largest colonizers and their displays of colonies that confirmed their sense of dominance.[41] Hence, observers most often focus on the approach of the British, French, Americans or Belgians to their subjects and the mechanisms used for turning them into anthropological and ethnographic exhibits. The superiority of their dominant power was often displayed through possessions of the colonies, the dependents and their visual and material culture. 'Typical' native villages and streets, 'typical' inhabitants performing various everyday and festival activities and making exotic products became an indispensable part of the process of reassuring the superior position of the white Europeans.[42]

Giovanni Arena's discussion of the 1938–39 exhibition in Naples turns attention to colonial displays of a nation that was relatively new to the overseas expansion and colony building. The author attends to the way colonial displays became an important element of exhibitions in cities that do not normally feature amongst the greatest exhibitors. Arena analyses the Triennial Exhibition of the Italian Overseas Territories of 1938–40 and its related imperial politics and takes it as an example of the evolution that took place in the official representations of Italian colonies at fairs both in Italy and abroad. The Italian colonial displays were also aimed at other colonizing nations such as Britain and France and, as the author claims, can be understood as a metaphorical compensation for the inferiority complex which Italy, as the most recent colonizing nation, brought into the colonial enterprise.

Finally, focusing on the 1903 National Industrial Exhibition in Osaka, Jeffer Daykin demonstrates that the exhibition medium, with its political and economic goals, quickly spread to countries relatively new in the colonial enterprise. The exhibition medium became an important promotional tool for Japan, an emerging power in the community of nations and a coloniser herself. As the fifth exhibition in Japan and only the first with truly international ambitions, this event was meant to bolster Japan's international standing, educate its public, and increase its share of global trade. Daykin points out that the Osaka exhibition did not appear as an isolated event but had close relationships with its more established model overseas, especially in the US. Similarly to the other contributions, the case of the Osaka exhibition shows that exhibitions in the geographical and political margins did not lack the determination to engage in a lively interaction with the wider exhibition network.

Conclusion

As the exhibition in Osaka and the many other case studies examined in this book show, one of the most important aspects in the study of exhibitions in the margins is the relationship between the metropolis, on the one hand, and, on the other, regional towns that strove for recognition by means of hosting a great exhibition. Influences between exhibition locations did not flow as a one-way traffic, for exhibitions of various sizes often engaged in mutual interchanges. Small, regional events in many cases did not simply copy the larger exhibitions but rather modified and adapted practices to suit local agendas.

From the range of studies of the different geographical locations in Europe, the Americas, Asia and Australia, it becomes clear that the medium of exhibition was indeed a global phenomenon that shared many features across regions, countries and nations of whatever size through the tangible objects and intangible ideas. However, the regional and case-specific issues raised by these examples of smaller exhibitions demonstrate the wide variety of original forms the medium took across cultural, ethnic or geographical regions. Indeed, it is not possible to provide an exhaustive list of 'marginal' exhibitions in the period in question – the volume, however, attempts to initiate further research in the aspects covered here and those that had to be left aside. The examples from 'the margins' are thus used to comment on and draw conclusions about the wider, global phenomenon of exhibitions and their cultures. The discussions turn attention to the existence of the global village in the way the world, however imagined, was displayed, in the universal embrace of modernity, or in the shared individual features of the exhibitions, including emphasis on popular entertainment and recognizable sculptural or architectural features.[43] Regional exhibitions can thus be understood as examples of the shared experience of modernity.

At the same time, exhibitions 'in the margins' highlight a number of novel and original issues within their contexts. The examinations of the struggles for cultural, political or economic recognition, as well as of the attempts to construct or reassure the host's identity using local historic, industrial or artistic heritage, contribute to the understanding of the multifaceted and flexible nature of exhibitions. As a medium closely linked with the modern age, an exhibition in the geographical, political or artistic margins informs us about how universal values, inherent to grand exhibitions, can be showcased alongside regional and national agendas. And being vehicles of regional or local identity, often complementing or even challenging the identity of a nation or empire, these exhibitions had the largest impact on the area from which they came. As such they help to shed light on the relationship between local, regional modernities and the global one. This book's novel, original look at exhibitions within their specific context *and* the context of universal modernity thus brings exhibition studies to the regions, provinces and margins and shows their importance for the understanding the exhibition medium as a complex, multilayered phenomenon.

Notes

1 George Wallis, 'Special Report of Mr. George Wallis', in Nathan Rosenberg, ed., *The American System of Manufactures: The Report of the Committee on the Machinery of the United States* (Edinburgh, 1969).

2 Wallis, 'Special Report', 294.

3 Wallis, 'Special Report', 207–208.

4 Nathan Rosenberg, ed., *The American System of Manufactures: The Report of the Committee on the Machinery of the United States* (Edinburgh, 1969).

5 *Oxford Dictionary of National Biography*, ed. R. C. Denis (Oxford, 2004), s.v. G. C. Boase, 'Wallis, George (1811–1891)', accessed 8 May 2013.

6 Rosenberg, 21–22.

7 Bibliographies related to international and universal exhibitions can be found in, for example, Alexander C. T. Geppert, Jean Coffey and Tammy Lau, *International Exhibitions, Expositions Universelles and World's Fair, 1851–2005: A Bibliography* (Berlin, 2006); John Allwood, *The Great Exhibitions, 150 Years* (London, 2001); Christian Beutler, *Weltaustellungen im 19. Jahrhundert: Ausstellung Katalog* (Munich, 1973); Robert Bordaz and Marc Vellay, *Le Livre des Expositions Universelles 1851–1989* (Paris, 1983); John E. Finding and Kimberley D. Pelle, eds, *Encyclopedia of World's Fairs and Exhibitions* (Jefferson, NC, 2008); Kira Homo et al., *World's Fairs: A Guide to Selected English-Language Resources* (Bloomington, 2005).

8 Most recently, Joseph Heathcott, 'Ephemeral City: Design and Civic Meaning at the 1904 World's Fair', *Journal of Design History* 26, no. 1 (2013): 25–46.

9 Paul Greenhalgh, *Ephemeral Vistas: History of the Expositions Universelles, Great Exhibitions and World's Fairs* (Manchester, 1990).

10 Robert W. Rydell, *The World of Fairs: The Century-of-Progress Expositions* (Chicago, 1993).

11 Perilla Kinchin, Julie Kinchin and Neil Baxter, *Glasgow's Great Exhibitions 1888, 1901, 1911, 1938, 1988* (Wendlebury, 1988); Jutta Pemsel, *Die Wiener Weltausstellung von 1873: das gründerzeitliche Wien am Wendepunkt* (Vienna and Cologne, 1989); Matthew Rampley, 'Peasants in Vienna. Ethnographic Display and the 1873 World's Fair', *Austrian History Yearbook* 42 (2011): 110–132; Serge Jaumain and Wanda Balcers, eds, *Bruxelles 1910. De l'Exposition universelle à l'Université* (Brussels, 2010).

12 Alexander C. T. Geppert, *Fleeting Cities: Imperial Expositions in Fin-de-Siècle Europe* (Basingstoke, 2010), 3 and 4.

13 Burton Benedict, 'Exhibitions and National Identity', *Anthropology Today* 7, no. 3 (1991): 5–9; Burton Benedict, *The Anthropology of World's Fairs* (London, 1983), 20; Zeynep Çelik and Leila Kinney, 'Ethnography and Exhibitionism at the Expositions Universelles', *Assemblage* 13 (1990): 35–58.

14 Paul Greenhalgh, *Fair World: A History of World Fairs and Expositions from London to Shanghai, 1851–2010* (Winterbourne, 2011), 156.

15 Edward Royle, 'Introduction', in *Issues of Regional Identity: In Honour of John Marshall*, ed. Edward Royle (Manchester, 1998), 2.

16 Cf. Paul Young, *Globalization and the Great Exhibition: The Victorian New World Order* (Basingstoke, 2009); Penelope Harvey, *Hybrids of Modernity: Anthropology, the Nation State and the Universal Exhibition* (London, 1996).

17 Tristram Hunt, *Ten Cities that Made an Empire* (London, 2014).

18 Rydell, *World of Fairs*.

19 Geppert, *Fleeting Cities*, 206; Robert W. Rydell, *All the World's a Fair: Visions of Empire at American International Expositions, 1876–1916* (Chicago, 1984), 4.

20 Jeffrey Auerbach, *The Great Exhibition of 1851: A Nation on Display* (New Haven and London, 1999); John R. Davis, *The Great Exhibition* (Stroud, 1999); Charles Harvard Gibbs-Smith, *The Great Exhibition of 1851* (London, 1981); Michael Leapman, *The World for a Shilling: How the Great Exhibition of 1851 Shaped a Nation* (London, 2001); Young, *Globalization*.

21 Geppert, *Fleeting Cities*, 204.

22 Cf. Greenhalgh, *Ephemeral Vistas*, 3–11.

23 For example Kenneth E. Carpenter, 'European Industrial Exhibitions before 1851 and Their Publications', *Technology and Culture* 13, no. 3 (1971): 465–486; Toshio Kusamitsu, 'Great Exhibitions before 1851', *History Workshop Journal* 9, no. 1 (1980): 70–89; Richard Daniel Altick, *The Shows of London* (London, 1978).

24 Tony Bennett, 'The Exhibitionary Complex', *New Formations* 4 (1989): 74.

25 Rampley; Daniel Unowsky, 'Staging Habsburg Patriotism. Dynastic Loyalty and the 1898 Imperial Jubilee', in *Constructing Nationalities in East Central Europe*, ed. Pieter M. Judson and Marsha L. Rozenblit (New York and Oxford, 2005).

26 Timothy Mitchell, 'The World as Exhibition', *Comparative Studies in Society and History* 31, no. 2 (1989): 217–236, at 221.

27 Çelik and Kinney, 36.

28 Rydell, *All the World's a Fair*, 4; Bennett, 73–102, at 95.

29 George F. Barwick, 'International Exhibitions and Their Civilising Influence', in *The Civilisation of Our Day: A Series of Original Essays on Some of Its More Important Phases at the Close of the Nineteenth Century. By Expert Writers*, ed. James Samuelson (London, 1896), 301–313.

30 George Wallis, *Special Report on the Local Manufactures of South Staffordshire and East Worcestershire as represented in the South Staffordshire Exhibition held at Wolverhampton, 1869* (Wolverhampton, 1869), 3.

31 Geppert, *Fleeting Cities*, 245.

32 Cf. Maurice Roche, *Megaevents and Modernity: Olympics and Expos in the Growth of Global Culture* (London, 2000).

33 Victor Hugo, 'Introduction', in *Paris-guide de l'exposition universelle de 1869* (Paris, 1867), reprinted in Victor Hugo, *Oeuvres complètes* (Paris, 1969), 561–609.

34 Patricia Mainardi, *Art and Politics of the Second Empire: The Universal Expositions of 1855 and 1867* (New Haven, 1987); Greenhalgh, *Ephemeral Vistas*, 198; Benedict, *Anthropology*.

35 Greenhalgh, *Fair World*, 157.

36 Mainardi, 46.

37 Greenhalgh, *Ephemeral Vistas*, 198.

38 Eric Hobsbawm and Terence Ranger, eds, *The Invention of Tradition* (Cambridge and New York, 1984), 1.

39 Barbara J. Smith, 'The Books of the Fairs. Materials about World's Fairs, 1834–1916, in the Smithsonian Institution Libraries', *Primary Source Media*, at http://microformguides.gale.com/Data/Introductions/10020FM.htm, accessed 8 May 2013.

40 Rydell, *World of Fairs*, 63. Cf. also Peter H. Hoffenberg, *An Empire on Display: English, Indian and Australian Exhibitions from the Crystal Palace to the Great War* (Berkeley and London, 2001).

41 Such as Rydell, *All the World's a Fair*; Mitchell; Raymond Corbey, 'Ethnographic Showcases, 1870–1930', *Cultural Anthropology* 8, no. 3 (1993): 338–369; Benedict, *Anthropology*.

42 William Schneider, 'Colonies at the 1900 World Fair', *History Today* 31 (1981): 31–36.

43 Cf. Young.

Bibliography

Allwood, John. *The Great Exhibitions, 150 Years*. London: Exhibition Consultants, 2001.

Altick, Richard Daniel. *The Shows of London: A Panoramic History of Exhibitions, 1600–1862*. Cambridge, MA: Belknap Press of Harvard University Press, 1978.

Auerbach, Jeffrey. *The Great Exhibition of 1851: A Nation on Display*. New Haven and London: Yale University Press, 1999.

Barwick, George F. 'International Exhibitions and Their Civilising Influence'. In *The Civilisation of Our Day: A Series of Original Essays on Some of Its More Important Phases at the Close of the Nineteenth Century. By Expert Writers*, edited by James Samuelson, 301–313. London: Sampson Low, Marston, 1896.

Benedict, Burton. 'Exhibitions and National Identity'. *Anthropology Today* 7, no. 3 (1991): 5–9.

Benedict, Burton. *The Anthropology of World's Fairs*. London: Scolar, 1983.

Bennett, Tony. 'The Exhibitionary Complex'. *New Formations* 4 (1989): 73–102.

Beutler, Christian. *Weltaustellungen im 19. Jahrhundert: Ausstellung Katalog*. Munich: Die Neue Sammlung, Staatliches Museum für angewandte Kunst, 1973.

Bordaz, Robert, and Marc Vellay. *Le Livre des Expositions Universelles 1851–1989*. Paris: Editions des Arts Décoratifs and Herscher, 1983.

Carpenter, Kenneth E. 'European Industrial Exhibitions before 1851 and Their Publications'. *Technology and Culture* 13, no. 3 (1971): 465–486.

Çelik, Zeynep, and Leila Kinney. 'Ethnography and Exhibitionism at the Expositions Universelles'. *Assemblage* 13 (1990): 35–58.

Corbey, Raymond. 'Ethnographic Showcases, 1870–1930', *Cultural Anthropology* 8, no. 3 (1993): 338–369.

Davis, John R. *The Great Exhibition*. Stroud: Sutton, 1999.

Finding, John E., and Kimberley D. Pelle, eds. *Encyclopedia of World's Fairs and Expositions*. Jefferson, NC: McFarland, 2008.

Geppert, Alexander C. T., *Fleeting Cities: Imperial Expositions in Fin-de-Siècle Europe*. Basingstoke: Palgrave Macmillan, 2010.

Geppert, Alexander C. T., Jean Coffey and Tammy Lau. *International Exhibitions, Expositions Universelles and World's Fair, 1851–2005: A Bibliography*. Berlin: Freie Universität Berlin, 2006.

Gibbs-Smith, Charles Harvard. *The Great Exhibition of 1851*. 2nd edn. London: Victoria and Albert Museum, 1981.

Greenhalgh, Paul. *Ephemeral Vistas: History of the Expositions Universelles, Great Exhibitions and World's Fairs*. Manchester: Manchester University Press, 1990.

Greenhalgh, Paul. *Fair World: A History of World Fairs and Expositions from London to Shanghai, 1851–2010*. Winterbourne: Papadakis, 2011.

Harvey, Penelope. *Hybrids of Modernity: Anthropology, the Nation State and the Universal Exhibition*. London: Routledge, 1996.

Heathcott, Joseph. 'Ephemeral City: Design and Civic Meaning at the 1904 World's Fair'. *Journal of Design History* 26, no. 1 (2013): 25–46.

Hobsbawm, Eric, and Terence Ranger, eds. *The Invention of Tradition*. Cambridge and New York: Cambridge University Press, 1984.

Hoffenberg, Peter H. *An Empire on Display: English, Indian and Australian Exhibitions from the Crystal Palace to the Great War*. Berkeley and London: University of California Press, 2001.

Homo, Kira, et al. *World's Fairs: A Guide to Selected English-Language Resources*. Bloomington: Indiana University, 2005.

Hugo, Victor. *Oeuvres complètes*. Paris: Le Club Français du Livre, 1969.

Hunt, Tristram. *Ten Cities that Made an Empire* (London: Allen Lane, 2014).

Jaumain, Serge, and Wanda Balcers, eds. *Bruxelles 1910. De l'Exposition universelle à l'Université*. Brussels: Racine, 2010.

Kinchin, Perilla, Julie Kinchin and Neil Baxter. *Glasgow's Great Exhibitions 1888, 1901, 1911, 1938, 1988*. Wendlebury: White Cockade, 1988.

Kusamitsu, Toshio. 'Great Exhibitions before 1851'. *History Workshop Journal* 9, no. 1 (1980): 70–89.

Leapman, Michael. *The World for a Shilling: How the Great Exhibition of 1851 Shaped a Nation*. London: Headline, 2001.

Mainardi, Patricia. *Art and Politics of the Second Empire: The Universal Expositions of 1855 and 1867*. New Haven: Yale University Press, 1987.

Mitchell, Timothy. 'The World as Exhibition'. *Comparative Studies in Society and History* 31, no. 2 (1989): 217–236.

Oxford Dictionary of National Biography. Edited by R. C. Denis. Oxford: Oxford University Press, 2004; online edn, October 2006. At http://www.oxforddnb.com/view/article/28570, accessed 8 May 2013.

Pemsel, Jutta. *Die Wiener Weltausstellung von 1873: das gründerzeitliche Wien am Wendepunkt*. Vienna and Cologne: Böhlau, 1989.

Rampley, Matthew. 'Peasants in Vienna. Ethnographic Display and the 1873 World's Fair'. *Austrian History Yearbook* 42 (2011): 110–132.Roche, Maurice. *Megaevents and Modernity: Olympics and Expos in the Growth of Global Culture*. London: Routledge, 2000.

Rosenberg, Nathan, ed. *The American System of Manufactures: The Report of the Committee on the Machinery of the United States* (Edinburgh: Edinburgh University Press, 1969).

Royle, Edward, ed. *Issues of Regional Identity: In Honour of John Marshall*. Manchester: Manchester University Press, 1998.

Rydell, Robert W. *All the World's a Fair: Visions of Empire at American International Expositions, 1876–1916*. Chicago: University of Chicago Press, 1984.

Rydell, Robert W. *The World of Fairs: The Century-of-Progress Expositions*. Chicago: University of Chicago Press, 1993.

Schneider, William. 'Colonies at the 1900 World Fair', *History Today* 31 (1981): 31–36.

Smith, Barbara J. 'The Books of the Fairs. Materials about World's Fairs, 1834–1916, in the Smithsonian Institution Libraries'. *Primary Source Media*. At http://microformguides.gale.com/Data/Introductions/10020FM.htm, accessed 8 May 2013.

Unowsky, Daniel. 'Staging Habsburg Patriotism. Dynastic Loyalty and the 1898 Imperial Jubilee'. In *Constructing Nationalities in East Central Europe*, edited by Pieter M. Judson and Marsha L. Rozenblit, 141–156. New York and Oxford: Berghahn, 2005.

Wallis, George. 'Special Report of Mr. George Wallis'. In *The American System of Manufactures: The Report of the Committee on the Machinery of the United States*, edited by Nathan Rosenberg, 201–328. Edinburgh: Edinburgh University Press, 1969.

Wallis, George, *Special Report on the Local Manufactures of South Staffordshire and East Worcestershire as Represented in the South Staffordshire Exhibition Held at Wolverhampton, 1869*. Wolverhampton: Steam & Blacket, Steam Printing Works, 1869.

Young, Paul. *Globalization and the Great Exhibition: The Victorian New World Order*. Basingstoke: Palgrave Macmillan, 2009.

EXHIBITION AS A CONCEPT

1

A Capital in the Margins: Concepts for a Budapest Universal Exhibition between 1867 and 1917

Miklós Székely

In the region of Central Europe, universal and international exhibitions shared many features with their Western precedents. At the same time, however, they adopted independent agendas, related to the specific political circumstances in which they were organized. Hungary, as part of the Dual Monarchy of Austria-Hungary, provides an especially pertinent example of such autonomous transformation of the exhibition medium, which was used to proclaim the Hungarian sovereignty, modernity and national identity.

In general, universal exhibitions were addressed to international audiences and in most cases were organized in national capitals with ever-increasing number of exhibitors. The evolution of universal exhibition can be traced back to national industrial exhibitions at the end of the eighteenth century, which displayed diversity of goods and artefacts created nationwide. In the course of the nineteenth century, small trade fairs and industrial exhibitions around Europe increasingly opened up to international exhibitors and audiences. The first industrial exhibitions adopted the name 'general exhibition' in the last quarter of the nineteenth century. In the geographical context that will be the subject of analysis here, examples of such events include *Országos Általános Kiállítás* (The General National Exposition) in 1885 in Budapest or the *Expoziţia Generală Română* (Romanian General Exposition) in 1906 in Bucharest. They focused on a wide range of exhibits, but especially those produced locally. Universality, seen in the character of the exhibited goods, objects and inventions, remained the leitmotif for great fairs that put on display the latest and diverse material culture of the everyday life, important technical inventions and industrial products which were collected from international exhibitors. As the century 'progressed' the need for exhibitions specializing in fewer products and inventions increased although these events retained their international character by inviting foreign participants. The first International Art Exhibition in Venice in 1895 (today's Venice Biennale) originally started as an international exhibition specializing in fine arts while the International Health Exhibition in 1884 in London attracted international exhibitors of a new kind of specialization of the modern casual life.

Between 1867 and 1917, a number of attempts were made in Hungary to organize an international exhibition. The history of proposals for a Budapest Universal Exhibition therefore reveals the political and constitutional contradictions that the modern Hungarian state had to face as a result of the Austro-Hungarian Compromise of 1867. This contribution examines the cultural and political circumstances of these ambitious projects and the social and cultural background of the main initiators in order to understand the variety of influences that shaped exhibitions and their proposals. The analysis contributes to the understanding of the vulnerable position of the Hungarian cultural policy at the turn of the century. The stages of the formation of a modern nation were reflected in its participation at international exhibitions and in the will of hosting such a show in Budapest. Hungary shared similar intentions of nation building with many other countries in Central Europe that lacked full national sovereignty, which also hosted exhibitions displaying the best that the respective nations had achieved. The example of Hungary's attempts for a universal exhibition shows a determination of a nation in the making in constructing its own image using the modern and popular medium of exhibition while negotiating its internal politics and ethnic minorities.

Identity of the Hungarian State after 1867

The European universal exhibitions of the 1850s and 1860s coincided with significant changes in Hungary's political status and the construction of its political and cultural identity. The Kingdom of Hungary was a part of the Habsburg Empire until 1867 when the compromise with Austria converted the Empire into the Austro-Hungarian Dual Monarchy.[1] The previously gradual modernization of the Hungarian economy and culture accelerated when Hungary became one of the two political and administrative entities of the Austro-Hungarian Monarchy. It was only within the legal and internationally recognized framework of the Dual Monarchy that Hungary gained this relative degree of self-government. Apart from a few joint ministries of finance, foreign affairs and war, Austria and Hungary were led by separate governments under the same ruler: Emperor Franz Joseph I, based in Vienna who was also the Apostolic King Franz Joseph, whose official seat was in Budapest. From the Hungarian historical and juridical perspective, Franz Joseph represented the continuity with the medieval kings of Hungary.[2]

The question of 'being Hungarian' did not in the least manifest aspirations for political independence as the Hungarian parliamentary parties, the Liberal and the Independence Party alike, unanimously accepted the rule of the House of Habsburg. After 1867, the modern nation-building process was based on the Hungarian constitution, which in the view of national politicians dated to 1222. It was the legal foundation for the modernization of the country, based on its ancient political, economic and cultural heritage. For contemporary Hungarians, the restoration of historic Hungary in politics, economics and culture was a constitutionally based process of reconstruction.

Between the 1867 Compromise and the First World War, the terms 'sovereign' and 'Hungarian' remained the central concepts of nation-building, reinforced by the conscious construction of the country's image abroad. Using the constitutional lines, the political elite, mostly of aristocratic origin, was continuously willing to construct and show the 'sovereign' and 'Hungarian' economy and culture – thus representing the country abroad beyond the administrative capacities ensured by the limits of the Compromise, which entailed a reinterpretation of some historical facts, as will be mentioned shortly.

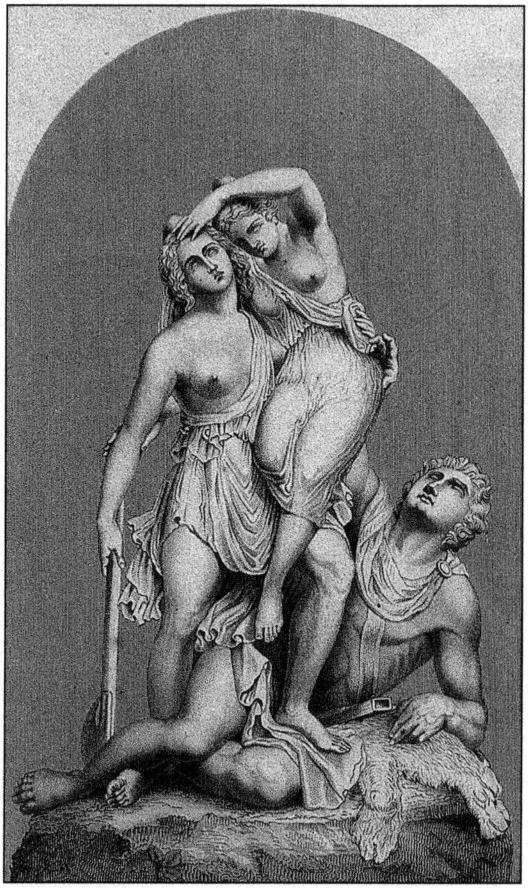

1.1 Joseph Engel, 'An Episode in the History of War Between the Amazons and the Argonauts', from John Tallis, *Tallis's History and Description of the Crystal Palace and the Exhibition of the World's Industry in 1851* (London: J. Tallis, 1851), p. 38

Such image making that Hungary engaged in can also be discerned in its politics of participation in universal exhibitions. Because of the Viennese political absolutism, the Hungarian presence was imbedded within the section of the Austrian Empire in the first three universal exhibitions of 1851, 1855 and 1862 (Fig. 1.1). At the London (1851) and Paris (1855) universal exhibitions, Hungarian exhibitors participated in a minor and fragmented way and objects of Hungarian origin were displayed among Austrian exhibitors' goods and pieces. The Hungarian economy was presented as being on the same level with the Austrian hereditary provinces and primarily as a source of agricultural products and raw minerals. Evidence of an improved political situation were the Hungarian sections at the 1862 Great London Exposition organized by the council of the governor-general in Buda, which was, nevertheless, still run under the strong authority of the Viennese court. In the economic sections, the country was primarily represented by its natural resources. However, a collection of applied arts objects and a small fine art collection were also exhibited and included a few works by contemporary Hungarian painters and sculptors: historical scenes and genre paintings by Gyula Benczúr, Mór Than, Károly Lotz and Bertalan Székely, the pupils of Karl von Piloty and Georg Ferdinand Waldmüller, who represented the latest artistic influences of the Munich Academy.[3]

The position of Hungary changed fundamentally after 1867. In culture and in internal politics, it became fully autonomous and this resulted in a sea change in its representation at fairs abroad. Beginning with the 1867 Paris *Exposition Universelle*, Hungarian thematic exhibition groups were organized by national commissars under the auspices of the Hungarian government. Thus, in the year of the political compromise with Austria, Hungary also made an independent debut on the international stage. An important sign of the new Hungarian approach to organization was the publication of a catalogue of the national exhibits in French, for the first time edited by a Hungarian exhibition committee and published in Paris.[4]

At the 1867 *Exposition Universelle* architecture was employed to represent national politics and thus the image of the sovereign and Hungarian state. The economic development of the country was articulated not only in the purely economic sections of industry, mining and agriculture but also in the industrial and applied arts sections. This division set a precedent and reappeared in Hungarian national exhibitions before the turn of the century.

The displays of Hungarian political, economic and cultural 'sovereignty' required visionary perspectives. This motivated the use of historic reference as a political tool: the revival of historic styles in Hungary has its roots in the Hungarian representational ideas beginning in the 1860s. An exhibition of historical forms outside the national borders was an echo of much stronger internal political connotation. One of the first manifestations of such an approach is contained in the memoirs of the Transylvania-born Countess Emma De Gerandó Teleki. In her description of the Hungarian section at the Paris *Exposition Universelle* of 1867, Countess De Gerandó paid special attention to the self-image of modern Hungary.[5] She also presented a plan for upcoming exhibitions, which included exhibition rooms entirely carved from salt from the Transylvanian town of Marosakna (today Ocna Mures) which would represent the mineral richness of Transylvania, a region she considered purely Hungarian. The reconstruction of the Hunyadi family's famous castle in Vajdahunyad as an architectural reference to the aristocratic Hunyadi family, linked with the fifteenth-century battles against the Turks and the introduction of quattrocento to Hungary, was another example of her attempt to construct a 'Hungarian' identity at future exhibitions.[6]

De Gerandó's vision complied with a trend that was popular at exhibitions in Europe and elsewhere at the time and which stressed the uniqueness of pavilions of individual nations. In the context of a universal exhibition, the architecture of national pavilions underwent important developments in the late nineteenth century, serving the representational needs of an increasingly secularized bourgeois society, the preservation of national memory and the entertainment of the masses. Nationally appropriated, the pavilion soon became the medium par excellence of national self-representation in universal exhibitions.[7] However, at the 1867 *Exposition Universelle*, which the Countess visited, a new type of exhibition pavilion appeared. Small-scale pavilions, showcasing gastronomy or small entrepreneurs, appeared here for the first time in considerable numbers and included a Hungarian display too.

The Hungarian wayside inn (*csárda*) was the only edifice aiming at representing Hungarian national characteristics here. Apart from serving as a political representation, proclaiming Hungarian sovereignty, it also promoted national economic interests: wine production and export were highly significant to the national economy of the 1860s and 1870s. The *csárda* building also reinforced a romantic perception of Hungary through the image of the Great Hungarian Plain (*Puszta*) with its ethnically diverse inhabitants wearing 'exotic' clothing.[8] The distinctive *csárda* building referred to the only internationally known notion of the *Puszta*, as the Great Hungarian Plain was one of the most widely acknowledged topoi of Hungary in the nineteenth century. On the other hand, the rural character of this display highlighted the weak industrial position of Hungary to international audiences.

Only six years later, at the 1873 universal exhibition in Vienna, organized not by the entire Monarchy but by Austria alone, Hungary was invited as a participant.[9] Taking advantage of this, Hungary appeared as a sovereign country with extensive and elaborate exhibition material. The forestry pavilion, for instance, served the interests of the state-owned forestry which provided significant state profit. As such, it became the first Hungarian pavilion to really meet its economic and marketing potential. Its design followed the traditional wooden architecture of the Greek Orthodox Church from the eastern Máramaros region which had mainly non-Hungarian inhabitants, especially the Ukrainian-speaking Ruthenians and Romanian-speaking Romanians. The emphasis on the *Puszta* romance prevailed in the 1870s too, in the form of a country house from the Great Plain which served as a wine bar.

As the references to the ethnically mixed origin of the Hungarian pavilion in Vienna indicate, the new national administration faced the challenge of the great linguistic, ethnic and religious diversity of Hungary after the Compromise.[10] During this period, the interest in ethnographic issues was as much political as it was scientific. The material objects and fleeting traditions that ethnographers documented served as fundaments on which the nation could be built.[11] Ethnography played a crucial role in the multi-ethnic politics of both sides of the Dual Monarchy and this was manifested in the insertion of most ethnic groups into national representations at the Vienna fair. Here, a range of objects at the ethnographic village represented ethnicities from the territories of the empire through their dwellings. The Hungarian committee also displayed Hungary as a multi-ethnic region, with a series of peasant houses representing the Germans, Hungarians, Szeklers, Rumanians and Croatians. Their cultural and political importance influenced the selection: the Germans, Szeklers and Romanians represented the centuries-old ethnic diversity of Transylvania, while the almost 900-year-long political union of Croatia with Hungary was made evident by the inclusion of a Croatian peasant house.[12]

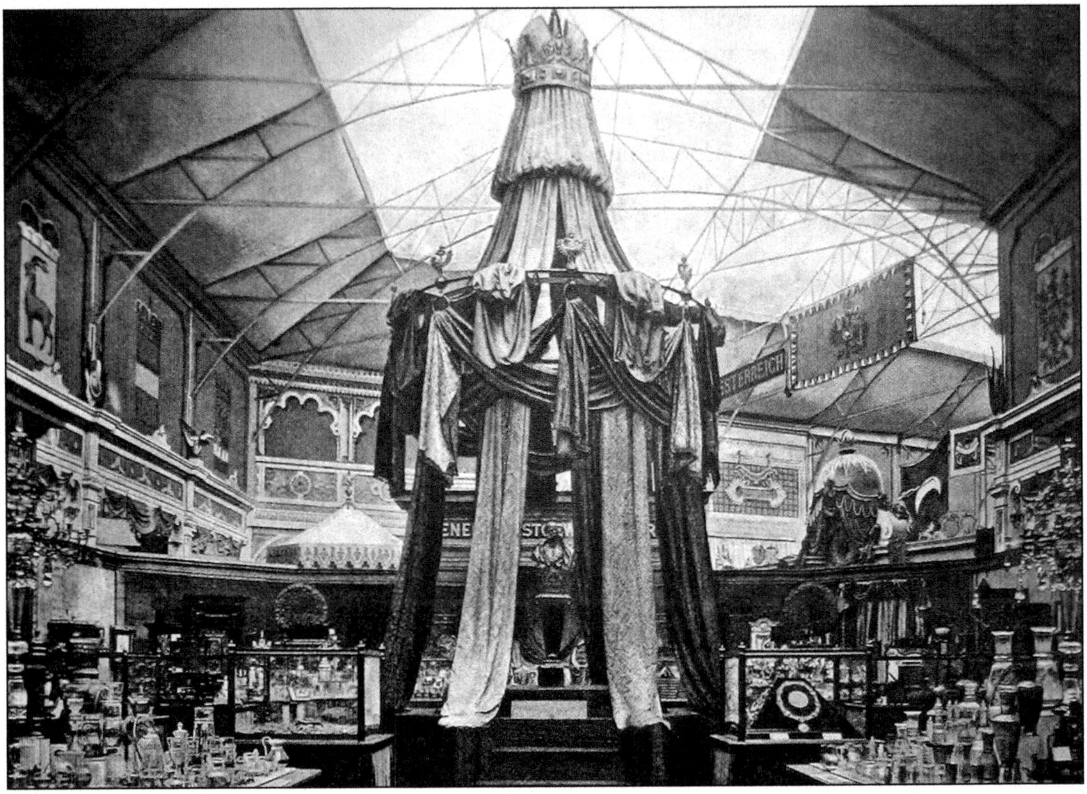

1.2 The Austro-Hungarian Exhibition at the Antwerp International Exhibition in 1885, from René Cornelli, *Antwerpen und die Weltausstellung 1885* (Leipzig: Pfau, 1887), p. 218

After this relatively extensive Hungarian involvement in Vienna in 1873, Hungarian presence decreased significantly at the 1878 Paris *Exposition Universelle*. Here, under the auspices of a Monarchic representation, all Austrian and non-Austrian exhibits were put together on common exhibition grounds. Serving as an important diplomatic act right after the Franco-Prussian war and the failure of the Paris Commune in 1871, this Monarchic representation buttressed the supranational and peaceful aspects the Dual Monarchy wished to assert. The unified representation of Austria-Hungary was also a part of a larger royalist representation in republican France: the German absence from Paris made the presence of the Dual Monarchy and Russia even more significant for the French diplomacy. In this joint exhibition the high-quality industrial products from Bohemia and Austria indirectly highlighted the weakness of the Hungarian economy (Fig. 1.2). As a consequence of the weak national representation in the economic sections the Hungarian government decided to abstain from further international exhibitions until the economic conditions attained European standards.

Plans for the Budapest Universal Exhibition of 1874

The idea to hold a universal exhibition in Budapest was encouraged by members of the nobility, industrial entrepreneurs and some members of the bourgeoisie and soon became popular among citizens of different social standing. One of first grand plans for a universal exhibition in Hungary was conceived by a talented individual and a patriot who remained more of an outsider and whose ideas did not come to fruition. The former army captain Szilárd Blána (1826–?), a political refugee who left Hungary after the 1848–49 War of Independence, visited the 1851 Great Exhibition in London and later, as a passionate patriot, took part in the preparatory works of the Hungarian section for the 1867 *Exposition Universelle* in Paris. It was his personal experience with international exhibitions that prompted Blána to publish his work promoting the idea of a permanent exhibition grounds in Budapest.[13]

In Blána's eyes, the foremost aim of a universal exhibition as envisaged by Prince Albert – a representation of the material and spiritual development of Mankind with the aim of a continuous comparison – had not been entirely fulfilled in the five previous large events (London 1851, Paris 1855, London 1862, Paris 1867, Vienna 1873). Blána's patriotism and his positivistic approach had a crucial impact on his concept. Combined with his idealistic and nationalistic ideas, it differed very much from the established dynamic economic concept of the universal exhibitions based on regularity and shifting locations among economic centres.

The permanent exhibition space in Budapest, Blána envisioned, would mostly serve Hungarian national economic development through the presentation of the latest universal trends in economy, industry and culture. Through the concentration of goods and products in one place over the course of a limited time, universal exhibitions answered the need of a modern society for universal publicity to learn about the latest achievements of urban modernity.[14] Exhibits displayed in the permanent venue, which would be constantly changing as Blána proposed, would ensure a never-ending influx of foreign visitors.

The exhibition area would include displays of six continents of the world classified according to their level of development, as Blána saw it: Europe, America, Asia, Africa, Australia and Oceania. The size of the exhibition spaces would not relate to the geographical dimensions of the continent, but would be proportional to the 'intellectual and cultural level of its inhabitants', Blána suggested.[15] The giant circular walled areas would present each of the nation's vegetation and mineral samples and the best artistic and industrial products, which would be constantly replaced by more recent inventions and innovations. The purpose-built permanent 'temple of modern life' was conceived more as a permanent place of display rather than a temporary exhibition – in Blána's proposal it would become an immense library of progress.[16]

In the last decades of the nineteenth century, universal exhibitions were often linked to the idea of memory and commemoration. The first such occasion was the 1876 Philadelphia World Exhibition, organized to commemorate the centennial of the signing of the Declaration of Independence. The conception of this fair was already made public by the time of the 1873 Vienna exhibition and most probably influenced Blána as well.[17] Due to the lack of original sources, though, a direct link cannot be established between the first commemorative type of universal exhibition and Blána's ideas. It is, however, quite plausible that Blána believed the best moment for the inauguration of the Budapest exhibition space should correlate with the celebration of Hungary's Millennium because he was captivated by the idea of the exhibition at the core of the commemoration act.

Once the occasion, the commemoration of the Millennium, was agreed upon, it became clear that the exact date of this important event was uncertain not only for Blána but for contemporary historians as well. The arrival of the Magyars, referred to as the Hungarian conquest, happened at the end of the ninth century. Determining one precise and historically justifiable moment for this crucial event had been the subject of intense speculation in historiography and, due to the tense political conditions in Hungary, in politics. The title of Blána's proposition clearly refers to its aim: the celebration of the 'thousands of years of existence of the Hungarian kingdom in the year 1883'.[18]

Blána's vision of the exhibition's organization, financial matters and the national contributions was undeniably original and ambitious. He wanted to divide the construction site for each nation based on its population and cultural development and shift the construction fees to the national governments. Further, his plan contained a waterpark which included an exhibition field with lakes, fountains, canals and its own transportation service. He also stipulated an obligation for all nations to send copies of their best industrial products and artworks. Apart from the temporary exhibition of the latest products, he considered the educational and museological aspects of the inventions and proposed to receive two copies of all new technical inventions so that they could be exhibited and published in the multilingual exhibition journal. While one artefact would be exhibited, the other would be displayed in a museum and placed in an international patent office to be set up in Budapest.

Blána also proposed that pavilions and national buildings should be erected in native materials and techniques by national professionals and that locals be sent to the exhibition to represent the diversity of national costumes and popular traditions. They would be in charge of the construction work, setting up the exhibition, hosting visitors, while they would be following (and displaying) their own local traditions. In brief, he imagined the foundation of national 'colonies' within the exhibition grounds. As there is only limited documentary evidence, the author's motivations for such colonial displays in the context of landlocked Hungary remain unclear. However, after several years abroad and his engagements in previous international exhibitions, Blána's vision must have been largely influenced by the colonialist approach to the non-Western world he encountered here. His perception of an international exhibition, based on his travels and services in London and Paris, were therefore different from those of other Hungarian politicians and compatriots in charge of organizing domestic exhibitions, all of whom were subjects of the Dual Monarchy, a non-colonizing European empire.[19]

Blána's idealistic and utopian plan was, however, unattainable partly due to the economic and financial limitations of Hungary and partly because of the unfeasibility of the exhibition's permanent character. It was, still, an important proof that the idea of an exhibition as a generator of modernization penetrated the entire Hungarian society. If implemented, the plan would have transformed Budapest into the spiritual and economic capital of Mankind, in which Blána believed. Its new role, according to Blána, would be to diffuse the achievements of the Enlightenment such as secularization, fraternity, progress and culture. Obviously such a highly ambitious aspiration was impossible to realize, not only for Hungary but for any other nation at that time. In Hungary, Blána's visionary patriotism and utopian technocracy were contrary to the nobility's collective memory of Hungarian medieval greatness. Although Blána's paper echoed a shared vision of his contemporaries, he was but a visionary patriot with no social capital or recognition – and this might explain the fact why Blána's exhibition plan remained unnoticed in future exhibition planning related to Hungary's efforts.[20]

Commemorating the Foundation of Hungary in 1896

The problem of establishing an exact conquest date to be commemorated by the Millennium festivities, that Blána had faced, was passed to the Hungarian Academy of Sciences by the government in 1882. A committee of scholars determined a 12-year period between 888 and 900 during which the event likely happened. For practical reasons the date of the festivities was set at 1894 but for organizational purposes it was later postponed to 1896. Not only the date but also the content of the Millennium was a point of discussion. This included talks about new monuments, a new parliament building, a National Pantheon in Budapest, festivities and commemorative acts.[21]

A key figure in the Millennium preparations was Count Jenő Zichy (1837–1906). In June 1890 he came up with a draft programme for the Millennium celebrations, which became the first and complete overview of the festivities; its printed version served as the basic for the future organization. Zichy, unlike Blána, was a prominent figure of the time, a promoter of industrial education, President of the National Industry Society, a generous patron of the fine arts and a passionate amateur of archaeology, national prehistory and historic monuments, but his most important activities were linked with domestic exhibitions. Starting in the 1840s, Zichy had organized a series of national industrial exhibitions, with the first held in Székesfehérvár, central Hungary, in 1872 and one in Budapest seven years later. The latter one, the National General Exhibition of 1885, although smaller in scale was similar in content to the planned Millennium Exhibition as it focused on the industrial and agricultural achievements of Hungary.

The Millennium soon turned into an event of great national enthusiasm. Intellectuals, politicians, priests, noblemen and even ordinary citizens promoted their ideas of how to commemorate this event. Zichy's proposal to organize a universal exhibition on the occasion of the Millennium was officially submitted to parliament by the Art Amateurs' Society in June 1891.[22] Combining the national, international and universal aspects, Zichy considered the arrival of the Magyars as an important event from a global perspective, so a universal exhibition would serve as an international frame for the commemorative festivities of the national jubilee. Zichy attributed a major role to the exhibition for a simple practical reason: opened for a period of six months, it would create a continuity among the various commemorative events, which included the joint jubilee session of the upper and lower chambers of the Parliament and the inauguration of a sculpture on Gellért hill in Budapest at the occasion of the closing of the commemorative year. Zichy planned the exhibition to be listed as the fourth item on a long list of festivities, which comprised seven components altogether: various national festivals, religious celebrations, a theatrical procession representing 'the ten centuries of Hungarian history', the exhibition, theatre and music performances, sports meetings and many other popular events.

The Exhibition was a proclamation of Hungary's historicity as well as modernity.[23] The contemporary aspect of the Exhibition was contained in the representation of the latest economic and cultural achievements of Hungary in the Main Contemporary Group, which consisted of, among others, industrial, ethnographic and art sections. The retrospective part of the Main Historical Group, housed in a romantic pavilion composed of replicas of 22 different historic buildings, focused on the historical development and culture going back to the coronation of St Stephen as King of Hungary in 1000. The interiors of the Main Historic Group displayed reconstructions of significant historic venues in the Hungarian history that were examples of a late inspiration in the museological concept of the French archaeologists Alexandre Lenoir and Alexandre du Sommerard in their creations of a lively atmosphere in the spirit of a Habsburg-oriented historiography.[24]

Starting in the 1870s, the public discourse had long been focused on the historical and political aspects of the Millennium, which included debates on the nation's origin but also the circumstances, the heroes and the possible timeframe of the conquest. This coincided with the spread of exhibitions – both of industrial, special and general profile – in Hungary and the rise in their popularity. The combination of a notable historical event and a large, national exhibition therefore proved successful.

Constructing a Modern Nation in Art and Architecture

The commemoration of Hungary's Millennium was not, however, limited to domestic displays in Budapest, but extended to exhibitions abroad. Hungary officially joined the 1900 *Exposition Universelle* in Paris as a participant and invested more financial, economic and intellectual effort into its national presentation than ever before. The Hungarian installations of various thematic groups in the great exhibition galleries employed in part the architectural style of Ödön Lechner (1845–1914) who had tried to develop a Hungarian 'national' language in architecture that applied folk patterns and motifs to facades and who was inspired by *Bekleidungstheorie*, the theory of dressing, of the German architect and architectural theoretician Gottfried Semper.[25] The Hungarian installations in Paris were designed by two of Lechner's students, Zoltán Bálint and Lajos Jámbor. The floral decoration of the immense galleries' interiors, inspired by peasant art, was conceived to highlight the visual distinctiveness of Hungarian culture. Beside the economic and cultural sovereignty exhibited in the galleries of the Hungarian historical pavilion in the Rue des Nations, the country's officially appropriated historical narrative was emphasized through a mixture of historic and vernacular architectural elements. Its decoration, the first example of the use of vernacular motives on ephemeral constructions, opened the way to the use of vernacular motifs and pre-modern tendencies in Hungarian pavilions during the forthcoming decades.

In contrast to the *csárda*-style Hungarian pavilions, which emphasized an idyllic *puszta* image of the country at the international exhibitions in the 1860s and 1870s, mentioned above, after about 1905 Hungarian self-definition radically changed. The Liberal Party which gained a majority in the Parliament for thirty years at eight consecutive elections lost its seats. The years 1905–1906 were a period of significant internal political tensions in Hungary, which affected presentation of Hungary at international exhibitions. The new image took vernacular arts and architecture as a source for the new culture of modernized Hungary.[26] As in many other parts of the Habsburg Empire, many young artists, architects and passionate amateurs understood peasant traditions as preserving national roots and the memories of the nation's past. As remnants of the mythic past, they interpreted peasant culture as the foundation of reinvented national myths and tales, and, what is more, important from the political point of view, encouraged attempts to revive national vocabulary in art and architecture.[27]

At the turn of the century, patterns and motifs inspired by peasant art – outsourced especially from the Hungarian-speaking region of Kalotaszeg in Transylvania, the Great Hungarian Plain and the Matyó land – decorated architectural elements as a common characteristic in the lands of Austria-Hungary and beyond. Architectural structures and peasant art objects were not considered as autonomous exhibits anymore. They were reinterpreted and served as the basis for new structures and ornaments. In political terms, this new Hungarian art and architecture reflected the new concept of Hungary as a modernized, large and powerful historic state. This ambitious new image was based on the memory of the political and economic power of medieval Hungary and combined with the new economic and cultural achievements the country had enjoyed since the Austro-Hungarian Compromise. The reinvention and popularization of traditions seemed historically, culturally and judicially justified, a natural consequence of the country's political and economic modernization.

The use of art and architecture for national representation became a major element of the official cultural politics after the 1896 Millennium exhibition and during the subsequent two decades. Part of the new and nationalistic paradigm of the national representation Hungarian pavilions reflected the image of a culturally sovereign country. Still, there was no attempt to proclaim political independence, rather, national life was envisioned within the framework of the Habsburg Monarchy and the political concept of being Hungarian and sovereign therefore did not exclude the acceptance of the 1867 Compromise.

The stylistic diversity of the turn-of-the-century international art and architecture tendencies had not only different aesthetic but also a different cultural and political background in each country. At the Paris Universal Exhibition in 1900 Hungary seemed to find its own voice in the mixture of vernacular traditions and pre-modern tendencies.[28] By the turn of the century the notion of culture had changed, and reflected a bounded nature of the Hungarian nation. All objects related to the inherited past were conceived to ensure the nation's historical grounding while the ethnological interest began to focus on ethnic Hungarians as heirs of an ancient heritage.[29] This was definitively not a new phenomenon; the nationalistic approach of the Hungarian political elite at this time had its origins already at the time of the Compromise of 1867 and in Countess De Gerandó's proposal for Hungarian exhibitions at the 1867 exhibition. Moreover, such attitude was not unique and had close parallels in the Czech, German, Finnish or Polish approaches which at this time used politically motivated references to ethnicity and the most ancient rural roots to construct their respective national identities.[30]

At the turn of the century, Hungarian folk art tradition was used following a new paradigm: it was promoted as a feature of modern national art and architecture and this became an important factor in pavilion architecture and decorative art objects.[31] Such use had not only a political but also an important economic motive: products decorated in the modern national style enhanced the country's international recognition and were successful on the market as part of a more general trend of the vernacular revival in Central Europe. The promotion of vernacular modernism – through the interpretation of folk traditions – was based on a more organic way of conceiving national architecture and art.[32] Between 1906 and 1911 vernacular modernism was an appropriated representational tool for the state; its promotion in the medium of exhibitions was strengthened by professionalization of the exhibition industry.[33]

In the 1910s, Hungarian art and architecture, which appeared in the international context of exhibitions, included also the use of ornaments borrowed from objects that dated back to the conquest period and the arrival

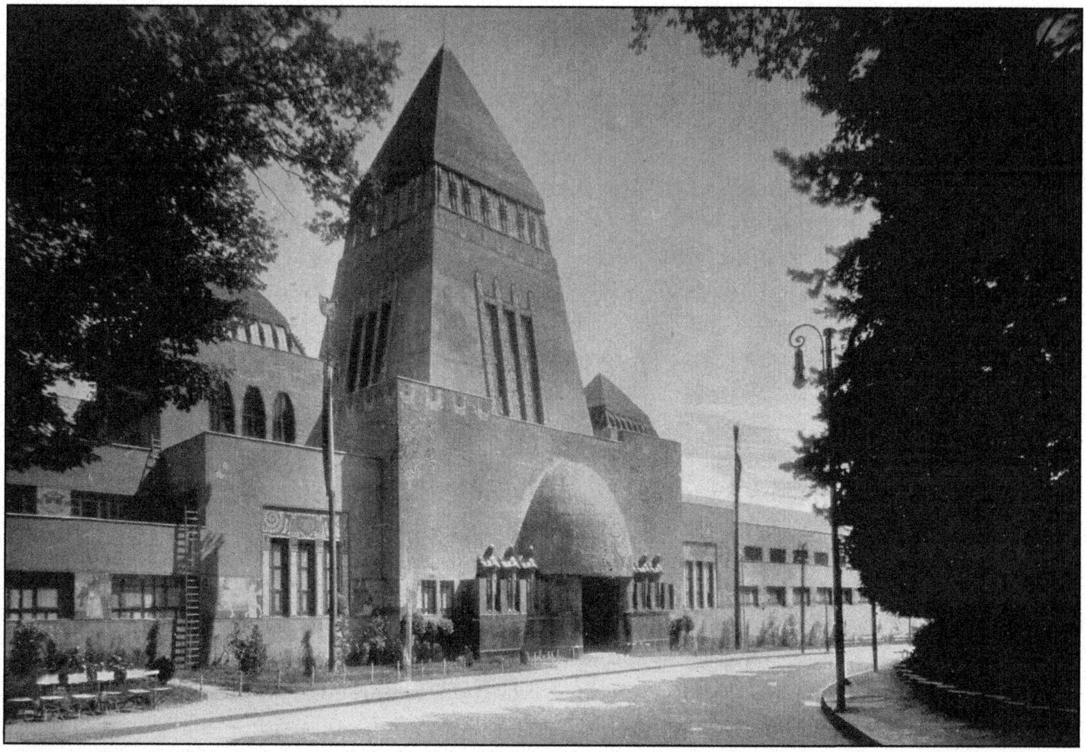

1.3 The Hungarian pavilion at the Turin Universal Exhibition in 1911, by Dénes Györgyi, Emil Tőry and Móric Pogány, from *Magyar Iparművészet* (1911), p. 261

of the Hungarians. In the spirit of the romantic nineteenth-century idea of peasantry, such objects were seen as proof of a historical national art vocabulary – they were tangible and much more concrete than one could have imagined some decades before. At the 1911 Turin *Esposizione internazionale dell' industria e del lavoro* (The International Exhibition of Industry and Labour), for instance, the display of Hungarian works and goods appeared to form a unity with the exhibition pavilion; the interior of the exhibition space was no longer merely a stylistically adequate frame, but became itself an exhibited object (Fig. 1.3). Wood as an important material of pavilion architecture beside plaster was used as a bearing structure as well as a visible, ornamented structural element, inspired by vernacular Transylvanian architecture, and one that proclaimed the connection between material and the structural solution.[34]

The Hungarian pavilion in Turin was designed by Dénes Györgyi, Emil Tőry and Móric Pogány, whose intention was to combine vernacular Hungarian architecture with modernism. The mostly wooden edifice combined the architectural approach of Károly Kós (1883–1977), a Transylvania-born architect of German origin and of Hungarian identity, with the architectonic language of the Young Ones, a group of young architects who began their careers around 1907 after graduating from the Budapest University of Technology. Their design was based on the Transylvanian-Hungarian vernacular traditions and combined late Secessionist, pre-modern architectural elements with heavy figural monumentality reminiscent of Franz Metzner's Battle of the Nations

Monument in Leipzig or the semi-cupola designed by Béla Lajta in the Charity Home of the Chevra Kadisha in Budapest. The exhibition displayed works of several well-known Hungarian designers, including Géza Maróti, Ede Telcs, Miklós Ligeti and Miksa Róth, gaining an outstanding importance from the perspective of domestic public opinion and foreign markets.[35] The interior of the pavilion and the installation were an exemplary summary of the official governmental position and of how the head of the art department of the Ministry of Religion and Education, Elek Koronghi Lippich (1862–1924), saw Hungarian industrial art based on vernacular traditions and, at the same time, preserving its alleged oriental (Asiatic) characteristics.[36] Art and architecture, with strong historical references, gained political and economic significance when displayed in domestic and international exhibitions.

Commemorating the Coronation of Franz Joseph

As pointed out before, most of the Hungarian sections at the international exhibitions between 1900 and 1911 attempted to represent Hungary as a modern nation. The pavilions in Turin (1902 and 1911), St Louis (1904), Milan (1906), Bucharest (1906), Rome (1911) and Dresden (1911) were conceived as 'true mirrors' of the modernized economy and cultural policy of Hungary for a mostly foreign audience. Even if Hungarian self-representation was defined by strong historical awareness of the political and financial elite devoted to national conventions, their main aim was to make the economy prosper and to refine and modernize culture generally.

In the pre-WWI period, the global exhibition enterprise underwent significant changes as the national organizations, commerce museums and exhibition centres ensured the professionalization of universal exhibition organization worldwide. These changes also influenced the last project for a universal exhibition in Budapest which closes an important chapter in the Hungarian exhibitions history. The grand plans were devised by Leo Lánczy (1852–1921), an assimilated Jewish member of the Upper House, president of the Budapest Chamber of Commerce and Industry and a liberal patriotic plutocrat. Lánczy's proposal was again linked to commemoration, this time to the fiftieth anniversary of Franz Joseph's coronation as King of Hungary and of the Compromise in 1867. Since this double political act was seen as the re-establishment of the (nearly) sovereign state of the Lands of the Crown of St Stephen, Franz Joseph appeared to be the 'second founder of the country', establishing a link to St Stephen's crowning as the first Christian king of Hungary.

Lánczy submitted his proposal to Károly Hieronymi, the minister of trade and commerce in February 1911, only a couple of months before the opening of the Turin exhibition.[37] He referred to Zichy's 1891 proposition which asked for the Millennium celebrations to be turned into a universal exhibition. Lánczy, a wealthy banker and a promoter of the latest industrial and financial innovations in Hungary and a patron of the fine arts, had much in common with Count Zichy. The national economic, industrial, and agricultural factors, which had contributed to the country's fast growth after 1867, formed an important part of his concept. He argued that the Millennium Exhibition was as an example of how a universal exhibition could put the country's economy and cultural life on an international map.[38] In his view, participation by Hungary in exhibitions abroad brought some 'beautiful results' for the economy but, as he stated, none of these shows ensured a conveniently high percentage in the exhibition section to represent the latest achievements of the economic and cultural modernization of Hungary after 1867.[39] So, as Lánczy had argued, the Budapest Universal Exhibition would fulfil the long-existing aspirations for the recognition of the economic, technological and cultural modernization of Hungary by the international audience. Due to the restrictions on the exhibition area of the invited counties as compared to the host, the 'abundance, diversity and excellent qualities' of the Hungarian products could not be presented in full.[40] Lánczy claimed that invited countries were altogether given the same area which was equal to the size of the space occupied by the host country, as was often the practice. The proposed Budapest Universal Exhibition of 1917 would therefore provide opportunities for a larger-scale Hungarian products, the full spectrum of presentation and comparison, since, as he stated: 'Here we see which of our products are competitive and where and in what field we are lagging behind ... It is clear that the Budapest Universal Exhibition would promote all aspects of our economic and cultural life and ensure new opportunities and new perspectives for our development.'[41] Lánczy also highlighted the direct and indirect advantages, which he saw in tourism and innovation and investment from which Budapest, with nearly one million inhabitants, could benefit. He also considered all direct political gains of such an event which would be the worthiest celebration of the jubilee in the year of 1917, a celebration that would serve as an occasion to invite foreign states and civilized nations and to present the country's achievements of the previous half-century.[42]

Lánczy maintained that at universal exhibitions abroad Hungarian economic and cultural policies were identified with *sovereign* (economically and culturally separated from Austria) and *Hungarian* (historically relying on its own constitution). Lánczy intended not only to raise the international reputation of Hungary but, more importantly, to construct an image of the country as a self-sufficient political entity. This was still a dream at the time, for in foreign policy and in military issues Hungary was not sovereign and decisions were made by Emperor Franz Joseph. Still, the Hungarian government made several efforts in the prewar period to increase the visibility of Hungary as a sovereign entity and not a mere province of Austria. 'It is a fact that even those foreigners who sympathize with us, in many cases, consider us from a romantic perspective and appreciate us from an ethnographically interesting perspective.'[43] As a passionate Freemason, Lánczy ultimately believed that the bloody rivalry between nations would be moved to the peaceful terrain of international trade and commerce, of which the universal exhibition would be the utmost tool.[44]

The official position of the Hungarians on Lánczy's proposal is unknown. After six decades of Hungarian participation at foreign universal exhibitions, his arguments were probably familiar to most Hungarian politicians, entrepreneurs and economists. Despite the fact that no parliamentary or ministerial records can be found in any reference to the discussion of the draft, Leo Lánczy printed and edited his proposal for the Budapest Universal Exhibition, which, ultimately, was the last attempt of this kind.

Generally, a universal exhibition requires five to ten years to get organized. Even though Lánczy's ideas had been published at the right time, they were left unanswered, first, because of the great inner political problems the country faced and, second, because of the outbreak of the First World War. In any case, Lánczy's plan, had it been realized, could not propose definitive solutions for the economic and social challenges Hungary had to face. The Budapest Universal Exhibition of 1917 might have indeed achieved at least some of those political, cultural and economic goals, which were described in the draft concepts of Blána, Zichy or Lánczy. The history of Hungarian proposals ends with the first coherent project of 1917.

Conclusion

In the period discussed here, hosting an international exhibition almost became a necessity for a modern country, a sign of political ambition, economic strength and cultural advancement. Capitals of leading colonialist countries (London, Paris), newly created nation-states (Brussels) or conservative monarchies (Vienna) played the lead in this process. The role of capitals in increasing the international visibility of a country through the medium of exhibition is even more evident in the case of Italy: the fiftieth anniversary of the country's unification had been commemorated in 1911 by three exhibitions, organized simultaneously in the former and actual national capitals: Turin, Florence and Rome. The concepts for the Budapest Universal Exhibition reflected this new role of a modern national capital: through the organization of such an event Budapest would have appeared in the competition of nations as a new political, economic and cultural hub. Apart of being a capital of a semi-sovereign country, the fast-growing Budapest would have been able to present many of the modern traits of a universal exhibition.[45]

The differences of the concepts for a Budapest universal exhibition throughout the period between the 1867 Austro-Hungarian Compromise and the dissolution of the historic state of Hungary in 1918 have indicated the transformation of a new modern political entity on the basis of its domestic and international circumstances. In the times of democracy and nationalism concepts for a Budapest universal exhibition are testimonies of its semi-sovereign statute. Hungary, perceived in 1867 as a newly created modern state by the international audience, had been ranked among other new nation states in way of modernization at the eastern and southern borders of the Monarchy. Due to its legal status within the Monarchy, the shared responsibilities with Austria, a continuous effort can be detected from the Hungarian administration to emphasize the nation's sovereign history, economy and culture in all communication towards the international audience.

The history of the ideas for a universal exhibition in Budapest reflects the changing approach to the image of Hungary and to the self-image of the Hungarians. On a more general level, the history of the plans for a Budapest universal exhibition also demonstrates the ambitions, the motivations and the expected results of a country in which nation-building process and the accelerated modernization was manifested simultaneously within the frame of a newly created national administration. From the 1870s, this coincided, with the redefinition of the universal exhibition, when this typical phenomenon of the nineteenth century was no longer a collection of artefacts for a certain period of time, aimed at celebrating the common achievements of the Mankind. Exhibitions faced the seemingly contradictory challenge of professionalization of the exhibition industry, the thematic specialization, the need for mass entertainment, and the apparition of ephemeral pavilions as means of national representation. The plans for a Budapest universal exhibition indicate that Hungary, as any other country in the process of modernization, saw itself through the optics of its glorious past while the latest achievements of its political, economic and cultural modernization were demonstrated in the exhibition galleries and pavilions abroad as well as in the nation's capital.

Notes

1 Bryan Cartledge, *The Will to Survive. A History of Hungary* (New York, 2010), 219–250.

2 Daniel Unowsky, 'Staging Habsburg Patriotism: Dynastic Loyalty and the 1898 Imperial Jubilee', in *Constructing Nationalities in East Central Europe*, ed. Pieter Judson and Marsha Rozenblit (New York and Oxford, 2004); Daniel Unowsky, *The Pomp and Politics of Patriotism: Imperial Celebrations in Habsburg Austria, 1848–1916* (West Lafayette, 2005).

3 *A londoni kiállítás ügyében működő Országos Központi Bizottmány 1862-dik april 26-ik napján tartott ülésének jegyzőkönyve* (Pest, 1862), 12.

4 *Catalogue special du Royaume de Hongrie* (Paris, 1867).

5 Emma De Gerandóné Teleki, *1867diki világtárlat* (Paris, 1868), 518–519.

6 Árpád Mikó, 'A reneszánsz művészet története Magyarországon' (doctoral thesis, Hungarian Academy of Sciences, Budapest 2011). Dániel Pócs, *A Didymus corvina. Hatalmi reprezentáció Mátyás király udvarában* (Budapest, 2012).

7 Pieter van Wesemael, *Architecture of Instruction and Delight. A Socio-Historical Analysis of World Exhibitions as a Didactic Phenomenon (1798–1851–1970)* (Rotterdam, 2001), 136–142.

8 Mónika Lackner, 'Az első magyar néprajzi gyűjtemény: a magyar népi kultúra prezentációja az 1873-as bécsi világkiállításon', in *Az áttörés kora. Bécs és Budapest a historizmus és az avantgárd között (1873–1920)*, ed. Katalin F. Dózsa (Budapest, 2004), 101–110.

9 Katalin Sinkó, 'Az Alföld és az alföldi pásztorok felfedezése a külföldi és a hazai képzőművészetben', *Ethnogaphia* 100 (1989): 121–154; Lackner, 101–110.

10 Sinkó, 'Az Alföld', 121–131.

11 Bjarne Stoklund, 'How the Peasant House Became a National Symbol. A Chapter in the History of Museums and Nation-Building', *Ethnologia Europaea, Journal of European Ethnology* 29, no.1 (1999): 5–18.

12 Matthew Rampley, 'Peasants in Vienna. Ethnographic Display and the 1873 World's Fair', *Austrian History Yearbook* 42 (2011): 110–132.

13 Szilárd Blána, *Egy terv Magyarországnak, illetve fővárosának virágzóvá tételére, s ennek folytán a világkereskedelem és ipar közvetítő főpontjává leendő emelését* (Budapest, 1874).

14 Alexander C. T. Geppert, *Fleeting Cities: Imperial Expositions in Fin-de-Siècle Europe* (New York, 2009).

15 Blána, 17–18.

16 Timothy Mitchell, 'The World as Exhibition', *Comparative Studies in Society and History* 31, no. 2 (1989): 222–230.

17 Pál Lővei, 'Magyarország és a világkiállítások (főleg építészettörténeti vázlat)', in *Pavilonépítészet a 19-20. században a Magyar Építészeti Múzeum gyűjteményéből*, ed. Zoltán Fehérváry, Virág Hajdú and Endre Prakfalvi (Budapest, 2000), 21.

18 Blána, 1.

19 Mitchell, 222–224.

20 Ferenc Vadas, 'Programtervezetek a Millennium megünneplésére (1893)', *Ars Hungarica* 24, no. 1 (1996): 3 –55, at 44–46.

21 Vadas, 4–11.

22 Vadas, 6–8, 12.

23 Even though Zichy used the Hungarian word *világkiállítás* (world or universal exhibition) in his draft, he referred to the event as *országos kiállítás* (a general national exhibition) in other printed versions which drew on the legacy of the 1885 event. It is highly probable that in Zichy's approach the universality of the exhibition would have been ensured by a high number of foreign (international) visitors for whom the new Hungarian economy and cultural life would have been presented.

24 Katalin Sinkó, 'A millenniumi kiállítás mint Gesamkunstwerk', in *A historizmus művészete Magyarországon*, ed. Anna Zádor (Budapest, 1993), 141.

25 József Sisa, 'The Beginnings of Art History and Museology in Hungary. Some Semper Connections', *Centropa* 2, no. 2 (2002): 128–135.

26 Katalin Keserü, 'The Transformation of Architectural Thinking in Central Europe at the Turn of the Twentieth Century', in *The Beginnings of Modernism in Central European Architecture*, ed. Katalin Keserü (Budapest, 2005), 17–24. Ákos Moravánszky, *Competing Visions: Aesthetic Invention and Social Imagination in Central European Architecture, 1867–1918* (Cambridge, MA and London, 1998), 217–284.

27 Eric Hobsbawn, *The Invention of Tradition* (Cambridge, 1983), 263–307; Benedict Anderson, *Imagined Communities* (London and New York 1983), 83–112.

28 Rebecca Houze, 'National Internationalism. Reactions to Austrian and Hungarian Decorative Arts at the 1900 Paris Exposition Universelle', *Studies in the Decorative Arts* 12 (2004–05): 55–97.

29 Zoltán Fejős, 'Nemzeti diskurzus és néprajz', in *XIX. Nemzet és Művészet. Kép és Önkép*, ed. Erzsébet Király, Enikő Róka and Nóra Veszprémi (Budapest, 2010), 110–111.

30 Cf., for example, Grace Brockington, ed., *Internationalism and the Arts in Britain and Europe* (Oxford, 2009); Elizabeth Clegg, *Art, Design and Architecture in Central Europe, 1890–1920* (New Haven, 2006); David Crowley, *National Style and Nation-State: Design in Poland from the Vernacular Revival to the International Style* (Manchester, 1992); Marta Filipová, '"National Treasure or a Redundant Relic". The Roles of the Vernacular in Czech Art', *RIHA Journal* 0066, at http://www.riha-journal.org/articles/2013/2013-jan-mar/filipova-national-treasure-or-a-redundant-relic, accessed 4 June 2013; Charlotte Ashby, 'The Pohjola Building: Reconciling Contradictions in Finnish Architecture around 1900', in *Nationalism and Architecture*, ed. Raymond Quek, Darren Deane and Sarah Butler (Farnham, 2012), 135–145.

31 Tamás Csáki, 'A finn építészet és az "architektúra magyar lelke". Kultúrpolitika, építészet, publicisztika a századelő Magyarországán', *Múltunk* 29, no. 1 (2006): 200–230.

32 Paul Stirton, 'From "The Stones of Venice" to "The Stones of Transylvania": Károly Kós, Ruskin and the English Arts and Crafts Tradition', in *Britain and Hungary: Contacts in Architecture, Design, Art and Theory*, vol. 3, ed. Gyula Ernyey, Hungarian University of Craft and Design (Budapest, 2005), 166–179.

33 Hungarian pavilions in Milan and Bucharest (both in 1906); Turin and Dresden (both in 1911) echoed this new concept. Cf. Eszter Magdó, 'Magyar részvétel az 1906-os Román Általános Kiállításon', in *Opus Mixtum 1/1 – Proceedings of the Conference: Progressive Tendencies in Ephemeral Architecture*, ed. Miklós Székely (Budapest, 2012), 36–47; Zsófia Hutvágner, 'Az 1911-es drezdai Nemzetközi Higiéniai kiállítás Magyar pavilonja', in *Opus Mixtum* 1, no. 1, 70–77.

34 Gyula Török, *A Turini világkiállítás magyar háza* (Budapest, 1912), 4–8.

35 Paolo Cornaglia, 'A magyar pavilon az 1911-es torinói világkiállításon', in *Pavilon építészet a 19–20. században a Magyar Építészeti Múzeum gyűjteményéből*, ed. Zoltán Fehérvári, Virág Hajdú and Endre Prakfalvi (Budapest, 2000), 79–88.

36 László Jurecskó, 'K. Lippich Elek – a hivatalos művészetpolitika irányítója – és a Gödöllőiek', *Studia Comitatensia* 10, no. 1 (1982): 10–28.

37 Leó Lánczy, *Előterjesztése Lánczy Leó főrendiházi tagnak, a budapesti kereskedelmi és iparkamara elnökének a Budapesten 1917–ben rendezendő nemzetközi világkiállítás tárgyában* (Budapest, 1911), 1.

38 Lánczy, 5.

39 Lánczy, 5.

40 Lánczy, 2.

41 Lánczy, 10.

42 Lánczy, 10.

43 Lánczy, 10.

44 Lánczy, 10.

45 Three new bridges over the Danube, new railway stations, the first electronic underground line, new university buildings, theatres, museums and many other investments were made for the Millennium festivities in 1896. Ilona Sármány-Parsons, 'Ungarns Millenniumsjahr 1896', in *Der Kampf um das Gedächtnis. Öffentliche Gedenktage in Mitteleuropa*, ed. Emil Brix and Hannes Stekl (Vienna, 1997), 273–291.

Bibliography

A londoni kiállítás ügyében működő Országos Központi Bizottmány 1862-dik april 26-ik napján tartott ülésének jegyzőkönyve. Pest: Herz János, 1862.

Anderson, Benedict. *Imagined Communities: Reflections on the Origin and Spread of Nationalism*. London and New York: Verso, 1983.

Ashby, Charlotte. 'The Pohjola Building: Reconciling Contradictions in Finnish Architecture around 1900'. In *Nationalism and Architecture*, edited by Raymond Quek, Darren Deane and Sarah Butler, 135–145. Farnham: Ashgate, 2012.

Blána, Szilárd. *Egy terv Magyarországnak, illetve fővárosának virágzóvá tételére, s ennek folytán a világkereskedelem és ipar közvetítő főpontjává leendő emelését*. Budapest, 1874.

Brockington, Grace, ed. *Internationalism and the Arts in Britain and Europe*. New York and Vienna: Peter Berg, 2009.

Cartledge, Bryan. *The Will to Survive. A History of Hungary*. New York: Columbia University Press, 2010.

Catalogue special du Royaume de Hongrie. Paris, 1867.

Clegg, Elizabeth. *Art, Design and Architecture in Central Europe, 1890–1920*. New Haven: Yale University Press, 2006.

Cornaglia, Paolo. 'A magyar pavilon az 1911–es torinói világkiállításon'. In *Pavilon építészet a 19–20. században a Magyar Építészeti Múzeum gyűjteményéből*, edited by Fehérvári Zoltán, Hajdú Virág and Prakfalvi Endre, 79–96. Budapest: OMvH, Magyar Építézeti Múzeum, Pavilon Alapítvány, 2000.

Crowley, David. *National Style and Nation-State: Design in Poland from the Vernacular Revival to the International Style*. Manchester: Manchester University Press, 1992.

Csáki, Tamás. 'A finn építészet és az "architektúra magyar lelke". Kultúrpolitika, építészet, publicisztika a századelő Magyarországán'. *Múltunk* 29, no. 1 (2006): 200–230.

De Gerandó neé Teleki, Emma. *1867diki világtárlat*. Paris, 1868.

Fejős, Zoltán. 'Nemzeti diskurzus és néprajz'. In *XIX. Nemzet és Művészet. Kép és Önkép*, edited by Király Erzsébet, Róka Enikő and Veszprémi Nóra, 83–111. Budapest: Magyar Nemzeti Galéria, 2010.

Filipová, Marta. '"National Treasure or a Redundant Relic". The Roles of the Vernacular in Czech Art', *RIHA Journal* 0066, 26 February 2013.

Geppert, Alexander C. T. *Fleeting Cities: Imperial Expositions in Fin-de-Siècle Europe*. New York: Palgrave Macmillan, 2009.

Hobsbawn, Eric, and Terence Ranger, eds. *The Invention of Tradition*. Cambridge: Cambridge University Press, 1983.

Houze, Rebecca. 'National Internationalism. Reactions to Austrian and Hungarian Decorative Arts at the 1900 Paris Exposition Universelle'. *Studies in the Decorative Arts* 12 (2004–05): 55–97.

Hutvágner, Zsófia. 'Az 1911-es drezdai Nemzetközi Higiéniai kiállítás Magyar pavilonja'. In *Opus Mixtum I – Proceedings of the Conference: Progressive Tendencies in Ephemeral Architecture*, edited by Miklós Székely, 70–77. Budapest: Centrart. 2012.

Jurecskó, László. 'K. Lippich Elek – a hivatalos művészetpolitika irányítója – és a Gödöllőiek'. *Studia Comitatensia* 10, no. 1 (1982): 10–28.

Keserü, Katalin. 'The Transformation of Architectural Thinking in Central Europe at the Turn of the Twentieth Century'. In *The Beginnings of Modernism in Central European Architecture*, edited by Katalin Keserü, 17–25. Budapest: Ernst Múzeum, 2005.

Lackner, Mónika. *Az első magyar néprajzi gyűjtemény: a magyar népi kultúra prezentációja az 1873-as bécsi világkiállításon*. In *Az áttörés kora. Bécs és Budapest a historizmus és az avantgárd között (1873–1920)*, edited by Katalin F. Dózsa, 101–110. Budapest: Budapesti Történeti Múzeum, 2004.

Lánczy, Leó. *Előterjesztése Lánczy Leó főrendiházi tagnak, a budapesti kereskedelmi és iparkamara elnökének a Budapesten 1917–ben rendezendő nemzetközi világkiállítás tárgyában*. Budapest, 1911.

Lővei, Pál. 'Magyarország és a világkiállítások (főleg építészettörténeti vázlat)'. In *Pavilonépítészet a 19–20. században a Magyar Építészeti Múzeum gyűjteményéből*, edited by Zoltán Fehérváry, Virág Hajdú and Endre Prakfalvi, 17–64. Budapest: OMvH, Magyar Építézeti Múzeum, Pavilon Alapítvány, 2000.

Magdó, Eszter. 'Magyar részvétel az 1906-os Román Általános Kiállításon'. In *Opus Mixtum I – Proceedings of the Conference: Progressive Tendencies in Ephemeral Architecture*, edited by Miklós Székely, 36–47. Budapest: Centrart, 2012.

Mikó, Árpád. 'A reneszánsz művészet története Magyarországon'. Doctoral thesis, Hungarian Academy of Sciences, Budapest 2011.

Mitchell, Timothy. 'The World as Exhibition'. *Comparative Studies in Society and History* 31, no. 2 (1989): 222–230.

Moravánszky, Ákos. *Competing Visions: Aesthetic Invention and Social Imagination in Central European Architecture, 1867–1918*. Cambridge, MA and London: The MIT Press, 1998.

Pócs, Dániel. *A Didymus corvina. Hatalmi reprezentáció Mátyás király udvarában*. Budapest: Institute of Art History, Research Centre of the Humanities, Hungarian Academy of Sciences, 2012.

Rampley, Matthew. 'Peasants in Vienna. Ethnographic Display and the 1873 World's Fair'. *Austrian History Yearbook* 42 (2011): 110–132.

Sármány-Parsons, Ilona. 'Ungarns Millenniumsjahr 1896'. In *Der Kampf um das Gedächtnis. Öffentliche Gedenktage in Mitteleuropa*, edited by Emil Brix and Hannes Stekl, 273–291. Vienna: Boehlau, 1997.

Sinkó, Katalin, 'A millenniumi kiállítás mint Gesa mkunstwerk'. In *A historizmus művészete Magyarországon*, edited by Anna Zádor, 132–147. Budapest: Institute of Art History, Hungarian Academy of Sciences, 1993.

Sinkó, Katalin. 'Az Alföld és az alföldi pásztorok felfedezése a külföldi és a hazai képzőművészetben'. *Ethnogaphia* 100 (1989): 121–154.

Sisa, József. 'The Beginnings of Art History and Museology in Hungary. Some Semper Connections'. *Centropa* 2, no. 2 (2002): 128–135.

Stirton, Paul. 'From "The Stones of Venice" to "The Stones of Transylvania": Károly Kós, Ruskin and the English Arts and Crafts Tradition'. In *Britain and Hungary: Contacts in Architecture, Design, Art and Theory*, vol. 3, edited by Gyula Ernyey, 166–179. Budapest: Hungarian University of Craft and Design, 2005.

Stoklund, Bjarne. 'How the Peasant House Became a National Symbol. A Chapter in the History of Museums and Nation-Building'. *Ethnologia Europaea, Journal of European Ethnology* 29, no. 1 (1999): 5–18.

Török, Gyula. *A Turini világkiállítás magyar háza*. Budapest, 1912.

Unowsky, Daniel. 'Staging Habsburg Patriotism: Dynastic Loyalty and the 1898 Imperial Jubilee'. In *Constructing Nationalities in East Central Europe*, edited by Pieter Judson and Marsha Rozenblit, 141–156. New York and Oxford: Berghahn, 2004.

Unowsky, Daniel. *The Pomp and Politics of Patriotism: Imperial Celebrations in Habsburg Austria, 1848–1916*. West Lafayette, IN: Purdue, 2005.

Vadas, Ferenc. 'Programtervezetek a Millennium megünneplésére (1893)'. *Ars Hungarica* 24, no. 1 (1996): 3–55.

Wesemael, Pieter van. *Architecture of Instruction and Delight. A Socio-Historical Analysis of World Exhibitions as a Didactic Phenomenon (1798–1851–1970)*. Rotterdam: 010, 2001.

2

Barcelona's Universal Exhibition of 1888: An Atypical Case of a Great Exhibition

Marina Muñoz Torreblanca

Introduction: Great Exhibitions Before 1930

Universal and international exhibitions were representations of what was believed to be progress and modernity, and were conceived as being miniature versions of the modern totality.[1] In the search for this 'universalism', world's fairs embodied the material representation of the principle promoted by encyclopaedias in the later eighteenth century, in that they were attempting to gather together a general image of the world.[2] The organization of a universal exhibition thus contained a mixture of economic, imperialist and exotic desires.

In general, such exhibitions could be defined as an organized public display of agricultural, industrial, scientific or artistic products aimed at promoting production, commerce and culture as well as stimulating public interest. They sometimes also communicated and showcased colonial achievements, incorporating ethnographic elements from cultures dominated by the colonial power or exhibiting native groups and individuals who inhabited the colonial territories. In these fairs, ethnographic objects from the remotest countries were displayed alongside the great technical advances, the latest works of art, local products and various spectacles. The first time that all the nations of the world were invited to participate in a Great Exhibition was in 1851, by Britain, seeking to throw open its markets at the famous Great Exhibition of the Works of Industry of all Nations held in London's Hyde Park. This occasion also marked the birth of the concept of the 'Great International Exhibition'.[3]

Exhibitions typically embraced a whole range of events, from the first acclaimed exhibition in London in 1851 to those that took place at the end of the nineteenth century, up to the point when various participating countries started to feel a certain unease as regards funding and the quality of the objects displayed.[4] This emerged particularly after the *Exposition Universelle du Paris* of 1902, due to the momentum generated by previous French exhibitions and the influence of committees such as the 'Comité Initiative des Expositions'. Active from 1895, this was a precursor of the 'Comité Français des Expositions' founded by decree in 1901 and which led to the creation of similar committees in other countries: Belgium (1903), Italy (1905), Germany and Holland (both 1906) and Hungary (1907). These committees met for the first time on 30 November 1907 in Paris, to align the parallel doctrines they had implemented in the field of international exhibitions. It was, however, during another meeting, held in Brussels in 1908, that these committees created a 'Fédération des Comités Permanents' whose work formed the basis of what would lead to the conventions that are still in place.[5]

It soon became necessary to establish rules for these events. The Diplomatic Conference of Berlin held in 1910 stipulated the basis of an international convention (Federation of Permanent Exhibition Committees) that would regulate international exhibitions. In October 1912, the German government took the initiative and called together the representatives of 16 states – Germany, Austria, Belgium, Denmark, Spain, France, Great Britain, Holland, Hungary, Italy, Japan, Norway, Portugal, Russia, Sweden and Switzerland – with the intention of coming to an agreement about the principles of future exhibitions.[6] However, things had to be put on hold because of the First World War and it was not until 1928 that some governments were invited through French diplomatic channels to send their delegates to Paris to establish a Convention on international exhibitions. The result was that 43 countries – Spain included – signed an agreement for the creation of the Bureau International des Expositions (BIE) to regulate the frequency, length and quality of exhibitions.[7]

This short history of exhibitions' institutionalization highlights the difficulty of analysing and comparing events that took place prior to 1928. That is certainly the case with the Universal Exhibition of Barcelona, which was held in 1888, forty years before the creation of the regulatory framework. One must also take into account that expressions such as *exposition universelle, great exhibition, world's fair, international exhibition* and *gran exposición* are commonly, but not correctly, used as synonyms. So it might seem that, for example, the *Exposició Universal de Barcelona de 1888* was comparable to the *Exposition Universelle du Paris du 1889,* despite their being thoroughly different cases in terms of objectives, scale, numbers of exhibitors and visitors, etc. Given the difficulty of classification of these exhibitions, it would appear that the academic world has opted not to categorize them under any particular headings and to speak about exhibitions before 1928 in general.

Spain, like its European neighbours, also followed the fashion for organizing exhibitions and held a number of events, which, nevertheless, had a lesser impact at the European level when compared with those of other countries, such as Great Britain or France. One of the main reasons for this lack of international impact is the nature of the exhibitions in Spain – the majority of them was national or sector-specific. The number of so-called universal exhibitions, as defined by the BIE, that have taken place in the country up to the present has been relatively low. In fact, there have only been two: the Universal Exhibition of Barcelona (1888), the universality of which is the subject of this study, and the Universal Exhibition of Seville, Expo '92. As regards international exhibitions, there have also been only two: the Barcelona & Seville, Spain International Exhibition of 1929 in Barcelona and the Water and Sustainable Development Exhibition of 2008 in Zaragoza.

The Universal Exhibition in Barcelona in 1888, the main focus of this chapter, took place at the same time as other large and smaller exhibitions across Europe and beyond. The name that it adopted was 'Exposición Universal', a universal exhibition, which would suggest – as mentioned before – that the organizing team aspired to compare it to other such events, including The Great Exhibition and the Great London Exhibition (London, 1851 and 1862 respectively), the *Exposition Universelles* (Paris, 1855, 1867, 1878), the *Weltausstellung* (Vienna, 1873), or the *Wereldtentoonstelling van Antwerpen* (Antwerp, 1885).

In this chapter, I shall attempt to establish whether the 1888 Barcelona Exhibition belongs among the universal exhibitions, such as those held in Paris and London before the twentieth century, or if it was the outcome of a series of good intentions that coalesced in an exhibition of little lasting significance. Examination of the origins of the project, the site and the main pavilions, the monuments and attractions linked to the exhibition, along with the recruitment of exhibitors and visitors will enable me to discuss what might constitute a 'typical' universal exhibition and, indeed, whether such a concept exists.

Planning a Universal Exhibition in Barcelona

The earliest records that mention an industrial exhibition in Spain date back to 30 March 1826 when a royal decree established that a Public Exhibition of Spanish Industry would be held annually.[8] However, after the first one took place in Madrid in 1827, subsequent exhibitions were discontinued.[9] Following the success of the Great Exhibition of 1851, the Spanish government tried – and failed – to organize a Great Spanish Exhibition of Industry and Arts in Madrid.[10] The idea first formed in 1852 and was mentioned in various decrees of 1859, 1872 and 1881 but, again, did not come to fruition.[11] In 1885, a royal decree was issued giving state protection to a private group who wanted to put on an exhibition of Industry and Arts, but this also came to nothing. Exhibitions in other sectors, however, did take place, all in Madrid: the General Agricultural Exhibition of 1857, the Exhibition of Viniculture Products, Machines, Apparatus, Tools and Utensils for Vinification of 1877, the Permanent Maritime-Industrial Exhibition on products related to the construction of boats in 1876, and the National Exhibition of Mining, Metallurgical Arts, Ceramics, Crystals and Minerals in 1883. All of these efforts came together in the first great exhibition with international participation that was held in Spain and which came to be called the Universal Exhibition of Barcelona of 1888. At the same time, we must take into account the experience the city of Barcelona had with holding Catalan industrial exhibitions from 1822 using the model that had began with the first of the French exhibitions in 1798.[12]

This first universal exhibition in the history of Spain took place in the industrial capital of the country that Barcelona was at the time. The municipal authorities were favourably disposed towards accepting the proposal for a spectacle of this type. Industrialists, businessmen and intellectuals in Barcelona wanted to show the world the splendour of the city after a period of economic prosperity known as the *Febre d'Or* (Gold Fever) of 1871–85 in Catalonia. This had been characterized by a boom in the wine sector (coinciding with the Phylloxera plague in France), an increase in investment in metallurgy and textiles, the export of iron, lead and copper, and the political stability of the Bourbon Restoration in Spain. This period of economic growth began to run out of steam with the arrival of Phylloxera in Catalonia (1879) and the free-trade turn taken by the restored Bourbon monarchy. However, it converted the Catalan bourgeoisie, thanks to speculative movements, into the motor driving that society. The Exhibition, which was held in the midst of what was already an advanced crisis, was seen, following the end of the *Febre d'Or*, as a second chance for the Catalan industrialists.[13]

The idea of creating an exhibition in Barcelona derived from the private proposal of Eugenio Serrano de Casanova. Many authors emphasize that it was an initiative of a non-Catalan, unconnected to Barcelona society but who had an in-depth knowledge of it thanks to long-standing contacts in the city.[14] A Galician ex-soldier, he had fought with the Carlists and become a health spa promoter in Paris. On his own initiative, he joined the Spanish contingent that went to the Philadelphia Exhibition of 1876 and entered into partnership with the architect Alexandre Sallé, who had designed some of the structures for the Spanish pavilion at the Universal Exhibition of Antwerp in 1884. Serrano de Casanova's advantage in proposing the exhibition lay in the fact that no Catalan had experience in the organization of industrial fairs.

In 1885 Serrano de Casanova presented his plan for the universal exhibition to the city council without asking the city for a single penny. Instead he asked for a small percentage of the takings as commission when the event was held. Serrano de Casanova was initially given 500,000 *pesetas* by the city council and six million *reals* by the state, amounting to another 1,500,000 *pesetas*. But as the work started, so the first difficulties arose. The official capital obtained from subsidies was scarce and, given the size of the project, Serrano de Casanova could not source enough private capital to match the scope of the plan (the costs in the end reached 8,534,441 *pesetas*). On top of this, the paperwork for demolishing old buildings in Ciutadella (where the exhibition was to take place) could not be obtained. For these reasons and a good number of other small difficulties, Rius i Taulet, the four-times mayor of the city, finally accepted that, once it was approved by the town hall, Serrano de Casanova would be paid some 800,000 *pesetas* for the construction materials he had already acquired and the city council would take over the project.[15]

To make up the necessary money, according to the chronicler Conrad Roure (a lawyer, publicist and theatrical writer, as well as federalist politician in the first wave of Catalanism), the government of Spain eventually approved a bill with an extraordinary credit to cover the costs of the Universal Exposition of Barcelona. The city council was committed by this law to reimburse this amount to the state with the profits from the exhibition and, in the event of their being insufficient, to return 75 per cent of the total amount in six payments over a period of six years following the end of the exhibition.[16] The role of the Spanish government was therefore secondary and reduced to that of a lender, which would make the Barcelona exhibition comparable to a regional expo, rather than a great universal exhibition planned by a central government. However, the involvement of the royal family was extensive during the months that the exhibition was open. It is also important to mention that the inauguration was presided over by King Alfonso XIII (who was then two years old), the Queen Regent María Cristina, the Princess of Asturias María de las Mercedes and the Prime Minister Práxedes Mateo Sagasta, in addition to the Mayor of Barcelona, Francesc Rius i Taulet. The very presence of the monarchy and the Prime Minister gave the event a decidedly national rather than regional profile.

Once the city council took over the exhibition project, they made the plan their own, forming an executive committee in 1886 to carry it forward. This committee consisted of the city's industrialists and conservative politicians, without any representation for the monarchy or Spanish government.[17] Serrano de Casanova was the project secretary, although as the work progressed this post was fulfilled by the journalist Carlos Pirozzini Martí.[18] Several historians have argued that the Barcelona exhibition, accidental in character despite its success, responded to the will of the most influential sectors of society who desired power and knew exactly what this type of event represented.[19] As early as 1851, the Catalan economist and lawyer Joan Illas i Vidal was sent to London's Crystal Palace as an observer and drew up an influential report.[20] The exhibition was for Barcelona a preparation for modern life.[21] Among the arrangements made prior to the exhibition, of particular note were the changes introduced to the design of the city made by the city planner Ildefons Cerdà (1855) in a large area that would be known as the Eixample district. His plan defined a city with large open spaces, buildings placed in wide streets and no differentiation between social classes as all the streets were equal in a square Cartesian grid. Barcelona was thereby converted into a modern metropolis even before the Universal Exhibition.

The Barcelona exhibition also took advantage of a particular moment in the urban development of the city after demolition of the old city walls had opened up the possibility of expansion towards Eixample. The Exhibition entailed major urban restructuring, as was also the case in many of the European capitals holding a grand event of this type.[22] In the case of Barcelona, this included locating the exhibition in Ciutadella Park, improving the general state of the 'Rambla' and 'Passeig Colón' streets leading to the Ciutadella – that is, everything previously enclosed by the city walls that had been demolished for the Exhibition.

In architectural terms it may be surprising, however, that the Exhibition did not attract greater participation by the Modernist artistic movement which would soon become so important for the image of the city of Barcelona. The only 'Modernista' architectural traces evident in the Barcelona Universal Exhibition are two buildings by the architects Lluís Domènech i Muntaner and Antonio Gaudí. The first one on the exhibition grounds was constructed as a *café-restaurant*, though it did not come to be used as such because of delays in its completion. The building, however, still standing today and known as the Castle of the Three Dragons, is considered to be one of Modernism's most outstanding works. The second work, the Palau Güell by Antonio Gaudí, was the private residence of Eusebi Güell i Bacigalupi, the first Count of Güell (1846–1918). Eusebi Güell was the son of a famous industrialist and indiano, Joan Güell (1800–72), and at the same time Antonio Gaudí's patron and a member of the Exhibition organizing committee.[23]

The reforms to Ciutadella Park, where the Exhibition took place, had been discussed for some time. Before deciding to locate the Exhibition there, a previous project had included a large park with a fountain and a museum. In 1883, the construction of the Hall of Shadows by the architect Josep Fontserè Mestre, used as an exhibition space in 1888, was well advanced; and in 1884 the Greenhouse, designed by another architect, Josep Amargós, and used as a party and conference hall during the Exhibition, had already been built. A large fountain and the water tank that fed it had been in operation from 1881. For this reason, it seemed logical to hold the exhibition in the Citadel (Ciutadella), since the city could take advantage of what had already been built when the goal of the original plan for this place was to create a park for the city. It is likely that the rush to put the Exhibition on and the absence of urban plans for the construction of a specific site for these purposes meant that the two plans – the original one to transform the old citadel into a city park and the new one to accommodate the Exhibition – had to be merged. This would generate considerable controversy in the Barcelona of the time.

Aspirations of the Barcelona Exhibition

As I have mentioned before, the Exhibition was developed on the former grounds (and in the vicinity) of the military citadel toppled by the revolution of 1868, covering a total area of about 450,000 square metres.[24] The entrance to the grounds was through a ticket office located in the 'Arc de Triomf' (Triumphal arch) monument. Then came a long walkway called St John's

Hall up to the first building, the Palace of Fine Arts, by the architect August Font i Carreras, where the opening and closing ceremonies, as well as art exhibitions, concerts and cultural events were held. On the other side stood the Palace of Science, the work of Pere Falqués, where pedagogical materials and pharmaceutical and chemical industry products for science teaching were exhibited. It was also one of the conference venues.

According to the exhibition plans, the right side included Domènech i Montaner's *café-restaurant*, Josep Amargós' Greenhouse and Antoni Rovira i Trias' Museo Martorell, which opened in 1882 as the first building in Barcelona to be built specifically to accommodate a public museum. There was also the Hall of Shadows for parties and conferences; a church, where liturgical materials for Catholic worship were exhibited; and the Pavilion of Colonies and Customs, by the French architect Alexandre Sallé.

The Palace of Industry, the exhibition's main building, initiated and designed by Alexandre Sallé and finished by the master builder Elies Rogent, dominated the grounds. Behind this building were the maritime section bridge, built by the architect Gaietà Buïgas i Monravà to link the citadel with the maritime section of the exhibition, followed by the fort of Don Carlos, a military building of the old citadel and two constructions by Gaietà Buïgas: the Pavilion of Shipbuilding Industry and the docks. Just to the north and beyond the perimeter of the park and former Citadel was the Gallery of Machines. Opposite the Pavilion of Industry was a lake and a fountain, designed by Josep Fontseré, and beyond this, also outside the Exhibition grounds, stood Pere Falqués' Palace of Agriculture. As with the Palace of Science, the land on which these buildings were erected was private and given to the city council only for the duration of the exhibition. For this reason the buildings were designed here from the outset as temporary structures.

Once the site and the pavilions were in place, the need for further attractive features, which would resemble those at other great exhibitions, arose. One option was to erect a recognizable monument. At this time, the plans for the Paris *Exposition Universelle* of 1889, for which Gustave Eiffel constructed his famous tower, were already underway. Work on the Eiffel Tower began in 1887 (although the plans date from 1884) and, coincidentally, a couple of similar projects were proposed for the Barcelona exhibition.[25] The first one, by the French designer Lapierre from Tolouse, aspired to be a pyramidal wooden tower, two hundred metres high, which was to be placed in the centre of the semicircle created by the Palace of Industry (Fig. 2.1). The second project, the Comtal Tower by Pere Falqués, was to be a stone, brick and iron construction, and of the same height as the competitive design. In the end, both of these designs were rejected and the 1888 exhibition had no prominent architectural feature comparable to the Eiffel Tower (Fig. 2.2).[26] The only monuments worthy of any mention were the statue of Columbus at the top of the Ramblas, a project which predated the Universal Exhibition but was inaugurated during the event, and the 'Arc de Triomf' which served as the entrance to the exhibition.

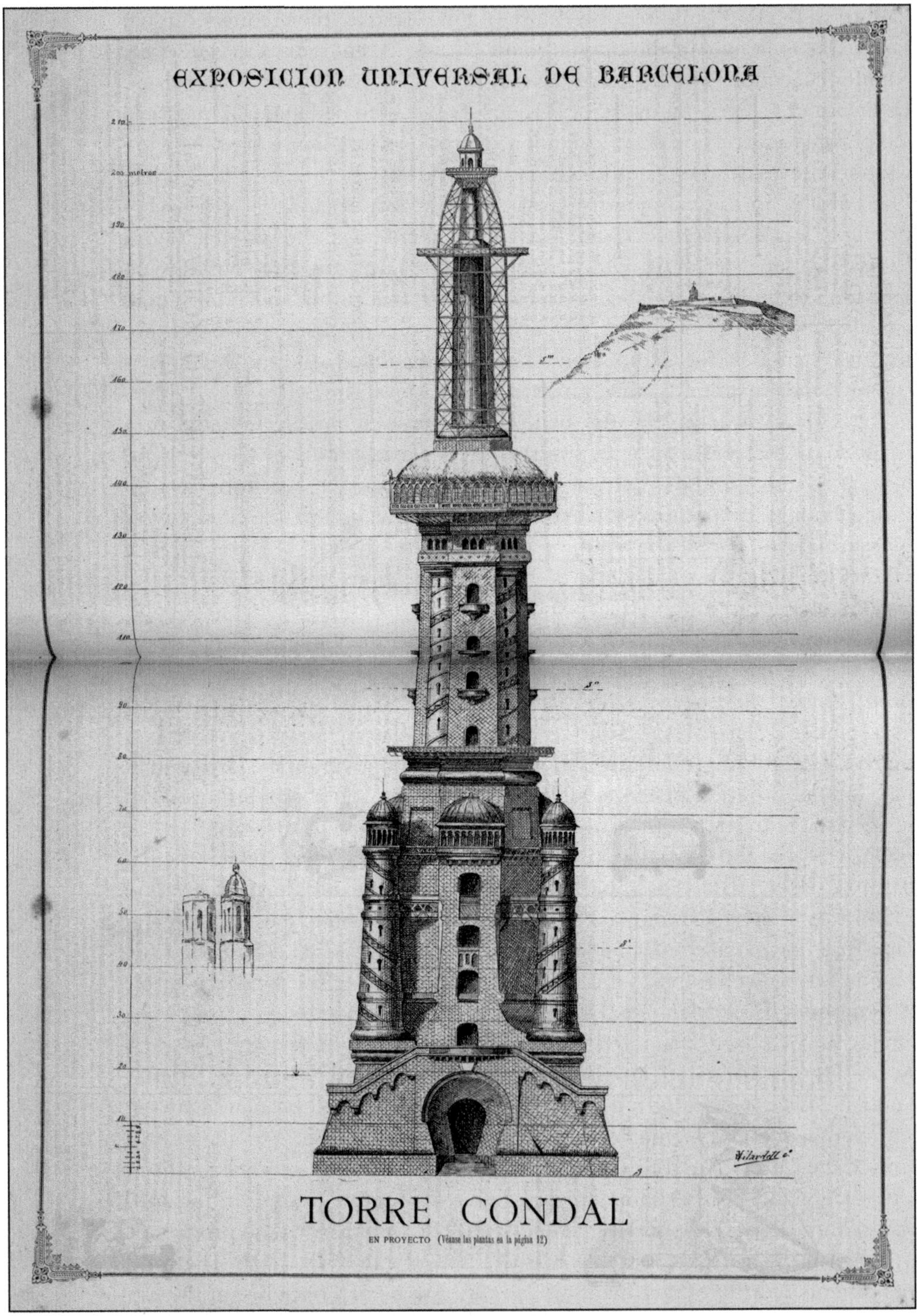

2.1 Lapierre's tower project reproduced in *La Exposición*, 10 March 1888, no. 45, pp. 8–9. Courtesy of Arxiu Històric de la Ciutat de Barcelona

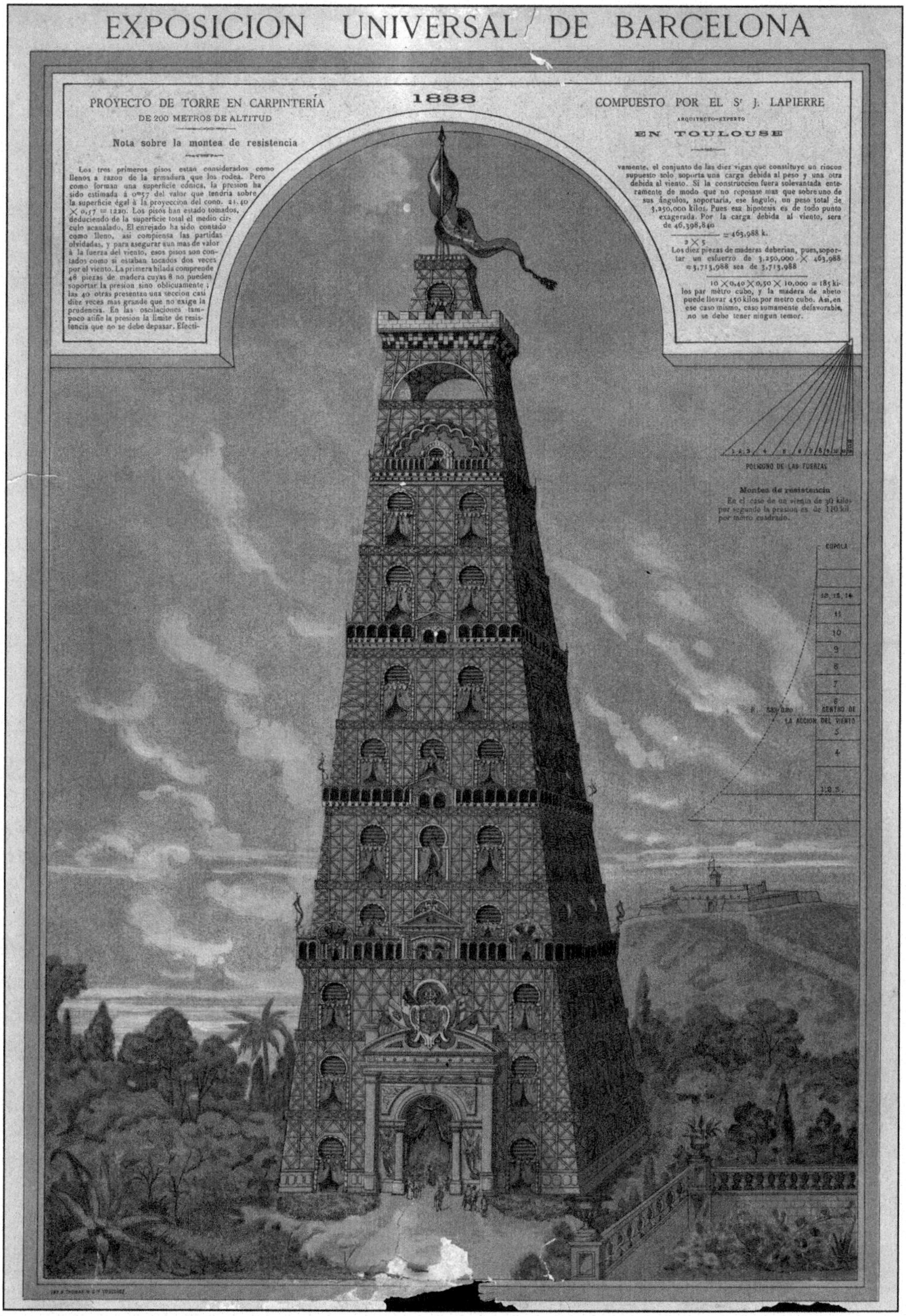

2.2 Comtal's tower project by Pere Falqués. Engravings collection of the Arxiu Històric de la Ciutat de Barcelona, AHCB 03496. Courtesy of Arxiu Històric de la Ciutat de Barcelona

In the context of universal exhibitions, visitors were also brought in by the attractions that were offered to them. The most remarkable entertaining feature in Barcelona was 'the magic fountain' placed in front of the current Catalan Parliament in which water was illuminated by voltaic arcs. This spectacle was popular with the public and had already featured at earlier exhibitions – the first such fountain was installed at the Exhibition of South Kensington in 1874. The work on the Barcelona fountain was planned by The Anglo-American Brush Electric Light Corporation under the direction of William Bates and carried out by the company J. Saint Noble.[27] It is not surprising that the fountain was success, as the contemporary press reported, partly because the use of electricity was still a relatively recent thing in Barcelona. Moreover, some other structures both within and outside the grounds, including the 'Arc de Triomf' monument, the park cascades, the entrance to the Pavilion of Industry and the waterfall at 'Plaça Catalunya' (a square in the Exhibition), were also illuminated at night. In this way Barcelona started a new tradition of lit fountains and the use of electricity that would be a fundamental element in its 1929 International Exhibition 40 years later.

Another important attraction for visitors was the aerostatic balloon, called 'España', and also known as the 'captive balloon' because it was tied to the ground by a rope (Fig. 2.3). It allowed visitors to fly up to the height of 300 metres to view the Exhibition and part of the city from above. The first balloon, however, lasted only ten days before it was struck and burnt down by lightning and a couple of months were needed before it could be replaced by a new one from France, called 'Catalunya'.[28] During the exhibition the most prominent figures, including members of the royal family, were invited to climb into the captive balloon.[29] And although again this was not the first time such an artefact had been presented at an exhibition – a look-out balloon had already featured at the *Exposition Universelle* of Paris in 1878 – the attraction in Barcelona became a symbol of modernity and progress.[30]

Another popular feature of great exhibitions of the time was what came to be known as 'exotic villages', displayed for instance at the *Exposition Universelle* of 1878 as the famous *Rue des Nations*. In these short-lived constructions, the host country would typically reproduce various architectural models of colonial territories and peoples from these countries, who were brought especially for the occasion and who took part in a variety of performances or the sale of hand-made products. Such artificial villages were meant to bring a certain degree of 'exoticism' to the presentation of scientific knowledge and universal advancement that the exhibition in question wanted to display.[31] No such model, however, was reproduced in Barcelona, even though a year before the General Exhibition of the Philippines (1887) had been held in Madrid, in which different types of habitats from this Spanish colony were built and individuals were brought over to participate. In Barcelona, the most 'exotic' thing a visitor could encounter was the pavilion of the Philippines Tobacco Company in which a group of Tagalog cigarette-makers rolled different types of cigarettes in front of the public. The Tagalogs were the same ones who had

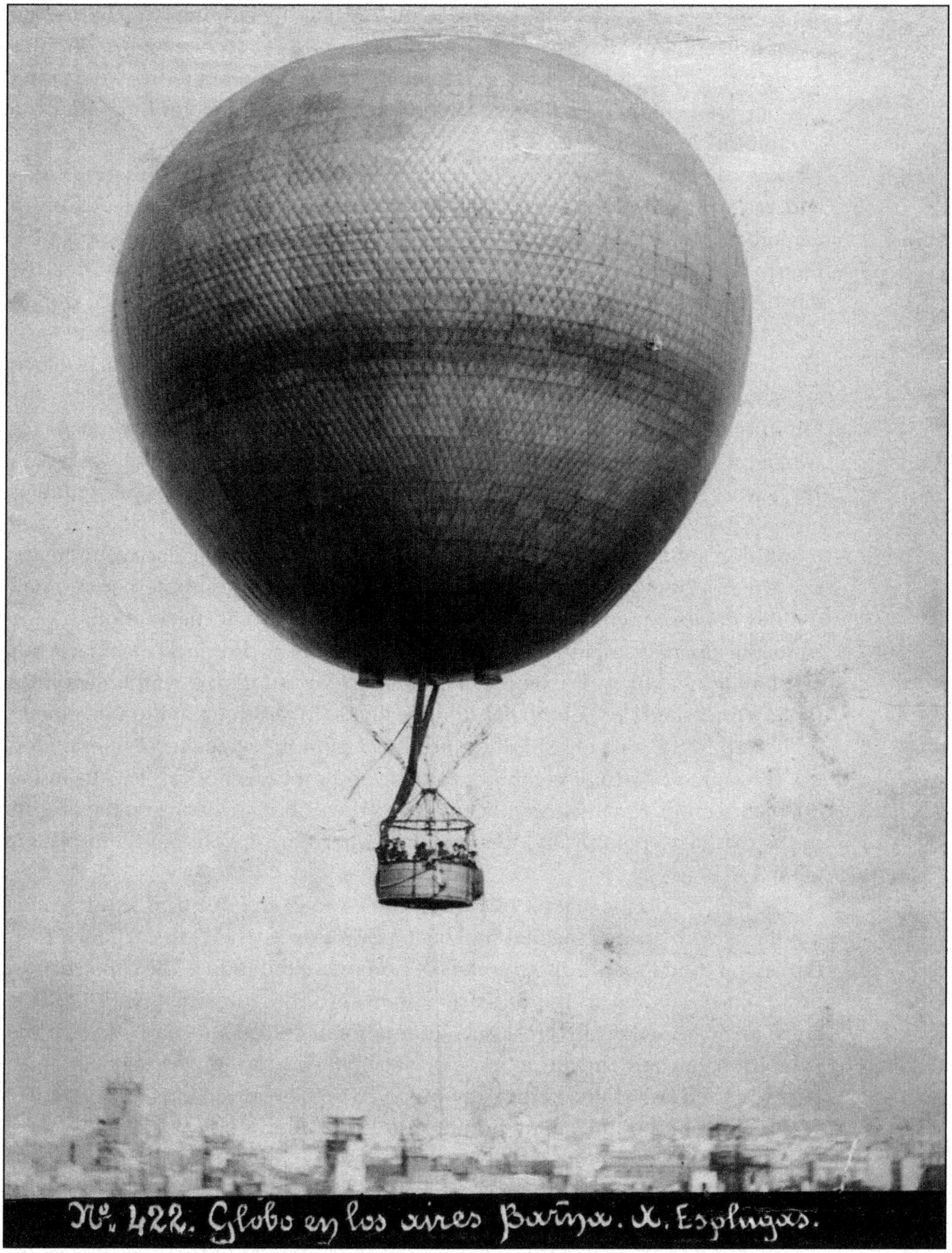

2.3 The aerostatic balloon, photograph by Antoni Esplugas.
Courtesy of Arxiu Fotogràfic de Barcelona

appeared in Madrid the year before but the rest of the Filipinos that had come to Madrid did not take part in Barcelona. The reason why the highly popular exotic village was understated in Barcelona can be ascribed to the weakening position that colonial Spain was experiencing at the time. Not long after the Exhibition closed, in 1898, Spain lost its last American colonies.

Nevertheless, the Pavilion of Colonies, 60 metres long and 27 metres wide, and covering a total area of 1,600 square metres, was constructed here to exhibit products of the Spanish overseas colonies, for example from Cuba, Puerto Rico and the Philippines. The chronicler Conrad Roure identified the most relevant installations to be those from the island of Puerto Rico due to the sheer diversity of the colony's products laid out in a long showcase.[32] The majority of these exhibits were manufactured luxury products (such as essences, oils and silks) and tobacco. The collection of objects from the Philippines, such as agricultural equipment, typical clothing, implements of war, zoological and art collections, which had all been exhibited in Madrid in 1887, were not brought to Barcelona either and as a result the representation of the Spanish colonies was not as rich as the initial plans of the Universal Exhibition envisaged. As Roure explained, 'annexed to the Palace of Industry was the Pavilion of Colonies, which was converted into a supplementary hall for the Palace, having, apart from the various colonial installations, some from the Spanish mainland ....'[33] Therefore, due to the lack of objects for display, the Pavilion of Colonies became little more than a warehouse for a motley miscellany of colonial and other items. In short, a pavilion originally dedicated to the idea of exhibiting products from the Spanish colonies ended up, through a dearth of colonial products and an excess of exhibits from the Spanish section, forcing organizers to open it without any focus on the origins of the exhibitors. Naturally, that did not help to bring value to the products exhibited.

There is no doubt that by including the various architectural features and pavilions with attractions that had been popular elsewhere, the organizers in Barcelona tried to reproduce models of previous exhibitions. The construction of a palace devoted to the fine arts and one to the industries were the two constant features of universal and international exhibitions and so were the exhibition gardens and attractions such as light sources or aerostatic balloons. Already tried and tested in the previous events, these attractions were popular with the general public, who did not visit the exhibition for business reasons, to attend conferences or specialized congresses, but to see the shows and varied entertainment on offer.[34]

The Exhibition was organized into five sections that were spread around a number of pavilions, but the focus was on the Palace of Industry as the centre for trade.[35] The large pavilion with 25 aisles was laid out in the shape of a fan and dedicated to various branches of industry and thus followed a patterned established at previous exhibitions. At the Paris *Exposition Universelle* of 1878, for instance, the Machine Gallery consisted of an industrial display under low transversal arches designed by the engineer Henri de Dion (1828–78), a design which later reappeared in the famous gallery constructed for the Paris *Exposition* of 1889. The fan design of the great pavilion in Barcelona, organized into semicircles, was also based on the 1878 model.

Seven of the pavilion's 25 aisles were occupied by exhibitors from the three largest Catalan cities of Barcelona, Terrassa and Sabadell and occupied a total area of 9,800 square metres. The province of Gerona took up approximately 1,000 square metres. The majority of the products exhibited here were related to the regional Catalan industries of the time: the production of refined sugars, spirits, chocolate, soup, pasta, cans, textiles and clothing.[36] The rest of the Spanish representatives came to take up some 3,000 square metres. The Spanish government had reserved the main gallery of 4,000 square metres, where they exhibited objects classified by the ministries, including those of war, marine, finance, development and overseas.[37] France and its colonies were given 6,600 square metres and Austria-Hungary 3,000 square metres.[38] Germany, Great Britain, the United States, Russia and Belgium appeared with smaller display areas. Lastly, there was little more than a token presence for Portugal, the Chinese, Japanese and Turkish empires, the Latin American republics of Ecuador, Bolivia, Honduras and Argentina as well as other European representatives, such as Switzerland, Sweden, the Netherlands and Denmark.

The distribution of space demonstrated the distinctively regional emphasis of the Exhibition. The majority of the Palace of Industry was occupied by local exhibitors, followed by the official Spanish presentations with less than half the space of the Catalan industrial capitals and, finally, with the exception of France, a discreet and symbolic international participation. The lack of a more extensive international presence outside the Palace of Industry, in for example national pavilions, contributed to the fact that Barcelona did not reach the status of other universal exhibitions.

Catalan National Sentiment

Although the international presence at the Exhibition was lower than expected, the organizers had been actively recruiting participants from both abroad and home. The first committee responsible for organization of the Exhibition was created in 1886 and changed its structure several times up to the actual event. In tandem with this, there were various other organizations, such as the Patronage Committee, which was something of a discussion forum on what the Exhibition should look like and what it would bring to the city. This Committee comprised a long list of members including influential municipal figures and representatives of various business sectors.

Among the great number of other committees, juries, commissions and sub-committees, the Commission for Encouraging Attendance of Exhibitors was one of the most important. It was responsible for promoting the Exhibition both abroad and in other provincial capitals in Spain, aiming to attract the greatest number of exhibitors possible and to increase awareness of the event. In Paris, a committee was formed in 1886 with the task of encouraging French industrialists to attend. Its success can be seen in the fact that a significantly large space was reserved for France in the Palace of Industry.[39] The significance of the Committee's impact can be seen in the considerable press coverage given to potential French exhibitors.[40] Conrad Roure mentioned more than 200 French exhibitors that could not register for the exhibition for capacity reasons and, consequently, were not included in the competition.[41]

Among the 'delegates', the individuals authorized to act as representatives for the Organizing Committee, one of the most proactive committees was based in Berlin, presided over by a Dr Jaunasch. He advertised the Exhibition and its advantages among German industrialists in the Berlin-based newspaper *Export*.[42] Delegates in other countries were not as successful as had been initially expected. Only nine European countries, mentioned above, finally attended the Exhibition, bringing the total number of foreign participants to 19.

In much the same way as on the international scale, there were also commissions and representatives organized on a regional level to promote and advertise the Exhibition. For instance in Terrassa, an industrial city close to Barcelona, a Committee for Initiative and Propaganda was created.[43] Different local and regional newspapers also advertised the Exhibition from the moment the idea for a universal exhibition in Barcelona was first born and assured the public of its importance. The regional paper for the Catalonian agricultural region of Lleida, for instance, claimed that

> We must give our readers the good news that on 15 September 1887 there will be a universal exhibition in the capital of the Principality. [The exhibition was originally planned to open on 15 September 1887 but the actual opening was on 8 April 1888.] Who has initiated this idea? Saying that it is not going to be held in Madrid is enough to show that it is out of an *individual initiative, an activity of the Barcelonans, a speculative enterprise and the vigour of the Queen of the Mediterranean* [Barcelona]. The realization of such a project is a triumph of decentralisation and Mr Almirall[44] insists it is a triumph of regionalism [Clarification and emphasis mine].[45]

The newspaper emphasized that, from the beginning, the Exhibition was planned to take place in Barcelona. The fact that it was initiated by an individual, Eugenio Serrano de Casanova, and by the powerful bourgeoisie of the city, emphasized its regional and municipal scope rather than official involvement of the government.

On the other hand, no one can doubt that the nationalist sentiments of this period accompanied the organizational process of the Barcelona Exhibition.[46] This patriotic sentiment around Catalan identity is distinct from the dynamics imposed by the tendency of the majority of the members of the Organizing Committee, who, apart from Mayor Rius i Taulet, were generally sympathetic to liberal-monarchic ideas and, therefore, nationalists as regards the idea of Spain. Against this group, the federalist republicanism of the Catalanist

Valentí Almirall stood out; he rejected the whole Barcelona exhibition project as linked too closely to the Court of Madrid. The Lliga de Catalunya party, created around Àngel Guimerà and Domenech i Montaner, took the initiative to present the Queen Regent with a message completely in Catalan, claiming rights for the Catalan nation and, for the first time, calling for the autonomy of Catalonia. The document was signed by more than 300 contemporary figures.[47]

In addition, discreetly embedded within the Universal Exhibition's programme of events right after the opening ceremony, came the *Jocs Florals* (Floral Games), the most important Catalan-language literary contest.[48] Curiously – and contrary to the wishes of the Catalanist groups – the 1888 games ended with the Queen Regent as 'Queen of the Floral Games'. However, the inaugural speech of these games ended up being a compliment to and a defence of the Catalan language. Finally, there was also a pedagogical Congress which propounded Spanish-Catalan bilingualism in schools, because the Catalan language had been completely excluded from the education law of 1857, despite the imposition of co-official status for the languages in schools in 1768 during the reign of Carlos III.[49]

At the same time, the press in Madrid recognized the fitness of the city of Barcelona to host an event of this type because of the particular character of the Catalan people, their geographical location and the facilities they could provide for port transport, and the city's economic relations with other capitals. The newspaper *La época*, for instance, referring to other major exhibitions held outside of capital cities observed:

> From the moment that whole of Catalonia supports such high-flown aspirations, it can no longer be in any doubt of success. Catalans, whose entrepreneurial, laborious and reflective character is justly celebrated, will now know to give a new proof of their unstinting perseverance. We have always believed that if a Universal Exhibition were to be held in Spain, Barcelona would be the place. Its excellent climate, its beautiful topographical situation, the importance of its port, which would both facilitate the transportation of objects and products that are to be included in the competition, the economic aspect of its relations with other capitals and, above all, Madrid, its essentially industrial spirit, all point to Barcelona for a competition of this nature. The United States held its Expo in Philadelphia; Holland in Amsterdam; Belgium in Antwerp.[50]

Impact and Success

Nevertheless, despite all the efforts of the Commission for Encouraging Attendance of Exhibitors, the international, regional and local delegates, and the different media that recognized the importance of attending the Barcelona Exhibition, it seems that the outcome was modest in comparison with other universal exhibitions. The Great Exhibition of London of 1851 attracted 13,917 exhibitors, while four years later, at the *Exposition Universelle* of Paris in 1855, the number increased to 23,954. In the 1862 London International Exhibition the figure increased further and the 1867 Paris *Exposition* saw the figures for the Great Exhibition more than double. The 1888 Universal Exhibition in Barcelona welcomed 12,223 exhibitors.[51] Barcelona thus saw only a quarter of the participants compared with the *Exposition Universelle* of 1867 and its numbers resembled those of the Great Exhibition held 37 years earlier.

Apart from the promotion of the Barcelona Exhibition by the delegates of the Commission, other advertising techniques were used to attract an international public. Advertising posters, for instance, were distributed throughout Spain and other countries, often placed at train stations. Such advertising was highest in France, where posters were distributed and displayed in 240 stations, and significant in Switzerland, Germany, Italy and Belgium. Furthermore, railway companies granted a 50 per cent reduction on their normal prices for all journeys to Barcelona during the Exhibition.[52] Even boat tickets for the journey between Cuba and Cadiz were offered with a 50 per cent discount so that the extensive Catalan community living in Cuba could visit the Exhibition.[53]

But despite all this promotion and effort, the final visitor numbers were lower than at other universal exhibitions. In comparison, over the 141 days of its duration, the Great Exhibition of 1851 received over six million visitors, while the 1867 *Exposition Universelle* in Paris welcomed over ten million people during its period of 210 days.[54] The Barcelona Exhibition, which lasted longer than its predecessors, 245 days, managed to attract just over 2 million people.[55]

At the same time, taking into account that in 1888, Barcelona had around 450,000 inhabitants, the number of visitors does not seem so small. For a city of this size, the Exhibition represented a considerable influx of people. Although there are no records of where the visitors came from, it is highly likely that a majority was from nearby cities in the region. The Catalan railway network was designed with the capital as its hub, linking Barcelona with the coastal and inland lines. Even other cities were well connected with Barcelona: for instance Valencia, Aragón and Madrid were all accessible by both rail and road. At the same time, though, the predominantly working classes of Barcelona were inevitably excluded from the Exhibition because of the ticket prices: 25 *pesetas* for a season ticket or one *peseta* for a single entry represented a considerable amount of money. However, at the Universal Exposition of Barcelona, free tickets for disadvantaged classes were offered, as happened in other major exhibitions, but it seems that mismanagement by the royal delegation delivered passes and free passes to people who could easily have paid for them.[56]

The Organizing Committee made various efforts to attract visitors and exhibitors and promote commercial exchanges. Discounts and special tours to the Exhibition from other countries were offered and a great number of city guides were available to facilitate tourists' visits.[57] The Exhibition was well prepared for the influx of exhibitors and visitors, which, nevertheless, failed to materialize and the pavilions were not always as busy as planned.

Conclusion

Referring to the 1928 Paris Convention, Greenhalgh identified official or officially recognized international exhibitions as all displays to which foreign countries are invited through diplomatic channels, which are not necessarily held periodically and the duration of which does not exceed six months.[58] The main object of these events is to showcase the progress of different countries in one or several branches of industry.[59] Applying such a definition in hindsight to an event that took place in 1888, the Barcelona Exhibition conforms with the majority of characteristics of the exhibitions: big urban changes, construction of pavilions in a space designated for that purpose, a big pavilion dedicated to showcasing industry, national and international products, a pavilion dedicated to the arts, different types of entertainment, transport facilities, tickets for disadvantaged people, everything except the invitation of foreign countries via diplomatic channels. In this sense, the Exhibition was more a regional than a great exhibition (remember that this project was initiated by a private individual and picked up by the city council). At the very start, therefore, this was not an official initiative or a project promoting Spain on the international scene. Probably the correct description would be the promotion of Catalonia on the international scene thanks to the initiative of the ambitious Catalan bourgeoisie.

The participation of the government of Spain was minor. While it eventually approved a bill with an extraordinary credit to cover the costs of the Universal Exhibition of Barcelona, it was only committed by this law to reimburse this sum to the state with the profits from the exhibition and, in the event of the profits being insufficient, to return 75 per cent of the total amount in six payments over the period of six years following the end of the exhibition.[60] Therefore the role of the Spanish government was secondary and reduced to that of a lender. However, the involvement of the royal family was important, especially in everything relating to the public character of the exhibition: the inauguration, the Floral Games, the visits of different members of the royal family during the Exhibition, etc. It seems that, in addition to serving to promote the exhibition, this strengthened the binomial of Taulet, the Mayor, and the monarchy.

Barcelona provided a large space for its universal exhibition, a quarter of which was taken up by buildings and the rest by gardens or leisure areas. It constructed a number of pavilions (which were not always completed) for the official sections and for all participants, but it did not manage to assemble a sufficient number of exhibitors to fill them with. The understated colonial dimension with its highly popular 'exotic' spectacles also reduced the 'universal' character of the Exhibition. And although there were some elements of entertainment for the public, such as the magic fountain or the aerostatic balloon, and special shows, such as a parade with floats on the occasion of the inauguration of the Columbus monument, which were widely followed by the press, the visitor numbers were the lowest of all the so-called universal exhibitions.[61]

Being based on aspirations and ambitions rather than actual delivery and results, the Exhibition became a transformational event for the city of Barcelona in the first place and an international event in the second. From the point of view of Catalan nationalism, the Exhibition served to vindicate their rights and make their cause visible through the document delivered to the Queen Regent, in addition to the contributions of the pedagogic Congress in matters of Catalan language. However, all the pre-Exhibition criticism stoked by anti-absolutist sectors converged at its close in a collective enthusiasm around the event and the city. In many respects, it was an exhibition that brought the Barcelona brand to life. Definitively, the Barcelona exhibition is an atypical case from any of the points of view that have been analysed, a magnificent regional exhibition with universal aspirations – or a universal exhibition with unexceptional results.

Notes

1 Mauricio Tenorio Trillo, *Artilugio de la nación moderna, México en las exposiciones universales, 1880–1930* (Mexico, 1998), 14.

2 Tenorio, 15.

3 Paul Greenhalgh, *Ephemeral Vistas: The Expositions Universelles, Great Exhibitions and World's Fairs, 1851–1939* (Manchester, 1988), 10 and 12.

4 See, for a historical perspective of the origin of the exhibitions, Chapter 1 'Origins and Conceptual Development', in Greenhalgh, *Ephemeral Vistas.*

5 Charles Piat, *Les Expositions Internationales relevant du Bureau International des Expositions* (Paris, 2001), 8.

6 Piat, 8.

7 Greenhalgh, 9.

8 Eduardo Rodríguez Bernal, *Historia de la Exposición Ibero-Americana de Sevilla de 1929* (Seville, 1994), 36.

9 Bernal, 36.

10 María Teresa Solano Sobrado, 'Antecedentes históricos de la Exposición Iberoamericana de Sevilla', *Cuadernos de Historia Moderna y Contemporánea* 6 (1986): 163–187, at 165.

11 Bernal, 36.

12 'Les primeres exposicions industrials de Barcelona', in *Exposició Universal de Barcelona. Llibre del centenari*, ed. Daniel Fernandez and Angels Prat (Barcelona, 1988), 113–119.

13 For an analysis of the crisis after the *Febre d'Or*, see Albert Carreras, 'Dels anys de crisi a l'eufòria collectiva. El context econòmic de l'Exposició de 1888', in *L'Exposició de 1888 i la Barcelona de fi de segle* (Barcelona, 1985), 92–98.

14 For example (1) *Exposició Universal de Barcelona. Llibre del centenari, 1888–1988* (Barcelona, 1988) 31 and 316; (2) Pere Hereu Payet, *Arquitectura i ciutat a l'Exposició de Barcelona* (Barcelona, 1988), 74; (3) Josep Ma. Garrut, *L'Exposició Universal de Barcelona* (Barcelona, 1976), 13.

15 Rius i Taulet was Mayor of the city of Barcelona on four occasions: from 1872 to 1873, when he resigned from the town hall because of the first Republic; in 1874; after the Restoration period from 1881 to 1884; and finally, from 1885 to 1889.

16 Conrad Roure, *Memòries de Conrad Roure, Recuerdos de mi larga vida. La restauració dels Borbons (II). L'Exposició Universal de Barcelona de 1888* (Vic, 1999), 68–69.

17 Honorary President: Francesc de P. Rius i Taulet; Effective President: Josep Pujol i Fernández; Vice President: Francesc López i Fabra; Consultant lawyer: Manuel Duran i Bas; Vocals: Manuel Porcar i Tió; Josep Ma. Nadal, Félix Macià i Bonaplanta, Roma Macaya i Gibert, Andreu de Sardi de Roselló, Ramon de Manjarrés, Francesc Sitjà; Secretary General: Eugenio Serrano de Casanovas.

18 Garrut, 21.

19 For example (1) Jordi Cassassas i Ymbert, 'La regeneració interior de la societat catalana', in *L'Exposició de 1888*, 87–91; (2) Garrut, 21; (3) *Exposició Universal de Barcelona. Llibre del centenari, 1888–1988* (Barcelona, 1988), 316.

20 Cassassas i Ymbert, 88. This fact and other actions carried out up to the holding of the exhibition are the arguments of some authors who insist that the project was conceived long before Eugenio Serrano de Casanovas' proposal.

21 Joan Ramon Resina, in his book *Barcelona's Vocation of Modernity: Rise and Decline of an Urban Image* (Stanford, CA, 2008), quotes the Universal Exhibition of 1888 when he develops the different concepts about the idea of the city and the imagined city within the context of modernity.

22 For the urbanization of the city of Barcelona as a consequence of the Universal Exhibition of 1888, see A. G. Espuche, M. Guardia, F. J. Monclús and J. L. Oyón, 'Modernization and Urban Beautification: The 1888 Barcelona World's Fair', *Planning Perspectives* 6, no. 2 (1991): 139–159; also another example about the city model can be read in Resina, *Barcelona's Vocation*.

23 An *indiano* is a Spaniard who returned to Spain after having made his fortune in Latin America. Many invested their money in Catalan businesses and became philanthropists. Some of them were the patrons of architects and artists during the Modernist period.

24 All the content related to the description of the pavilions of the exhibition of 1888 has been taken from Garrut, 27–32, and Roure, 77–110.

25 Although some websites and blogs refer to this fact and it is even intimated that Eiffel came to offer his project to the city of Barcelona, the historian of the city Lluís Permanyer denies this historical myth in an article dedicated to the issue; see Lluís Permanyer, 'Eiffel jamás ofreció su torre a Barcelona', *Vida*, at http://www.lavanguardia.com/vida/20090601/53714791930/eiffel-jamas-ofrecio-su-torre-a-barcelona.html, accessed 20 April 2015.

26 Frenchman Lapierre's Tower was rejected at the end of 1887. Falqués' Tower is first mentioned by the press in January 1888. In February, there is a change of location and in May, Falqués withdrew from the construction. According to Hereu Payet, the Municipal Commission ruled that its placement and technical direction were not in compliance with its location within the enclosure. For more information see Hereu Payet, 108.

27 Garrut, 31.

28 *La Vanguardia*, no. 301, 25 June 1888 (late edition), 2.

29 *La Vanguardia*, 30 October 1888, no. 517, 2; and *La Vanguardia*, 31 December 1888, no. 627, 2.

30 See illustration 'Le Ballon Captif de la Cour de Carousel, from l'Album de l'Exposition 1878' in Anna Jackson, *Expo: International Expositions 1851–2010* (London, 2008), 82; and about the history of ballooning in Catalunya see Ròmul Brotons, *Més lleugers que l'aire, globus i dirigibles al cel de Barcelona (1783–1937)* (Barcelona, 2012).

31 The PhD dissertation of the author of this chapter examined the exhibition of the so-called 'primitive' people and objects at the Spanish exhibitions. The publication of the work is in progress. Some examples that examine this topic are: (1) Peter H. Hoffenberg, *An Empire on Display: English, Indian and Australian Exhibitions* (Berkeley and Los Angeles, 2011); (2) Raymond Corbey, 'Ethnographic Showcases, 1870–1930', *Cultural Anthropology* 8, no. 3 (1993): 338–369; (3) Zeynep Çelik and Leila Kinney, 'Ethnography and Exhibitionism at the Expositions Universelles', *Assemblage*, no. 13 (1990): 35–59; (4) Nicolas Bancel, Pascal Blanchard, Gilles Boëtsch, Eric Deroo and Sandrine Lemaire, *Zoos humains et exhibitions coloniales: 150 ans d'inventions de l'Autre* (Paris, 2011).

32 Roure, 102.

33 Roure, 101.

34 The first exhibitions held in London (1851) and Paris (1855) received large numbers of visitors for their novel character. Those held after these needed to find varied entertainment to attract the public. The first were the French who incorporated an amusement park and fairground into the plans for the *Exposition Universelle* of 1867. See Greenhalgh, 41–42.

35 First section: Teaching, Beauty and Medicine. Second section: Public Works, Defence. Third section: Marine, Commerce, Colonization. Fourth section: Agriculture, Mining, Quarrying. Fifth section: Industry. Sixth section: Decorative and Industrial Arts. Each section was divided into groups and each group into classes.

36 *La Vanguardia*, 11 August 1888, no. 373, 2.

37 *Exposición Universal de Barcelona 1888. Catálogo Oficial Especial de España* (Barcelona, 1888).

38 *Exposició Universal de Barcelona. Llibre del centenari, 1888–1988*, 350.

39 Garrut, 23.

40 *La Vanguardia*, 17 January 1888, no. 26, 2; and *La Vanguardia*, 28 January 1888, vol. 8, no. 43, 1.

41 Espuche et al., 75.

42 Greenhalgh, 14; and specifically about Mr Jaunash promoting activities see *La Exposición, Órgano oficial de la Exposición Universal de Barcelona*, 29 December, 7.

43 Garrut, 14.

44 Valentí Almirall (1841–1904) was a Catalan politician, considered one of the fathers of modern Catalan nationalism. As a republican, he opposed the project of the Barcelona Universal Exhibition, although by the end of it he had repeatedly recognized the success of the event.

45 *Diario de Lérida*, 12 December 1886, quoted in Garrut, 14. 'Una buena noticia tenemos que dar nuestros lectores y es que el 15 de septiembre de 1887 [fecha

en que se proyectó la inauguración en un principio, finalmente celebrada el 8 de abril de 1888] se va a celebrar una Exposición Universal en la capital del Principado. ¿Quién a iniciado esta idea? Con decir que no va a celebrarse en Madrid, huelga manifestar que es debida a la iniciativa individual, a la actividad de los barceloneses, a la especulación y al poderío de la reina del Mediterráneo [Barcelona]. La realización de semejante proyecto es un triunfo de la descentralización, y hasta si se empeña el señor Almirall un triunfo del regionalismo.'

46 To expand this issue see Stephen Jacobson, 'Interpreting Municipal Celebrations of Nation and Empire: The Barcelona Exhibition of 1888', in *Nationalism and the Reshaping of Urban Communities in Europe, 1848–1914*, ed. William White and Oliver Zimmer (Basingstoke, 2011).

47 Josep M. Ainaud de Lasarte, 'El missatge a la Reina Regent' in *L'exposició del 88 i el nacionalime català* (Barcelona, 1988), 92.

48 Josep Faulí, 'Els jocs florals del 1888, final de l'època de lluita pel prestigi social', in *L'exposició del 88*, 86.

49 Josep González Agapito, 'El congrés pedagògic. La reivindicació de la llengua en el marc de l'Exposició' in *Exposició Universal de Barcelona*, 100.

50 'Desde el momento en que Cataluña entera apoya tan levantado pensamiento, no puede ya dudarse del éxito. Los catalanes cuyo carácter emprendedor, laborioso y reflexivo, es justamente celebrado, sabrán dar ahora un nuevo testimonio de su incansable perseverancia. Siempre hemos creído que si algún día había de celebrarse en España una Exposición Universal, debería verificarse en Barcelona. Su excelente clima su hermosa situación topográfica, la importancia de su puerto que tanto facilitaría el transporte de los objetos y productos que hayan de figurar en el certamen, loe económico de la vida en relación con otras capitales y, sobre todo, con Madrid, su espíritu esencialmente industrial, todo designa a Barcelona par aun certamen de esta naturaleza. Los Estados Unidos celebraron su Exposición Universal en Filadelfia; Holanda en Ámsterdam; Bélgica en Amberes.' *La época*, 12 August 1886 cited in Garrut, 15.

51 Garrut, 49.

52 Garrut, 25.

53 *La Vanguardia*, 9 July 1888 (late edition), no. 326, 2.

54 At the *Exposition Universelle* four years later, the Paris Universal Exhibition of 1855 (200 days long), the number of visitors dropped significantly to 5,162,000 visitors. London in 1862 almost equals its figures from 1851, although the days of exposure increase to 171; the visitor numbers differ very little: 6,211,000. Six years later (1873), Vienna celebrates its exhibition in 176 days and reaches a lower number than its predecessor in Paris, with 7,250,400 visitors.

55 Garrut, 49.

56 Manuel Arranz, 'El rerefons d'un endeutament', in *L'exposició de 1888*, 115.

57 María Teresa Sala, *Barcelona 1900* (Barcelona, 2007), 25.

58 Greenhalgh, 15.

59 Greenhalgh, 15.

60 Roure, 68–69.

61 *La Vanguardia*, 6 July 1888 (late edition), no. 321, 2. In the author of this article's PhD thesis published in March of 2009, this aspect was widely treated; for a discussion of this event from the nationalist point of view, see Jacobson.

Bibliography

Printed Primary Sources

Catálogo General Oficial de la Exposición Universal de Barcelona, 1888. Barcelona: Imprenta de los sucesores de N. Ramírez, 1888.

Exposición Universal de Barcelona 1888. Catálogo Oficial Especial de España. Barcelona: Imprenta de los sucesores de N. Ramírez, 1888.

La Vanguardia, 1888.

Roure, Conrad. *Memòries de Conrad Roure, Recuerdos de mi larga vida. La restauració dels Borbons (II). L'Exposició Universal de Barcelona de 1888*. Vic: Eumo, 1999.

Secondary Sources

Bernal, Eduardo Rodríguez. *Historia de la Exposición Ibero-Americana de Sevilla de 1929*. Seville: Servicio de Publicaciones del Ayuntamiento de Sevilla, 1994.

Cassassas i Ymbert, Jordi. 'La regeneració interior de la societat catalana'. In *L'Exposició de 1888 i la Barcelona de fi de segle*. Barcelona: Ajuntament de Barcelona, 1985, 87–91.

Espuche, A. G., M. Guardia, F. J. Monclús and J. L. Oyón. 'Modernization and Urban Beautification: The 1888 Barcelona World's Fair'. *Planning Perspectives* 6, no. 2 (1991): 139–159.

Exposició Universal de Barcelona. Llibre del centenari 1888–1988. Barcelona: Avenç i la Comissió Ciutadana per la Commemoració del Centenari de l'Exposició Universal, 1988.

Fernandez, Daniel, and Angels Prat, eds. *Exposició Universal de Barcelona. Llibre del centenari*. Barcelona: L'Avenç, 1988.

Garrut, Josep Ma. *L'Exposició Universal de Barcelona*. Barcelona: Ajuntament de Barcelona, Delegació de Cultura, 1976.

Greenhalgh, Paul. *Ephemeral Vistas: The Expositions Universelles, Great Exhibitions and World's Fairs, 1851–1939*. Manchester: Manchester University Press, 1988.

Grau, Ramon, and Marina López. *Exposició Universal de Barcelona, Llibre del centenari, 1888–1988*. Barcelona: L'Avenç, 1988.

Hereu Payet, Pere. *Arquitectura i ciutat a l'Exposició de Barcelona* 1888. Barcelona: Universitat Politècnica de Catalunya, 1988.

Hoffenberg, Peter H. *An Empire on Display: English, Indian and Australian Exhibitions*. Berkeley and Los Angeles: University of California Press, 2011.

L'Exposició de 1888 i la Barcelona de fi de segle. Barcelona: Ajuntament de Barcelona, 1985.

L'exposició del 88 i el nacionalime català. Barcelona: Fundació Jaume I, 1988.

Permanyer, Lluís. 'Eiffel jamás ofreció su torre a Barcelona'. *Vida*, at http://www.lavanguardia.com/vida/20090601/53714791930/eiffel-jamas-ofrecio-su-torre-a-barcelona.html, accessed 20 April 2015.

Piat, Charles. *Les Expositions Internationales relevant du Bureau International des Expositions*. Paris: Bureau International des Expositions (BIE), 2001.

Resina, Joan Ramon. *Barcelona's Vocation of Modernity: Rise and Decline of an Urban Image*. Stanford, CA: Stanford University Press, 2008.

Solano Sobrado, María Teresa. 'Antecedentes históricos de la Exposición Iberoamericana de Sevilla'. *Cuadernos de Historia Moderna y Contemporánea* 7 (1986): 163–187.

Tenorio Trillo, Mauricio. *Artilugio de la nación moderna, México en las exposiciones universales, 1880–1930*. Mexico: Fondo de Cultura Económica, 1998.

3

A Marginal Exhibition? The All-German Exhibition in Berlin, 1844

John R. Davis

Introduction

The history of international exhibitions has rested upon a set of general assumptions. Consequently, the notion has existed of a recognizable canon or sequence of events. Many exhibitions have been excluded from consideration and academic analysis. This chapter will seek to approach and question these assumptions from a variety of angles. By focusing on the Berlin exhibition of 1844, it uses an event that falls outside the standard exhibition canon to demonstrate both the artificiality of the distinctions made, and the benefits to research of looking more widely.

Until twenty years ago, in English-language histories at least, the Crystal Palace and the Great Exhibition of 1851 were seen as the starting point and even yardstick of exhibitions. The event in Hyde Park formed the intersection of the axes, chronologically and geographically. Chronologically, the Great Exhibition was viewed as the first such event, and later displays were presented as following in some sort of apostolic succession. Promoters of later exhibitions aspired to recapture the glory of 1851, if not its profit margins, with varying degrees of success. Geographically, the Anglo-centric historian could, and did, look at later exhibitions in Continental Europe, and in colonies and former colonies, and view them as aspiring to follow the British example. Progress, this implied, really was a project launched in benign Britain. The Crystal Palace was presented as the real thing, with later exhibitions diluting the message, becoming overinflated caricatures of the original, eventually bogged down by the weight of their rhetoric and removal from the real world.[1]

The Great Exhibition and Expos have also been analysed in overwhelmingly national terms. 1851 is seen a high point in British history. Exhibitions are also, however, part of the mapping of national decline: the reaction of British industrialists to their showing at late nineteenth-century exhibitions is a set piece within this narrative.[2] Expos are discussed with respect to rising national self-consciousness, national rivalry, imperial pretension and the onset of war. More recently, exhibitions have been looked at in terms of discourses regarding national and international transformation and modernization – the *Nation on Display*, as it were, to itself and to others.[3] If we set aside terminological dispute regarding what constitutes a nation, and whether we mean nations or states, the point remains that nations and states have defined a great deal of the way we think about exhibitions, just as they define much of our thinking about history as a whole. Interestingly, the title 'Great' was appended to the exhibition of 1851 to distinguish it from the local, institutional or national exhibitions up to that point. The Crystal Palace in Hyde Park housed – officially – 'the Great Exhibition of All Nations'. It was titled 'Great' because it was international and placed nation alongside nation.[4]

Even in 1851, this official perspective on the exhibition – part of the exhibition promoters' rhetoric – was highly problematic. If this was a representation of nations, then it was a highly curious one. Imperial territories were not represented, they were portrayed. The level of democratic involvement, or even popular participation, in defining representation was widely variable. For the German states, the idea of national representation in an exhibition was extraordinarily complicated.[5]

Some general points, then: as with most history, research regarding exhibitions has been dominated by nation-based approaches. On the one hand, this has been highly useful and productive, and recently interest has produced valuable research regarding how exhibitions contribute to nation-building and modernization. On the other hand, it can be argued that this leaves much about exhibitions untouched. In relation to the Great Exhibition, for example, local committees, regional rivalries and popular participation were arguably a key to the event's success. It can also be argued that these nation-based approaches are chronologically and geographically conditional. Local and regional exhibitions have a history that reaches back beyond and outside the nation-based tradition. They constituted a rich well of practice and thought from which the organizers of national exhibitions drew.

Berlin 1844 and Marginality

The Berlin exhibition of 1844 can be considered as part of a discussion of marginality in several respects therefore. This was an exhibition in a city which was not, at that point, a national capital. It took place in 1844, on the margins, if not outside, the orthodox canon. The exhibition was the result of collaboration between states and extending beyond one nation's boundaries. It might be said, therefore, to be on the margins of the move towards international exhibitions. The Berlin exhibition of 1844 was organized within the context of technical education and market expansion. Yet it showed signs of the propaganda of industrialization and modernization, the historicism of industrial development, meta-narrative and bombast of Expos. In this sense too it might be seen as on the margins, or cusp, of the Great Exhibition tradition.

Exhibitions before 1851 might also be said to be marginal historiographically speaking, with few historians entering into this territory. Kenneth Carpenter wrote an important, if very general article on pre-1851 exhibitions in 1972.[6] Toshio Kusamitsu studied the exhibitions at the British Mechanics' Institutes before 1851.[7] There has been research on the related subject of German trades fairs.[8] Thomas Groβbölting's excellent book on German exhibitions makes an important contribution to this area.[9] Still, none of these quite maps the number, regularity and geographic density of industrial exhibitions held on the Continent prior to 1851.

Background

In order to understand the 1844 exhibition more fully, a few points need making regarding its wider context. First, politically, the Vienna Acts of 1815 had established 39 states – so-called German states – between France in the west and Russia in the east. These were real states, with separate capitals, financial systems, legal and political constitutions, foreign services and so forth. However, they were loosely combined in what we would today call an intergovernmental organization – the German Diet, in Frankfurt am Main. German nationalism had been encouraged up to 1815 in the face of French occupation and oppression. One point to make about the political settlement of 1815 was that it was a frustrating outcome for nationalists, and borders reflected security priorities as well as monarchical claims, rather than cultural or linguistic commonality. From 1815 on, nationalist agitation made itself felt, but was firmly suppressed. There was not one capital city – not one centre around which margins could be defined – but 39. Moreover, 39 was a huge simplification of the situation prior to 1806, when there had been some 300 separate political entities in the same geographic area.

The states after 1815 were in many ways new amalgamations of older regions and states, new units which now faced serious challenges of internal integration and questions regarding their right to exist at all. Such was particularly, but not exclusively, the case for Prussia. Here, the older lands of the Prussian monarchs had now been added to with part of Saxony (which had backed Napoleon for too long, and was now made to pay for its mistake) as well as – more significantly – by the Rhineland and Westphalia, an addition which was viewed as ensuring that Prussia was committed to meeting any revival of the French threat.[10]

Second, the economic context: after years of war, and with the radical changes in borders, the German states after 1815 were forced to alter their position on economic questions and, to some extent, explore innovative new solutions to their problems. On the one hand, and generally speaking, governments began to favour more keenly the growth of industry – both artisan and mechanized – and the mercantile sector as sources of income. Old-style patrimony was displaced by new-style state-promoted industrialization. Within this was an interest in the promotion of technical education, which grew outwards from interested bureaucrats and politicians to become an educational movement, supported by industrialists themselves, but also by nationalists, who increasingly called for the creation of a German national economy, rather than one characterized by fragmented markets. Exhibitions – inspired by the French example – were at the heart of these developments, and featured for example in the writings of the most prominent and well-known national economist, Friedrich List, whose agitation was at its most active between 1819 and his death in 1846.[11]

The new borders of 1815 created serious economic disruption. They were often drawn with little view to economic realities. Customs controls were now introduced for the first time on state borders – again following French models. The solution, after twenty years of complex negotiation and friction, was the German Customs Union – *Zollverein* – founded in 1834. This united the markets of states, got rid of customs dues, and located them instead at one common external border. The Customs Union is an important factor to consider when studying the exhibition of 1844. So, too, is the fact that several German states had yet to join it.

This, then, was the context of the many local and state exhibitions that took place in the German states between 1815 and 1844. Generally speaking, their purpose was technical education and the dissemination of market information. They were organized by state governments, but also by a mixture of enlightenment-style or polytechnical societies inspired by the British Society of Arts and mechanics' institutes. Such societies were located at almost all the towns and cities throughout the German states, and their existence and activities are another great under-explored area of historical research. Though carefully observed and circumscribed by the state, they were highly active, and – despite the absolutist atmosphere – often implicitly, if not explicitly, inspired by the national economists. This mixture of state and non-state action had already brought about exhibitions in Berlin in 1822, 1827 and 1840, largely focused on technical education and restricted to Prussian manufacturers.[12]

1844: A New Stage

Exhibitions hitherto had contained goods from one state or locality. The lack of good transportation, and localized, mainly artisan forms of production were one reason for this. Another, however, was the complexity of taking goods from one state to another when one considers the multiplicity of customs and commercial laws in existence. The exhibition of 1844, however, was the first exhibition to apply to all the German states. When one considers the complexity of the political and economic backdrop, and the challenge of reaching agreement between 39 sovereign states, the achievement of the exhibition's organizers becomes apparent. Furthermore, while the Great Exhibition was the first truly international exhibition, the Berlin exhibition of 1844 was nevertheless the outcome of negotiation and cooperation between many states.

How did the step-change in 1844 come about? First, national economists had repeatedly suggested an all-German exhibition, and it was no surprise, therefore, that suggestions along these lines were repeatedly made to several of the German governments, particularly on the occasions that local exhibitions took place. After an exhibition in Kassel in 1839, the Prussian envoy to the court there reported back to Berlin that manufacturers had proposed to him that Prussia put on an all-German exhibition.[13] The response in Berlin in spring 1840, however, was cool. There was some debate about the practical complexities involved, not least the fact there was no suitable space for such a large event. There was also scepticism regarding the event's usefulness and worries about how it might be financed. Financing exhibitions via subscriptions and housing industrial exhibitions in temporary constructions lay in the future.

In less than two years, however, a reversal took place in the Prussian position. One major factor in this was the sudden possibility that emerged that Bavaria, and not Prussia, would take the initiative. Rivalry between the German states was rife and complicated, and particularly so between Prussia and Bavaria. Nevertheless, one benefit of rivalry was that it encouraged reform and innovation. After reporting on the Paris exhibition of 1839 and that in Nuremberg in 1840, one of Bavaria's leading commercial politicians and economists, Dr Friedrich Hermann, pointed out the growing popularity of such events and underlined the advantages France and Britain gained from a united national market. He advocated a German national exhibition, not least to meet the commercial threat felt by German producers from British and French industry.[14] While Prussia was ambivalent about the idea, and indeed regarding industrialization, the Bavarian government pursued a more decisive policy of economic reform and education, characterized by new museums, galleries and art schools. The Bavarian Interior Ministry supported Dr Hermann's recommendations and proposed the matter be taken up by Bavaria at the biennial conference of the German Customs Union.[15]

With its King's backing, Bavaria drew up a proposal.[16] This outlined the arguments in favour of a Customs Union exhibition. Three reasons put forward for the expanded exhibitions stand out: first, that consumers and industrial producers would be educated regarding each others' needs. Second, that after the foundation of the Customs Union in 1834 an exhibition of all its members' produce would help integrate their markets and bring the states closer together. Third, that such an exhibition would help promote the status of industry as a whole. An important rationale for exhibitions was that they would help tip the balance of forces in favour of industry.[17] In this respect, Bavarian thinking presages the modernist rhetoric of progress surrounding the Great Exhibition.

Bavaria's proposal also developed the exhibition idea further in practical terms. It suggested the Customs Union hold exhibitions every ten years. Of interest within the context of discussion of marginality, as well as within debates among economic historians regarding the importance of regions to industrial development, is the suggestion that four, regional economic groupings could be identified within the German states, and that the exhibitions should circulate between the cities at the hearts of these regions: Nuremberg within a Bavarian region; Leipzig or Dresden within a Saxon region; Cologne within a Rhenish-Westphalian region; and Berlin within a Prussian region.[18]

The Bavarian proposal was laid before the Customs Union in early 1841.[19] Berlin remained cool, however, and the proposal was, for now, simply put on record.[20] However, between 1841 and 1842 things changed quickly. First, governmental changes in Berlin meant the tide turned yet again there in favour of industrializing reforms.[21] Second, one of the many business societies across the German states, based in the Grand Duchy of Hessen in Mainz, in light of Bavaria's suggestions, proposed extending what was originally envisaged as a local exhibition in 1842 to all German states. This unofficial and somewhat impromptu project provided a lightning rod, however, for public discussion about a German national exhibition and was a great success.

Across Europe interest in exhibitions was strengthening – as expressed in Paris (1839), Nuremberg (1840), and now in Mainz in 1842. Industrialists encouraged governments to become involved. Official support for industrial projects certainly offered tantalizing rewards. The Hessian government had given the Mainz Society tacit backing. For Prussia, one of these rewards was the possibility that by strengthening its support base among industrialists, it might also put itself more firmly in the position of economic leadership of Germany. While nationalism was still a revolutionary ideology at all the conservative German courts, Prussian rivalry with Austria over leadership of Germany was now intensifying, and economics provided one arena where Prussia could assert itself.

Bavaria resubmitted its proposal to the Customs Union congress in 1842. After internal discussion between ministries, Berlin instructed its delegate to point to the severe difficulties that would be associated with any Customs Union exhibition, but now backed Bavaria's proposals in principle, and even engaged with its practical suggestions. Bavaria had suggested such exhibitions should be paid for from the revenues of the Union. Prussia, however, proposed that hosting states should carry the costs. Prussia noted Bavaria's proposals regarding four regional exhibitions, and even allowed that the first might be in 1844. Expressing its continued mixed feelings on the matter, however, it suggested a lottery decide where the first should happen.[22]

Agreement was nevertheless finally reached to hold periodic exhibitions, as well as regarding a set of stipulations on the matter. Given the difference of governments' positions, it is no surprise these remained general. Rather than the regional proposal, any state was allowed to propose an exhibition, as long as it communicated its intentions and the criteria of the exhibition with six months' notice. The costs of exhibitions would be sorted out and agreed as they arose. The host state would set up a commission to choose exhibits. The other states would appoint commissioners to assist sending of exhibits.[23] Official Customs Union exhibitions could not be held any more frequently than at five-year intervals. Still, while general, this set of agreements was significant. They constituted a powerful framework within which not just the exhibition of 1844 but also that of 1854 in Munich would be held. For the first time in history a considerable number of sovereign states had united to hold joint exhibitions. Exhibitions were thus becoming at least marginally international.

The Customs Union had also established its authority regarding multi-state exhibitions in Central Europe. The organizers of the Mainz exhibition were forced to state that their undertaking was not part of the Customs Union framework and did not prohibit a full-blown Customs Union exhibition.

Despite the Customs Union agreement, the initiative still seemed to rest with Bavaria. In November, however, reflecting particularly now on the success and publicity generated by the Mainz exhibition of 1842, the Prussian government began to take the lead. Prussia, rather than any other state, should host the event. Public opinion now obviously valued exhibitions, and Customs Union governments – rather than being threatened by exhibitions – could only benefit from them. If Prussia did not act speedily, meanwhile the prize would go to Munich. There were still concerns about where such an exhibition might be held, as Berlin did not seem to have any kind of space big enough. Bavaria had already stated that it had large enough halls at Nüremberg. By December 1843, however, the government had identified room at the Academy of Art in Berlin. Prussia communicated its general intentions of hosting an exhibition in Berlin to the other Customs Union governments on 9 January 1844 via diplomatic channels.[24]

Over the next two months, plans were drawn up and the actual nature of the event was determined. The formal declaration on this – the so-called *Publikandum* – was transmitted to the other governments in mid-February. The exhibition was to last eight weeks from 15 August onwards. Goods had to arrive by 22 July. The conditions of goods to display were very loosely set: fine arts were excluded, but otherwise any article could be displayed which found a wide market in the area of the Customs Union and was characterized by a high level of production in relation to its price. Domestic producers in Prussia and outside Berlin should register their desire to exhibit with their local government official. A central commission in Berlin would do the same for Berlin but also recommend to the government which goods should be accepted. Prices of goods would be displayed, as would the prices of raw

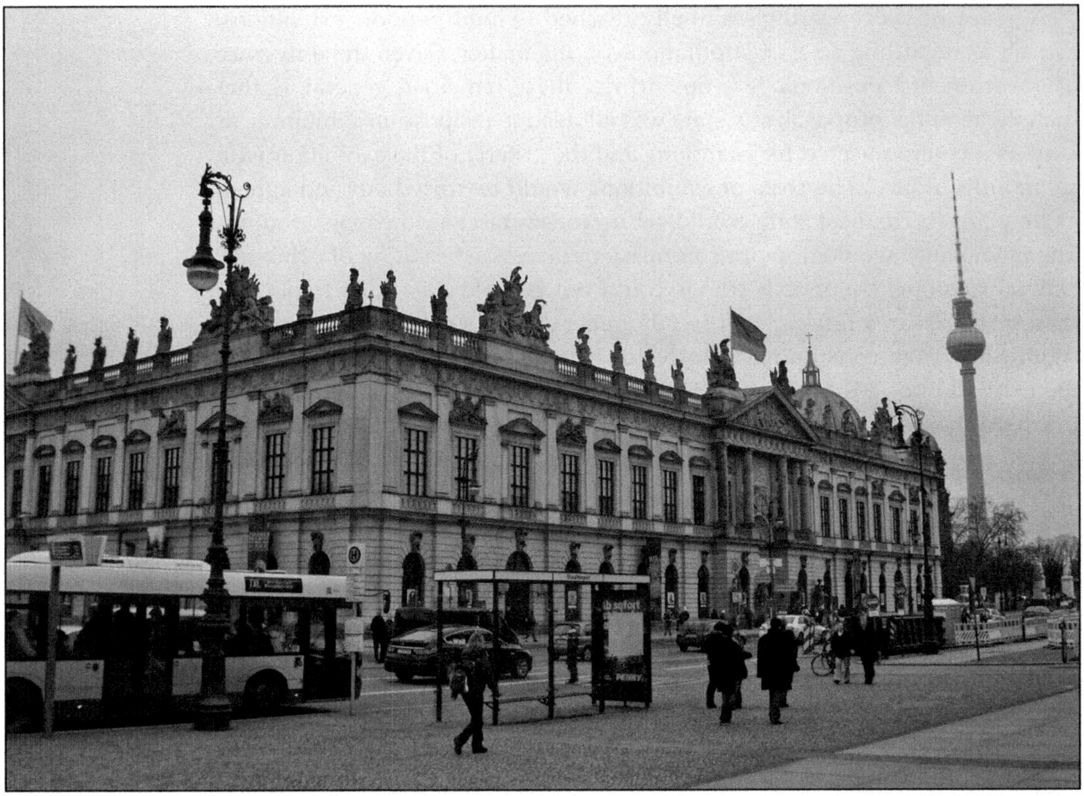

3.1 The *Zeughaus* (Armoury) in Berlin, viewed from the front, 2012, photograph. Author's collection

materials used for production, and goods could be bought, but not removed until the end of the exhibition. There would be entry tickets and a catalogue. The money made from these would pay for fire insurance and any excess then used to contribute to transport costs.[25]

A significant change was the decision to locate the exhibition in the Royal Armoury, the *Zeughaus*, built in 1736, instead of the Academy of Arts (Figs 3.1 and 3.2). This turned out to have many advantages, among them the strength of the building, its ability to take great weight, and its resistance to fire. There were unforeseen consequences, however. The *Zeughaus* was – and still is – covered in military emblems and even during the exhibition it remained filled with the accoutrements of war. A juxtaposition would do nothing to diminish Prussia's militaristic reputation. On the other hand, the *Zeughaus* seemed far more connected with the heart of the Prussian state and in particular with the King. The patriarchal imprimatur on an industrial enterprise – it might be argued – was important at a time when governments were still finding their way towards full support for industry and exhibitions. The triumphs of industry, it would be argued, were replacing the triumphs of war.

Probably the most significant change to the plans for the exhibition, one that transformed its meaning utterly, came at around this time. Even before the *Publikandum* had been sent out, reports had reached Berlin that Austria

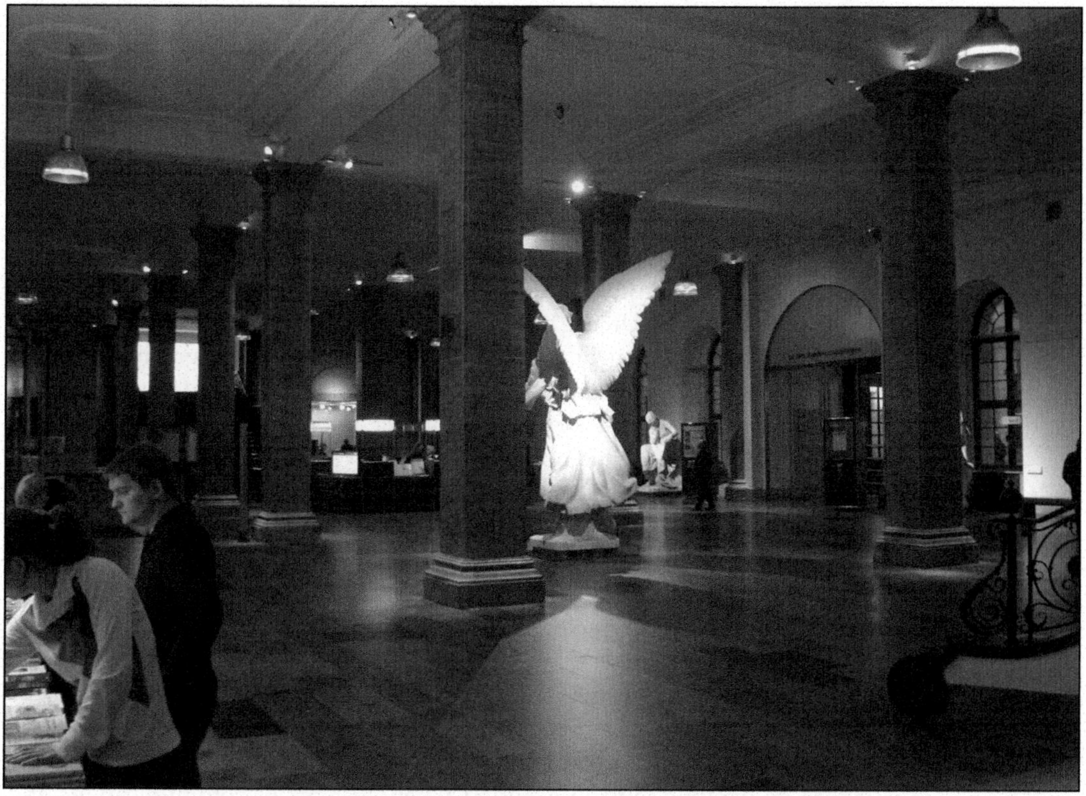

3.2 The interior of the *Zeughaus* in Berlin, 2012, photograph. Author's collection

was considering organizing an exhibition of all Germany; that is, not just of the Customs Union – of which Austria, significantly, was not a member – but all states within the Germanic Confederation as a whole. Nation-based German historical literature has downplayed the importance of economic development in the Austrian Empire, yet important steps were taking place there too, while the Austrian government was also aware that it needed to enter the industrial fray. Indeed, it seemed to hope that an all-German exhibition, as opposed to a merely Customs Union one, would pave the way to its own entry into the Customs Union – a somewhat naive expectation given the *Realpolitik* of Prussian politicians, but certainly not without great attraction to many German nationalists who saw Austria as part of Germany. A deal was offered to Berlin, whereby Austria would promote Prussia's 1844 exhibition among its own manufacturers if Prussia would do the same for an Austrian exhibition the following year.[26]

While Prussia did not want Austrian interference in the Customs Union, it recognized that the proposal for an All-German exhibition had its attractions. The Prussian King, unlike some of his ministers, had Romantic ideas of an ancient Germanic nation including Austria, and even less sentimentally included observers could appreciate the benefits of such an exhibition in Berlin to Prussia's status as the leading German state. The educational advantages of a wider exhibition were recognized. More significantly, however, Prussia might not have wished for Austrian entry to the Customs Union, but it was certainly interested in the accession of other German states in 1844, particularly that of Hanover and the Hanseatic States, which, apart from widening the Prussian market in important respects, would also secure it direct access to the North Sea.[27] The inclusion of all the states of the Confederation might be – to use a more modern term of international relations – an important confidence-building measure bringing the extension of the Customs Union in north Germany that much closer.

As the Customs Union had agreed a framework for such exhibitions, an extension to other states involved a further separate agreement on the matter.[28] It was also now necessary to agree via the Customs Union measures for handling goods free of charge coming into the Customs Union from outside – an issue that had not hitherto needed to be discussed. Nevertheless, the states unanimously agreed to the invitation to non-Customs Union states. This not only meant that Austria, Hanover and the Hanseatic states would take part, but also the two Mecklenburg states on the Baltic and – interestingly – the Grand Duchies of Luxembourg and Limburg, ruled by the Dutch King, and the principality of Schleswig under Danish rule. Though in the Confederation, these areas were marginally German. The complexity and practicalities associated with all this multi-lateral negotiation via diplomatic channels were immense. The schedule was also incredibly ambitious. States and manufacturers were only being invited to the exhibition four months before its opening, and in a period when systems of communications and transport were rudimentary and highly variable.

Still, arrangements continued in Berlin. A complex series of negotiations was undertaken relating to transport costs which could not be completely covered by the money from entry tickets. Prussia agreed to provide free postage and transport on its own railways for any goods going to Berlin, and encouraged other states to do the same or cover their exhibitors' costs. The costs of transport would then be settled as far possible with the state governments. Things were inordinately complicated as not all railways or even postal systems were state owned. Moreover, one was dealing in several different currencies in a period before bank transfers were easy to arrange. As per the agreements of 1842, a commission was set up in Berlin, and each of the states then appointed their own commissioners. Reporting to these was a series of provincial commissions in Prussia, and commissions in other states either for the state as a whole, or for provinces of the states, depending on size.

A structure for organizing the exhibition was formed, combining central and local elements. This structure was a fusion of state and non-state interests. On the one hand, the state retained the upper hand; for example a Ministerial Councillor – von Viehbahn – was put in charge of the Commission in Berlin, while in the Prussian provinces the provincial commissions were chaired by local government officials, usually those most closely involved with economic affairs. It was also accepted without much comment that the exhibition would be organized through the Finance and Foreign Ministries and via diplomatic channels. In this way, some of the patriarchal, state-led, centralist tradition of industrial education was able to assert itself. It was also notable that the Berlin Commission consisted solely of politicians and businessmen drawn from Berlin – a matter excused later in the catalogue on the basis of the expense of foreigners residing in Berlin.

On the other hand, however, the various state and provincial commissions were constructed to include important manufacturers and businessmen in each locality, and to this extent served as an extensive platform for the bringing together of state officials with industrial interests, often at a local level. Furthermore, outside Prussia, and especially outside the Customs Union, where the agreements of 1842 did not apply, states frequently devolved the role of commission partly and even fully to the local business societies, quasi-state authorities, patriotic societies and chambers of commerce.[29] Here again, one might argue, the state, and particularly the Prussian state, was putting down roots in business circles across the Germanic area. A final point, in connection with this, was that the Berlin committee was – to make arranging the exhibition less complicated – eventually given powers to communicate directly and without central government control with the committees and societies of other German states. This may appear unimportant today, but in the context of pre-1848 Germany, where non-governmental communication across borders had been severely controlled, this represented the first time such communication had been officially sanctioned.[30]

A pro-industry alliance was forming itself, therefore, in the same way as would happen in Britain in the run-up to the Great Exhibition. Just as in Britain, however, there was opposition to the exhibition, an opposition that would follow through into criticism of the exhibition's display once it had opened. Protectionism, rooted in support for localized and artisan economic structures, as well as in fear of competition, made itself felt, not least in the cool response to Prussia's invitation in the early summer of 1844. Such opposition severely limited the representation of some states, and completely prevented that of others.[31] There was resistance – even within Prussia from the Rhineland, which was both newly acquired and also had its own proud authority in industrial matters – to what were felt were overly centralized and bureaucratic instructions emanating from Berlin.[32]

It is not always easy to extrapolate from such discussions what was valid and what was not valid criticism, however. Certainly, in the Prussian government's defence, the exhibition was organized at high speed, and was completely unprecedented with regard to its complexity. As the *Aachener Zeitung* (a newspaper in the Rhineland) argued, a significant difference between exhibitions in France and this one now being organized in Germany was that when Paris officials sent out an instruction to the Departments, the provinces had to obey. In the German states, however, one was dealing with a multiplicity of sovereignties.[33] Nevertheless, even if the Prussian government had been centralist to start with, it had made compromises and concessions on arrangements in order to secure support. For example, though not originally intended, Prussia had quickly recognized that the awarding of prizes, medals and honourable mentions would attract industrialists to the event more strongly.[34] This was more significant than obvious at first sight: the Prussian King, after all, was now rewarding the citizens of foreign countries. Moreover, once the Prussian government had decided to put on the event, it had unreservedly and with a great deal of generosity supported an all-German exhibition.[35]

Comparisons were unavoidably drawn with the generosity of the Mainz exhibition's organizers, who had paid all transport costs, but this was hardly possible given the size and scope of the Berlin exhibition.[36] There was a thorny discussion with Bavaria on this, as well as on whether or not the Berlin commissioners would cover insurance for damaged goods, and also on whether or not producers should also show the price of their goods and their raw materials – requirements that Berlin also conceded could be voluntary rather than essential.[37] But it is hard to tell whether these grievances in Munich were real, a cover for protectionism among manufacturers, or fuelled by resentment that Bavaria's exhibition project had been hijacked by Berlin rather than by real concerns.[38] Whether excusable or not, there certainly were shortcomings in the way the first all-German exhibition was organized, and these would make themselves obvious in the appearance of the exhibition once it opened and in its public reception.

The Display

The exhibition opened, as planned, on 15 August. It was held on two floors of the *Zeughaus*, with heavier goods downstairs, and lighter ones in the upper storey (Fig. 3.3). After entering through the militaristic portal, and passing the ticket desks and barriers on the left and the offices of the Commission on the right, one was guided immediately upstairs via a purpose-built staircase in the courtyard and was greeted on the first floor by a bust of the King, Frederick William IV, surrounded by a display of flowers. Taking up the front half of the building – the back still contained the military artefacts – were textiles, clothing, jewels, chandeliers, light leather goods, flowers, feathers, dolls, lacework, typographical equipment, musical instruments, as well as lighter mechanical equipment for surgical, chronological and other purposes. Following a one-way system, one was then guided back downstairs where one found printing machinery and looms, locomotives, brass, knives, needles, heavier metal work and leather goods, stoneware and granite.[39] Outside the exhibition, on the river Spree which ran alongside the exhibition building, were several steam ships on display.

If any principle for the layout was to be identified, it appeared an aesthetic and commercial one rather than one which allowed comparison between

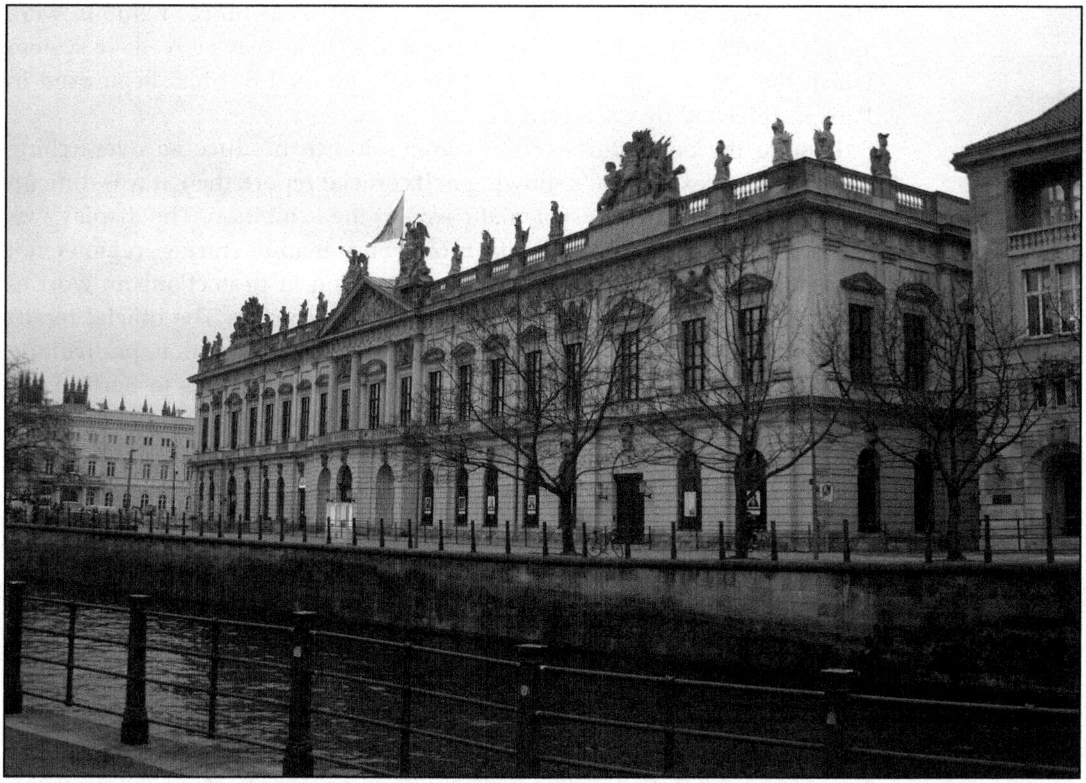

3.3 The *Zeughaus*, viewed from the side, across the river Spree, 2012, photograph. Author's collection

categories of goods or even between countries. The central commission had been divided into eight categories to look at different kinds of produce. However, such a division was not obvious in the exhibition. Rather, goods appeared in general groupings of category, arranged to look decorative. Where possible – as for example with silks – goods were separated according to country. Often, however, companies would display all their goods together. Furthermore, by arranging goods aesthetically, and moulding local displays into a national whole, the Berlin exhibition began to explore – accidentally and on purpose – a new, popular mode of arrangement that had less to do with education, and more with the kind of iconographic displays of later Great Exhibitions.

A crucial problem for the commission, however, was that a large number of goods arrived late for the exhibition, partly in consequence of the hurry with which it had been organized and complexity of sending goods there, but also because it seemed enthusiasm grew strongly for the event in the days running up to its opening. Rather than turn these goods away, out of misguided enthusiasm and, as the official report accepted, in light of the higher national goal of the exhibition, they were accepted, and goods moved aside to make way for them. This however, also made it impossible to produce a definitive catalogue on time for the exhibition. While a catalogue was available at the opening, it had to be added to with regular additional sheets – available at no cost from the commission office. Exhibits were simply numbered in the catalogue, and added to as they arrived – a system which the official report rather defensively pointed out had been used in Paris in 1839 and Brussels in 1841.

Despite the best efforts of the commission to produce an overarching analysis of the exhibition's showing in its official report, then, it was difficult to relate this back in any systematic way to the exhibition. The display was patchy, with important manufacturers and manufacturing regions not present. This no doubt found some explanation in protectionism, worries about displaying new products, and mutual suspicion. The official report was probably right to point to the difficulties of transportation, particularly as those areas that already had good railway or road links to Berlin were better represented. Still, the image of the latest standing of German industry was somewhat less reliably conveyed than had been hoped.

Reception

Much of this was reflected in the reception of the exhibition, though the public's reaction was harder to gauge in a country where censorship limited newspaper publication, and where political and territorial boundaries were many. It is obvious that the fact that this was the first all-German exhibition, indeed the fact it had come about at all, created a strong impact, particularly in official circles, but also among Prussian patriots, German nationalists, and national economists. Great praise was given to the exhibition in official newspapers. The idea that the exhibition conveyed a picture of German national economic development, and also supported the promotion of German national industry, was received positively – if guardedly – in official publications, enthusiastically in the exhibition's own report, and emotionally in more popular publications. The exhibition, it was believed, had begun to make German industrial potential visible. Its invisibility had been a long-standing grievance of nationalists and national economists alike. This first opportunity of looking at German industry as a whole unleashed a huge outpouring of comment on the reflection seen in the exhibition, a self-reflexive discourse that of course was seen again in 1851.

On the other hand, national economists, while delighted that such an exhibition had taken place, could not deny the exhibition's technical weaknesses when it came to enabling study and improvement of technical development and market forces. Prussia was also criticized for not having sacrificed its own particular interests more in favour of the national cause. Given that this exhibition still stood in the context, if on the margins, of exhibitions for the purpose of technical education, its weaknesses in this regard even had to be recognized by Prussian officials and in official Prussian newspapers.[40]

Meanwhile, in a context of economic and social transformation, such an aesthetic display of industrial produce caused comment in many quarters. As Karl Marx is quoted on the Great Exhibition of 1851, Ferdinand Lassalle, a contemporary and founder of German socialism, was similarly forced by the Berlin exhibition to reflect on the meaning of what was on show there. He acknowledged that the exhibition had captured the attention of the masses – as he put it the 'thoughtless masses who now push themselves into the halls of the *Zeughaus* with mouths gaping open', but he clearly recognized the modernist rhetorical message conveyed by the display.[41]

The Berlin exhibition also had an impact. For Prussia, it was an important milestone in its ascendancy as the economic leader of the German states. Given its timing it was also an important contribution to the integration of the German Customs Union. Economically, though difficult to quantify, and despite its weaknesses, the exhibition will have been important in terms of technology transfer. Commercially, manufacturers and merchants became better known across state borders, and some companies – for example Alfred Krupp – owed their later success largely to Berlin.[42] Politically, the exhibition and its commissions fostered all sorts of personal contacts between commercial politicians and industrialists at the local, regional and German level. It began to create a class and a political grouping that would be highly significant in later German history – the national liberals that would be so important to facilitating Bismarck's unification of Germany in 1871.

Two examples of impact are particularly worth noting: when the Society of Arts in London decided to explore the possibility of a Great Exhibition, its Secretary, John Scott Russell, undertook a research trip to Germany, furnished with introductions from Prince Albert, partly to find out whether the German states would collaborate in 1851, and partly to find out more about exhibitions there. He reported:

> he found the liberality, wisdom and advantages of the great plan everywhere appreciated – He found every disposition in the most influential quarters, to afford cooperation; and he found at Berlin, where already the greatest advances had been made in a similar direction, an organization existing, well calculated to secure effective cooperation, and which the Minister expressed his readiness to put in motion for so important and interesting a purpose.[43]

The Berlin exhibition of 1844 was an important influence on the organization of the exhibition of 1851, not least because – though national from the viewpoint of most historical discourse – it was the first exhibition to come about through multilateral negotiation.

Second: the exhibition of 1844 was important in establishing a set of procedures in the German Customs Union for the holding of all-German exhibitions. It also established an expectation. Despite all the turbulence of 1848, and despite the strong influences now of the London exhibition of 1851, the Munich exhibition of 1854 sat squarely within the organizational framework of the Customs Union established in 1844, and was able to make use of its established structures and procedures. Meanwhile, an important visitor to the exhibition in Berlin in 1844 had been, once again, Dr Hermann of Munich. Like many others, he would see clearly the failings of the exhibition in terms of industrial education and these obviously left a deep impression. While borrowing from Berlin, and learning from London, Hermann also ensured that the mistakes made in Berlin were not repeated this time round. His preparatory reports clearly express his intentions in this regard.

Conclusion

The Berlin exhibition of 1844 was on the margins in a variety of respects, therefore. It was part of that huge number of local and regional exhibitions that took place before and after in Europe, but which rarely receive focused attention. It was on the cusp of the step towards international exhibitions. In terms of exhibition curatorship it stood at the crossroads between technical education, market forces and conceptual display. In historical terms, it was national and non-national, political and economic, pre-modern and modern. For all these reasons, as with other marginal exhibitions, it has not received the historical attention it deserves.

Notes

1 For an overview of the historiography of exhibitions, see Jeffrey Auerbach, *The Great Exhibition of 1851: A Nation on Display* (New Haven and London, 1999), esp. ch. 8; and John R. Davis, *The Great Exhibition of 1851* (Stroud, 1999), Introduction. See also John F. Findling and Kimberly D. Pelle, eds, *Encyclopedia of World's Fairs and Expositions* (London, 2008).

2 For example, J. M. Blackman, 'The Woollen and Worsted Industries', in Derek H. Aldcroft, ed., *The Development of British Industry and Foreign Competition, 1875–1914* (London, 1978), 146.

3 See Auerbach. Also see: Burton Benedict, 'International Exhibitions and National Identity', *Anthropology Today* 7, no. 3 (1991): 5–9; Abigail Green, 'Representing Germany? The Zollverein at the World Exhibitions, 1851–1862', *Journal of Modern History* 75, no. 4 (2003): 836–863.

4 Note the terminology used in the official catalogue of the Great Exhibition: *Official Catalogue of the Great Exhibition of 1851* (5 vols, London, 1851).

5 See Green. For a wider perspective see Jeffrey A. Auerbach and Peter H. Hoffenberg, eds, *Britain, the Empire and the World at the Great Exhibition of 1851* (Aldershot, 2008).

6 Kenneth Carpenter, 'European Industrial Exhibitions before 1851 and Their Publications', *Technology and Culture* 13 (1972): 465–486.

7 Toshio Kusamitsu, 'Great Exhibitions before 1851', *History Workshop Journal* 9, no. 1 (1980): 70–89.

8 Nils Brübach, *Die Reichsmessen von Frankfurt am Main, Leipzig und Braunschweig (14.–18. Jahrhundert)* (Stuttgart, 1994).

9 Thomas Großbölting, *'Im Reich der Arbeit.' Die Repräsentation gesellschaftlicher Ordnung in den deutschen Industrie- und Gewerbeausstellungen 1790–1914* (Munich, 2008). See also one of the most detailed histories of exhibitions: Utz Haltern, *Die Londoner Weltausstellung von 1851. Ein Beitrag zur Geschichte der bürgerlich-industriellen Gesellschaft im 19. Jahrhundert* (Münster, 1971).

10 For a recent overview of Prussian history, see Christopher Clark, *Iron Kingdom. The Rise and Downfall of Prussia, 1600–1947* (London, 2006).

11 For the position of exhibitions in technical education, see: Ingeborg Cleve, *Geschmack, Kunst und Konsum. Kulturpolitik als Wirtschaftspolitik in Frankreich und Württemberg (1805–1845)* (Göttingen, 1996); Großbölting.

12 *Amtlicher Bericht über die allgemeine deutsche Gewerbe-Ausstellung zu Berlin im Jahre 1844* (3 vols, Berlin, 1845), 1: 5–7.

13 6.5.1840, Ministry of Foreign Affairs to Alvensleben and 16.5.40 Alvensleben to Foreign Affairs. Geheimes Staatsarchiv Preußischer Kulturbesitz, Dahlem, Berlin (GSPK).

14 11.5.1840, Excerpt from the report of Dr Hermann, MH9217 Industrie-Ausstellungen des gesamten deutschen Zollgebiets Bd.1 1839–44, Bavarian Hauptstaatsarchiv, Munich (BH). See also 19.9.1840 Graf Dönhoff to Ministry of Foreign Affairs, GSPK. Dönhoff reports an article on the subject in the *Allgemeine Zeitung* purportedly authored by Hermann.

15 1.4.1840, Antrag des Königlichen Ministeriums des Innern, MH9217 Industrie-Ausstellungen des gesamten deutschen Zollgebiets Bd.1 1839-44, GSPK. See also W. O. Henderson, *The Zollverein* (London, 1968), 141.

16 4.4.1840, Ludwig to Ministries of Foreign Affairs and Interior, MH9217 Industrie-Ausstellungen des gesamten deutschen Zollgebiets vol. 1, 1839–44, GSPK.

17 15.6.1840, von Abel (Minister of Finance), Voltz and Dr Hermann, MH9217 Industrie-Ausstellungen des gesamten deutschen Zollgebiets vol. 1, 1839–44, GSPK.

18 For the impact of exhibitions on regions, see Claus H. Boerner, 'Messen und Ausstellungen. Impulsgeber und Indikatoren für die Wirtschaft einer Region', *Wirtschaft und Standort* 10 (1978): 7–9.

19 Das Hauptprotokoll der 4.ten General-Conferenz in Zollvereins-Angelegenheiten. Contained in 2.1.1841. Ministry of Foreign Affairs to Alvensleben. GSPK.

20 13.1.1841. Alvensleben to Ministry of Foreign Affairs, MH9217, GSPK.

21 For the political background to this, see Eric Dorn Brose, *The Politics of Technological Change in Prussia. Out of the Shadow of Antiquity, 1809–1848* (Princeton, 1993).

22 9.7.1842, Bülow and Beuth to Pommer-Esche, MH9217, GSPK.

23 Paragraph 25, Hauptprotokoll of the General Conference of the *Zollverein*, MH9217, GSPK.

24 9.1.1844. Bülow to Representatives in Munich, Dresden, Stuttgart, Carlsruhe, Cassel, Darmstadt, Frankfurt am Main, etc. MH9217, GSPK.

25 *Publikandum*. Contained in 10.2.1844, Bodelschwingh to Ministry of Foreign Affairs, MH9217, GSPK.

26 The Austrian proposal came from Prof, Reutter, the Austrian Commissioner appointed as official reporter on the Mainz exhibition of 1842. 27.1.1844, Diergardt to Bülow, MH9217, GSPK.

27 14.2.1844, Bülow to Bodelschwingh, MH9217, GSPK.

28 31.3.1844, Ministry of Foreign Affairs to Munich, Stuttgart, Dresden et al., MH9217, GSPK.

29 Hannover used its Patriotische Gesellschaft. Baden used its Centralstelle für die Landwirtschaft. Luxembourg used its chamber of commerce. See the list of local commissions set up contained in 31.5.1844, Philipsborn to Ministry of Foreign Affairs, MH9217, GSPK.

30 See *Amtlicher Bericht*, 1: 30.

31 The state of Waldeck, for example, was forced to admit to Prussia that no manufacturers at all had registered interest in the exhibition: 19.6.1844. Hagemann to Ministry of Foreign Affairs, MH9217, GSPK. There were also no goods received from Hessen-Homburg.

32 23.5.1844. Diergardt to Ministry of Foreign Affairs, MH9217, GSPK.

33 23.5.1844, *Aachener Zeitung*.

34 *Amtlicher Bericht*, 1: 20.

35 9.6.1844, Bülow to Diergardt, MH9217, GSPK.

36 14.7.1844, Frankfurter Journal, MH9217, GSPK.

37 28.5.1844, *Allgemeine Preussische Zeitung*.

38 6.6.1844, Küster to Ministry of Foreign Affairs, MH9217, GSPK.

39 *Amtlicher Bericht*, 1: 55.

40 See, for example, the analysis in 7.11.1844, *Vossische Zeitung*.

41 6 September 1844, Ferdinand Lassalle to his father, at http://www.historische-kommission-muenchen-editionen.de/lassalle/anzeige/seite.php?id=L-001-0125, accessed 15 April 2015.

42 Particularly significant is Krupp. See Wilhelm Berdrow, ed., *The Letters of Alfred Krupp, 1826–1887* (London, 1930); *Krupp. A Century's History of the Krupp Works. 1812. 1912. Translated from the Commemorative Volume by the Krupp Works*. At http://www.archive.org/stream/cu31924030704880/cu31924030704880_djvu.txt, accessed 15 April 2015.

43 1.8.1849, Notes of a Meeting at Osborne House. Private and Confidential. RSA Archive, London.

Bibliography

Manuscript Sources

Gewerbeausstellung in Berlin. 2.4.2. Handelspolitischer Abteilung. Geheimes Staatsarchiv Preußischer Kulturbesitz, Dahlem, Berlin, Nr. 1287–1290.

Industrie-Ausstellungen des gesamten deutschen Zollgebiets, Volume 1, 1839–44. Trade Ministry. Bavarian Hauptstaatsarchiv, Munich, MH9217.

John Scott Russell Papers, Volume 1. Royal Society of Arts Archive, London.

Printed Primary Sources

Amtlicher Bericht über die allgemeine deutsche Gewerbe-Ausstellung zu Berlin im Jahre 1844. Berlin: Reimarus, 1845.

Official Catalogue of the Great Exhibition of 1851. London: Spicer, 1851, 5 volumes.

Secondary Sources

Auerbach, Jeffrey. *The Great Exhibition of 1851: A Nation on Display*. New Haven and London: Yale University Press, 1999.

Auerbach, Jeffrey, and Peter H. Hoffenberg, eds. *Britain, the Empire and the World at the Great Exhibition of 1851*. Aldershot: Ashgate, 2008.

Benedict, Burton. 'International Exhibitions and National Identity'. *Anthropology Today* 7, no. 3 (1991): 5–9.

Berdrow, Wilhelm, ed. *The Letters of Alfred Krupp, 1826–1887*. London: Victor Gollanz, 1930.

Blackman, J. M. 'The Woollen and Worsted Industries'. In *The Development of British Industry and Foreign Competition, 1875–1914*, edited by Derek H. Aldcroft, 128–158. London: George Allen and Unwin, 1978.

Boerner, Claus H. 'Messen und Ausstellungen. Impulsgeber und Indikatoren für die Wirtschaft einer Region'. *Wirtschaft und Standort* 10 (1978): 7–9.

Brose, Eric Dorn. *The Politics of Technological Change in Prussia. Out of the Shadow of Antiquity, 1809–1848*. Princeton: Princeton University Press, 1993.

Brübach, Nils. *Die Reichsmessen von Frankfurt am Main, Leipzig und Braunschweig (14.–18. Jahrhundert)*. Stuttgart: Franz Steiner, 1994.

Carpenter, Kenneth. 'European Industrial Exhibitions before 1851 and Their Publications'. *Technology and Culture* 13 (1972): 465–486.

Clark, Christopher. *Iron Kingdom. The Rise and Downfall of Prussia, 1600–1947*. London: Allen Lane, 2006.

Cleve, Ingeborg. *Geschmack, Kunst und Konsum. Kulturpolitik als Wirtschaftspolitik in Frankreich und Württemberg (1805–1845)*. Göttingen: Vandenhoek und Ruprecht, 1996.

Davis, John R. *The Great Exhibition of 1851*. Sutton: Stroud, 1999.

Findling, John F., and Kimberly D. Pelle, eds. *Encyclopedia of World's Fairs and Expositions*. London: McFarland, 2008.

Green, Abigail. 'Representing Germany? The Zollverein at the World Exhibitions, 1851–1862'. *Journal of Modern History* 75, no. 4 (2003): 836–863.

Großbölting, Thomas. *'Im Reich der Arbeit.' Die Repräsentation gesellschaftlicher Ordnung in den deutschen Industrie- und Gewerbeausstellungen 1790–1914*. Munich: Oldenbourg, 2008.

Haltern, Utz. *Die Londoner Weltausstellung von 1851. Ein Beitrag zur Geschichte der bürgerlich-industriellen Gesellschaft im 19. Jahrhundert*. Münster: Aschendorff, 1971.

Henderson, W. O. *The Zollverein*. London: Frank Cass, 1968.

Krupp. A Century's History of the Krupp Works. 1812. 1912. Translated from the Commemorative Volume Edited by the Krupp Works. At http://www.archive.org/stream/cu31924030704880/cu31924030704880_djvu.txt, accessed on 16 March 2012.

Kusamitsu, Toshio. 'Great Exhibitions before 1851'. *History Workshop Journal* 9, no. 1 (1980): 70–89.

CONSTRUCTING IDENTITIES

4

'Witness to the Momentous Significance of German Labour in Bohemia': Exhibitions in the German-Speaking Regions of Bohemia before the First World War

Tomáš Okurka

Introduction

Before the First World War, the northern, north-western and western parts of Bohemia formed the industrial heart of the entire Habsburg Empire. The region was home to thriving textile, chemical, food and glassmaking industries, porcelain production, coal mining and other branches of industry. Its ethnic composition was also noteworthy. Within the multi-ethnic Habsburg Monarchy, Bohemia had a prevailingly ethnic Czech population, but the northern, north-western and western border regions were inhabited mostly by Germans. Over the course of the nineteenth century, a dynamic process of national self-determination in Czech culture, politics and economics took place, during which many Germans in Bohemia felt they lost their hitherto unshakeably dominant position in society. They increasingly began to see their status of an ethnic minority within Bohemia as a threat to their interests and perhaps even to their very existence. As ethnic tensions between Czechs and Germans grew, the political programmes of the representatives of Bohemian Germans began to be centred on calls for the administrative separation of mainly German-speaking districts in Bohemia or for the creation of autonomous provinces in the border regions where Germans could form a majority population. This was rejected by Czech national political officials, who came out unequivocally against any division of territory belonging to the historical Kingdom of Bohemia.[1]

This text concentrates on the national and political aspects of German industrial and general exhibitions held in Bohemia between the 1890s and the First World War. National and political aims were common features of exhibition trends in the late nineteenth and early twentieth centuries and became particularly visible at international exhibitions. World fairs in this period began to be staged under an increasingly confrontational atmosphere and from the 1880s onwards a particularly prominent theme at these events and at special colonial exhibitions was the demonstration of colonial, and therefore economic and political, power. However, even provincial and regional exhibitions were marked by a strong nationalist and political tenor.[2] This contribution therefore offers a closer examination of the specific national and political aims of regional exhibitions in a climate profoundly marked by conflict between two ethnic groups living in one land.

The First Exhibitions in the German-Speaking Regions of Bohemia

I shall begin, however, with a brief, general introduction to the early stages of exhibition activities in the German-speaking regions of Bohemia in order to outline the main reasons and motivations behind these events. One of the earliest industrial exhibitions was held at a hall of a shooting range in the north Bohemian town of Liberec (known as Reichenberg in German), a major centre of textile production. Taking place even before the middle of the nineteenth century, in 1841, the majority of participants were manufacturers from Liberec and the town's immediate vicinity. Another regional type of exhibition was held in Liberec in 1852, this time held in a school building with approximately 240 exhibitors attending.[3]

From the 1870s the number of exhibitions organized in Bohemia soared, driven by the rapid development of activities initiated by new associations, which followed the introduction of the Act on Associations in 1867, and by the institution of municipal self-government which followed the introduction of the Act on Municipal Self-Government in 1862. Some of the most prominent organizers and supporters of regional exhibitions in Bohemia between the 1870s and the First World War were:

a) trade associations and federations, which were usually the initiators and in some cases also the organizers of exhibitions;

b) bodies of the municipal and district authorities, which usually provided the exhibitions with financial support, the use of land or buildings as exhibition sites free of charge, and whose members tended to sit on exhibition committees and oftentimes occupied the most influential posts;

c) chambers of commerce and manufacturing, which were the representative bodies of business circles and sought to promote the interests of industry, trade and business; the chambers supported many of the exhibitions either with funding or by sponsoring prizes at the events, their representatives usually held honorary positions in the exhibition committees, and in exceptional cases a chamber would even organize an exhibition itself (for example, in Cheb (Eger) in 1881 and 1892).

State authorities also supported exhibitions either with subsidies or national prizes, but they were not among the major actors in this area. For instance, the provincial government played an important role in organizing the Jubilee Exhibition that took place in Prague in 1891, which became an important event for the Czechs of Bohemia and which will be examined later. The support of regional exhibitions was limited mainly to the provision of subsidies, but even that was far from customary.

In the 1870s and 1880s many smaller industrial and agricultural exhibitions were held in northern regions of Bohemia and the organizers were for the most part local trade, industrial and agricultural associations or chambers of commerce and manufacturing. Some of these exhibitions were intended exclusively for members of the local trade association, but in most cases they were open to a wider circle of exhibitors. Regional producers were nevertheless always the dominant presence at these events. As exhibition venues they tended to use school buildings, athletics halls, pubs or some other extant building and, when necessary, several temporary exhibition halls were built on the grounds around the selected venue. An exhibition usually lasted for around one month, and sometimes for a shorter duration. The organizers frequently cited the goals of such exhibitions as being to provide information on the state of the art of production and technical innovations, to raise the level of taste, to mediate contacts between producers and consumers, to provide opportunities to make new business contacts, and to generally revive regional production and trade; these early exhibitions, including events in Cheb in 1871 and 1881, in Litoměřice (Leitmeritz) in 1877, in Česká Lípa (Böhmisch Leipa) in 1880, in Teplice (Teplitz) in 1875, 1879 and 1884, in Ústí nad Labem (Aussig) in 1874 and 1880, did not yet possess the nationalist tone that surfaced later on. And it is worth noting that no major exhibition was held in Liberec in the late nineteenth century, despite it being one of the main industrial centres of northern Bohemia. From 1873, however, the Museum of Applied Arts in Liberec largely fulfilled the same function.[4]

Czech Exhibitions in Prague

The Jubilee Provincial Exhibition in Prague in 1891 was an event of fundamental significance for the development of exhibition activities in Bohemia. It was initially conceived as a joint undertaking of both Czech and German exhibitors in Bohemia and intended to provide evidence of how well the two ethnic groups were able to come together on the 'neutral field of work' during a period of rising nationalistic tensions, and by joint effort work to improve the economic conditions in the land.[5] However, the political representatives of Bohemian Germans and the German-dominated chambers of commerce and manufacturing in Liberec and Cheb took a sceptical view of this exhibition project. They pointed to the ethnic and political situation in Bohemia as a factor that would prevent any such undertaking from succeeding. A temporary change in the Germans' attitude was brought about by negotiations in Vienna over a Czech-German settlement in early 1890, which offered Germans hope of achieving their political aims. When this failed, owing to strong resistance from many Czech political officials to accepting the negotiated agreement in the Bohemian diet, the political representatives of Bohemian Germans issued a statement recommending German artisans and industrialists not to take part in the exhibition. The leaders of the chambers of commerce and manufacturing in Liberec and Cheb soon added their voices to this exhortation. The vast majority of German producers heeded this call and stayed away from the exhibition.

Contrary to the many sceptical predictions, the Jubilee Provincial Exhibition was an enormous success for the Czech initiators. Although some areas were under-represented owing to the absence of the majority of German exhibitors, many fields, most notably the food industry, mechanical engineering and the electrical engineering industry were impressively showcased. An important section at the exhibition was the arts exhibit (in particular the successful displays of Czech fine arts) and the Czech village house (Česká chalupa) which epitomized the importance of ethnography and folk culture in contemporary Czech nationalism. The exhibition inevitably turned into a great celebration of the Czech nation, its economic and cultural maturity, and its autonomy from Germans.[6]

The Jubilee Provincial Exhibition served as an inspiration for regional exhibitions in Czech-speaking regions that, to the best of their means, wished also to demonstrate the sophisticated level of 'national' culture and the economy of the ethnic Czech population, to advance the Czech national tradition, and thus to strengthen Czech national identity. The upswing in Czech exhibition activities after 1891 followed the ethnographic efforts of Czech professionals and amateurs that materialized in the aforementioned Czech village house. They culminated in the Czechoslavic Ethnographic Exhibition held at the exhibition grounds in Prague in 1895, which offered a detailed showcase of what was conceived as Czech folk culture from across Bohemia and Moravia, including folk architecture, traditional costumes, customs and so forth.[7] The ethnographic element, embedded in the belief that folk culture was the genuine and traditional expression of the Czechs, was what gave the general and industrial exhibitions their Czech national character, which was further amplified by the addition of accompanying events, such as reconstruction of historic events or places, that were nationalistic in tone.

Nationalization of German Exhibitions

These ideologically tinted exhibition activities of Czechs in Bohemia affected events of Bohemian Germans that took place between the 1890s and the First World War. This was especially true of the 1891 Jubilee Provincial Exhibition, the success of which certainly came as quite a surprise to local Germans.

This success only deepened fears raised by the all-round mobilization of the Czech national sentiments. As I shall analyse below, some of the later German exhibitions were construed as a 'response' to the 1891 Exhibition in Prague and became strongly nationalistic in tenor.

However, German exhibitions did not acquire this strong nationalist tone immediately following the Prague exhibitions. Until 1897, nothing in the name of exhibitions in northern Bohemia indicated their national orientation and these events continued to bear neutral titles like 'The Exhibition of Trade, Industry and Agriculture' (Ústí nad Labem 1893), sometimes with a regional attribute like 'The North Bohemian Exhibition of Trade and Industry' (Teplice 1895) and so forth. Even the official documents of these exhibitions, such as invitations to exhibitors, exhibition guides, catalogues, and other materials, contained nothing to suggest anything nationalistic in tone at these exhibitions. If the introductions to the catalogues or guides to the exhibitions in Cheb in 1892, in Ústí nad Labem in 1893, in Teplice in 1895, or in Vrchlabí (Hohenelbe) in 1896 is scrutinized, they cite as the motivations for holding the event the desire to provide a picture of the economic situation in the given region, to support local producers, to stimulate advances in the fields of technology and art, and to boost trade. None of them contains any mention of the German character of the exhibition.[8]

Exhibitions in the German-speaking regions of Bohemia only began to assume a more nationalistic character towards the end of the 1890s. The above-noted development in the exhibitions organized by the Czech-speaking population after 1891 may have had influence on nationalizing the tenor of German exhibition activities in this period. While the Czech national character of the Jubilee Provincial Exhibition in 1891 could be ascribed to the deliberate absence of Germans at what was originally intended as an ethnically inclusive event, Czech exhibition projects that followed developed an overtly Czech national character. The increasingly German nationalist character of exhibition activities organized by Bohemian Germans was, however, primarily due to the generally increasingly nationalistic character of German society in Bohemia, especially after the so-called Badeni crisis of 1897. This occurred in connection with language ordinances issued by the Prime Minister of Austria, Count Badeni, which were intended to put Czech and German on more equal standing as official administrative languages in Bohemia and Moravia. According to these ordinances, all administrative proceedings should take place in the language in which an administration matter was initially filed with the authorities. Consequently, all officials in Bohemia and Moravia would thereafter have to be adequately fluent in both the Czech and German languages. Badeni's language ordinances sparked unprecedented resistance amongst German politicians and the German public throughout Cisleithanian Austria. Under pressure from them Badeni ultimately resigned and his ordinances never came into effect. Nevertheless, as a result, the year 1897 saw the further intensification of German nationalism in the land and an escalation of the Czech-German ethnic conflict. Following this crisis, politicians representing Bohemian Germans in the lands intensified their efforts to achieve the administrative separation of German-speaking settlements in the land.[9]

The nationalization of German activities after 1897 is apparent for instance from the titles given to the exhibition. None of the large industrial or general exhibitions in the German-speaking regions failed to stress the word 'German' (*deutsch*) or 'German-Bohemian' (*deutschböhmisch*) in their titles. The first large exhibition to adopt such a name was the Exhibition of German Industry, Trade and Agriculture in Northwest Bohemia (*Nordwestböhmische Ausstellung für deutsche Industrie, Gewerbe und Landwirtschaft*) held in the town of Most (Brüx) in 1898 and organized by the Most Trade Association. The introduction to the exhibition guide spoke about 'the glory and honour of the labour and industriousness of German entrepreneurship', and described the exhibition as 'a noble tournament of German products'.[10] The guide then proceeded in the same tone to offer a comprehensive history of the town and of the region of north-west Bohemia, which was full of descriptions of the 'civilizing' influence of German culture and devoted considerable attention to the 'age-old' nationalist conflicts between Czechs and Germans.[11]

Other large exhibitions in German-speaking regions put the same emphasis on the German character of the particular event. In the foreword to the catalogue for an exhibition organized in Děčín in 1902 by the local Trade Association, for instance, the very first sentence stresses three times that the exhibition is a *German* project and that it is intended to show 'what German skill and creativity are striving for and have achieved in various areas' and to provide 'the most comprehensive picture of German labour in our era', and finally states that participants shall in a united effort contribute 'to the honour and glory of this exhibition that is called "German"'.[12]

Similarly, in the information brochure for the German General Exhibition (*Allgemeine deutsche Ausstellung*) planned in Ústí nad Labem in 1903, the author voiced the explicit aim to turn the exhibition into a provincial exhibition for Germans in Bohemia (Fig. 4.1).[13] Later on, however, the organizers of this event, most of them members of the local Trade Association or top public office holders in the town, refrained from mentioning such an explicit objective. Ultimately the Ústí exhibition became a multi-regional event, as the majority of the exhibitors came from Vienna and, as a result, the organizers employed a relatively moderate nationalist tone compared to other large exhibitions organized by Bohemian Germans in this period.[14]

The strongest calls for a separate provincial exhibition for Bohemian Germans were heard in Liberec in 1906, which will be the subject of further discussion shortly, and in Chomutov (Komotau) in 1913. The latter was called the German-Bohemian Provincial Exhibition (*Deutschböhmische Landesschau*), though this title was somewhat exaggerated. The exhibition was centred primarily on craftwork and agriculture and featured very little in the way of industry and no art at all. Nevertheless, the organizers in Chomutov, led by the town's mayor, Ernst Storch, who was also chair of the exhibition's

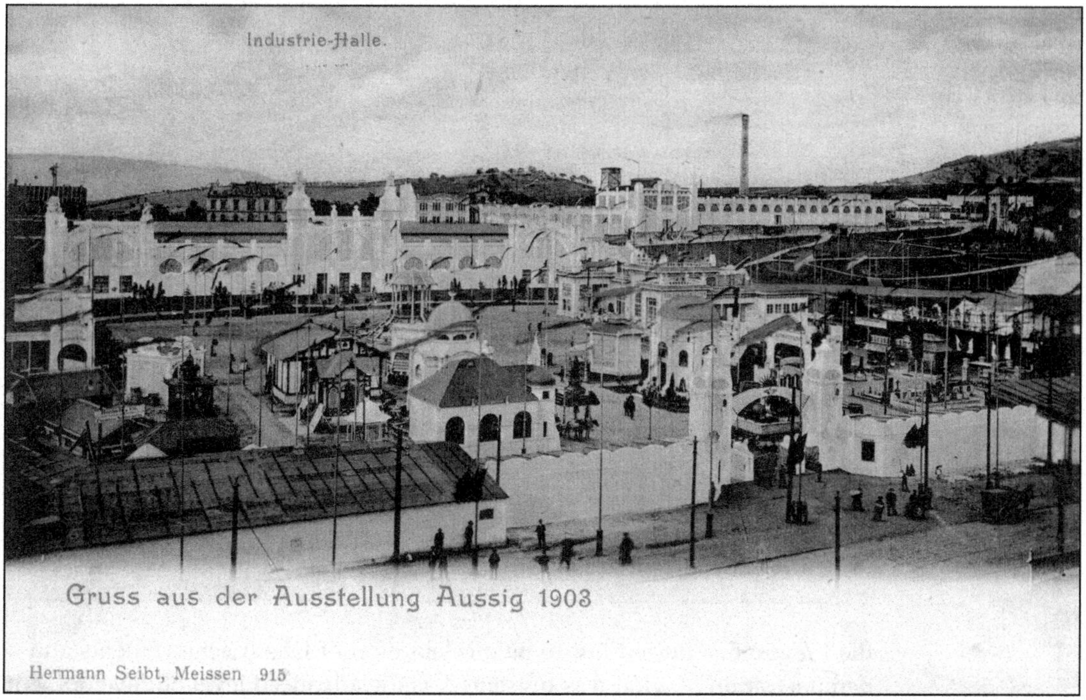

4.1 The German General Exhibition in Ústí nad Labem in 1903, postcard. © Museum of the City of Ústí nad Labem

executive committee, spoke of the exhibition as a 'great national act' and deemed participation in it an 'imperative' for all Bohemian Germans.[15]

The most ambitious undertaking in the German-speaking regions of Bohemia before the First World War was the German-Bohemian Exhibition (*Deutschböhmische Ausstellung*) in Liberec in 1906 (Fig. 4.2). Its organizers, dominated by the town's top officials, with the most influential post held by the mayor, Franz Bayer, emphasized the exhibition's political aims from the very start. The invitation that the executive committee sent out to potential exhibitors contained the following proclamation:

> Germans in Bohemia! We call on all Germans in Bohemia to attend this great and important undertaking – it will be an extraordinary economic and national event. The 'German-Bohemian Exhibition in Liberec in 1906' shall bear witness to the momentous significance of German labour in Bohemia and display proof that it is the German people in Bohemia who are the foremost agents of the empire's economic prosperity … This exhibition shall present in tangible form what has already been presented in numbers – evidence of the importance of German work in Bohemia, and thus confirm the legitimacy of the political and national claims thereby entailed of we the German people in Bohemia.[16]

The invitation further called on producers to take part in as large numbers as possible, but rather than citing the economic benefits to be had from attending, it referred to the national interests of Germans in Bohemia.[17] The organizers also openly declared their exhibition to be the German response to the Jubilee Provincial Exhibition in Prague of 1891. While in its ultimate form

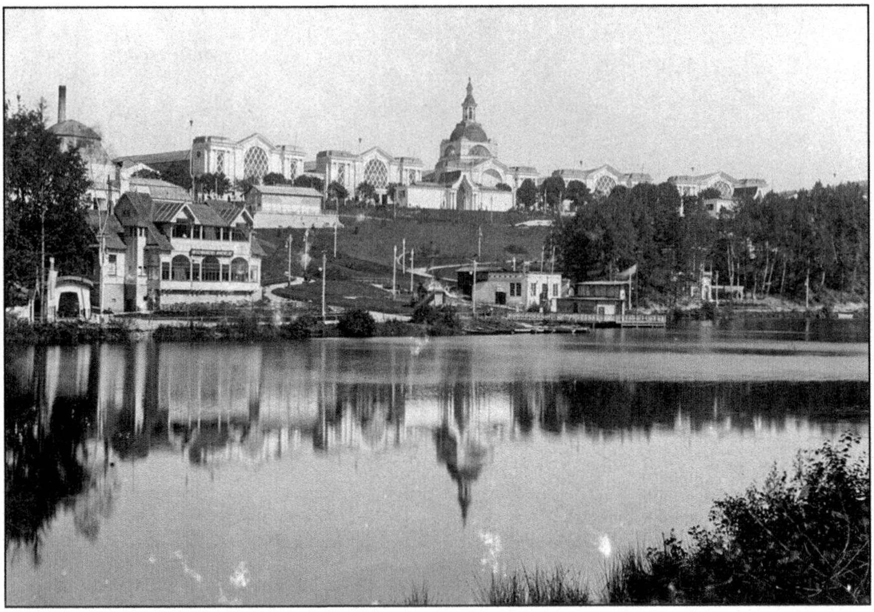

4.2 The Main Building of the German-Bohemian Exhibition in Liberec in 1906, postcard. © The North Bohemian Museum in Liberec

the Prague fair turned out to be a celebration of Czech achievements and a demonstration of what was presented as the advanced level of development of the Czech people in many areas, in a similar way the Liberec exhibition was to present German achievements in all their complexity and advanced level of development.[18]

The nationalistic focus that defined the event was already reflected in the preparatory work. The organizers, for example, issued an order that all of the exhibition's employees and supervisors (not, however, ordinary labourers) had to be Germans.[19] The committee also requested that because of the German orientation of the exhibition, the staff on all trains running through Liberec were only Germans.[20] The print matter accompanying the exhibition emphasized the role firms owned by Bohemian Germans in the province played in the preparation and content of the exhibition, even though many businesses from the Austrian Alps and from Germany were also invited to take part in this.[21]

Organization of this exhibition had also much to do with Liberec's long-term efforts to make itself the German-Bohemian capital. Prague, with its majority Czech population and Czechs-only government, was viewed by Bohemian Germans less and less as their capital. Radical German nationalists had been demanding since the 1880s that the German university and polytechnic institute be relocated from Prague to the border regions. A movement called *Los von Prag* (Out of Prague) emerged, whose ideology was driven not just by the proposition of leaving Prague, which – in their view – had lost its German character, but also by the liberal orientation of the German political leaders in Prague.[22] By contrast, from the 1880s the town hall in Liberec was controlled by German nationalists. Communal politicians in Liberec continuously came out with proposals for an administrative division of Bohemia and tried to assert a uniform position among German towns in terms of their political demands. In late 1897, the local government in Liberec passed a demand for the creation of a province called German Bohemia (Deutschböhmen), of which Liberec was to be the capital city.[23] Towns with German majorities in Bohemia, however, were plagued by rivalry, and they took a lukewarm view of Liberec's efforts to acquire this status for itself.

The grand exhibition of 1906 represented a good opportunity for Liberec to cast itself in the role of the voice for German nationalist demands and to bolster its ambition of being accepted as the capital city of Germans in Bohemia. This was a reason why soon after the event was announced by the local Trade Association, top politicians in the municipal government, foremost amongst them the mayor of Liberec, seized control of the organization process. They presented the Liberec fair as a provincial exhibition of Bohemian Germans, and they created the conditions enabling the advanced economic and cultural level of Bohemian Germans to be showcased in one location to lend weight to their political claims.

New exhibition grounds were built at great expense next to the lake in Liberec and stretched to approximately 400,000 square metres in size, which was roughly four times larger than the site of the largest German-Bohemian exhibition organized in Ústí nad Labem in 1903.[24] The Liberec exhibition grounds were even larger than some famous international exhibition and world fair grounds. For instance, the first three World Fairs in London and Paris were held on smaller sites, as was the World Fair in Melbourne in 1880, the colonial exhibition in Amsterdam in 1883, and the international exhibition in Antwerp in 1885.[25] Although most of the structures on the grounds were built of light construction materials and erected solely for the purpose of the exhibition, the grounds were nonetheless very impressive. The main exhibition building was designed by the renowned Viennese architect Max Fabiani (1865–1962), a student and colleague of Otto Wagner and one of the foremost representatives of the Art Nouveau within the Habsburg Empire.[26] The main pavilion was 240m in length and its centre aisle was crowned by a dome from which two spotlights shone at night.[27]

Approximately 1,500 exhibitors participated in the exhibition, not including the agricultural section, and that was by far the largest number ever to take part in any of the German-Bohemian exhibitions held before the First World War. There were 26 categories of exhibits, the majority of them representing crafts and the industrial arts. Agriculture was divided into two categories, and separate categories were devoted to fine art, education, health care, social services, the military and cultural history. Understandably, a very large proportion of the exhibitors were producers from Liberec and its vicinity, but there were also many from areas farther away. The chemical industry category, for example, was dominated by producers from Ústí nad Labem (such as the Association for Chemical and Metallurgical Production and the Schicht Plant); mining and metal-working were represented by large mining companies based in north-west Bohemia; the section on glass and porcelain production included the porcelain factories in north-west and western Bohemia (for example, the Thun Porcelain Factory in Klášterec nad Ohří (Klösterle)) and manufacturers from other parts of Bohemia, including the Pilsen-based Škoda Plant).[28] There was an important section at the Liberec exhibition consisting of contemporary German art from Bohemia, where considerable space was devoted to the sculpture of Franz Metzner, while the paintings of Alfred Kubin, Emil Orlik, Richard Teschner and other artists were also on show.[29]

The exhibition pavilion of the host town of Liberec was also notable. Its exhibits reflected, as was the custom at similar exhibitions, the historical, demographic, and structural development and activities of various of the town's institutions (Fig. 4.3). Uniquely, though, one part of the town's exhibit used information from a book titled *Deutschböhmen als Wirtschaftsgroßmacht* (German Bohemia as an Economic Hegemon) to demonstrate German economic superiority over Czechs.[30] One graph on display, for instance, showed the number of Czech and German factories in individual branches of production, their estimated value and the value of the goods they produced annually. Another presented the amount of German and Czech capital in banks. And yet another illustrated the amount of taxes paid by the Czech and German populations in Bohemia. The sums claimed that there were 5,159 German factories in Bohemia with a worth of 4.785 billion crowns and an annual production with a value of 1.721 billion crowns, while there were only 1,234 Czech factories with a value of 434 million crowns annually producing goods with a value of 184 million. Germans were allegedly paying 253 million crowns annually in taxes, while Czechs were paying just 130 million crowns.[31] Such calculations of taxes paid and the economic performance were frequently used by Bohemian Germans at that time to underscore their dominant position in the economic sphere and their significance for the state, whereby they hoped to strengthen the basis of their political demands. The Czech side responded with its own calculations, which were very different from the German ones.[32]

An Exhibition of German Towns in Bohemia (*Städteausstellung der deutschböhmischen Städte*) was organized as part of the fair, inspired by the recent German Exhibition of Towns (*Deutsche Städte-Ausstellung*) in Dresden in 1903, but the nature of the Bohemian event was quite different. While the Dresden exhibition aimed mainly to present various approaches to addressing the problems of the modern city,[33] the dominant idea behind the Liberec exhibition of towns was to symbolize the unity of Germans in Bohemia and to offer proof that the German-Bohemian Exhibition was not just a local undertaking but rather represented Germans right across Bohemia. The Exhibition of Towns was intended to be the representative centre of the entire exhibition, and for this reason it was located in the middle of the main building. The organizers strove to impart as uniform a character as possible to the exhibits of individual towns by selecting similar display

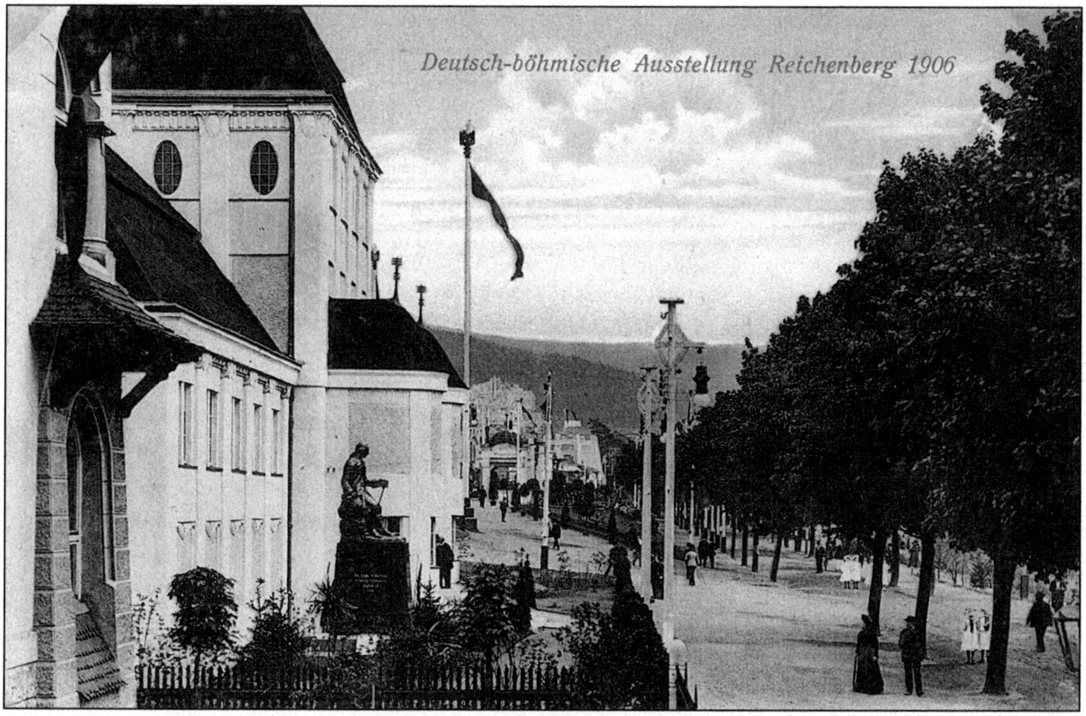

4.3 The Pavilion of the Host Town at the Exhibition in Liberec, postcard. © The North Bohemian Museum in Liberec

items and by employing one architect for the final design of the layout of the exhibition of towns. They asked the town authorities to send to Liberec the town's symbol and a wall plan of the town, as well as a plaster bust of the current mayor, a photograph of members of the town council, and photographs and paintings of earlier mayors and prominent figures. Other exhibits were to include population statistics, historical objects, such as guild flags and cups, and important historical documents.[34] This ambitious plan, however, was met with less interest from the other towns than the organizers would have wished. Only 15 towns contributed exhibits to the Exhibition of German Towns (alongside towns not far from Liberec, such as Jablonec nad Nisou (Gablonz) or Frýdlant v Čechách (Friedland), mainly the major German-Bohemian centres of Ústí nad Labem, Cheb, Teplice, Litoměřice and others). However, other towns that were invited refused to take part, citing the high financial expense it involved, a lack of interesting exhibition material, or insufficient time to prepare a decent exhibit.[35] Most towns were thus ultimately represented only by tables and statistical data.

The Liberec exhibition did not envision participation of Czech exhibitors. All the invitations sent out by the Exhibition Committee and the circulars inviting participants spoke explicitly to German producers and on every occasion the organizers stressed the German character of the exhibition and its German nationalist political aims. Nationalist Czech producers never considered attending this German-oriented event either and Czech producers in the Liberec chamber of commerce and manufacturing organized their own

Czech-oriented exhibition six years later in Mladá Boleslav. The introduction to their catalogue explicitly stated that it represented the Czech counterpart to the German exhibition in Liberec in which Czech exhibitors did not participate:

> The invitation extended by the Liberec exhibition explicitly stated to the invitees that it was an exclusively German exhibition, intended for German exhibitors in Bohemia ... The invitation and the subsequent conduct of the Exhibition Committee made clear that the exhibition would host nothing Czech, that no Czech was allowed to take part in it. And Czech members of the Liberec chamber truly were not able to participate, not unless they wished to wound their own national pride.[36]

However, it must be noted that some large enterprises, such as big joint-stock companies, consistently avoided any kind of nationalist inclinations. In the case of some there was a question whether the firm was actually Czech or German because their boards of directors, shareholders and employees were not always entirely German or entirely Czech. These firms regularly took part in both the Czech and the German exhibitions. On one side they were regarded as proof of the advanced level of Czech industry, and on the other side of German industry. Such examples were, for instance, mechanical engineering firms like the Joint-Stock Machine Works (formerly Breitfeld, Daněk & Co.), the Prague Joint-Stock Mechanical Engineering Works (formerly Ruston & Co.), and the Pilsen-based Škoda Plant.

The Liberec exhibition received extensive coverage in the Czech press, but the commentaries were primarily critical and sarcastic in tone. The Czech daily *Národní listy*, for example, stressed that the Liberec exhibition fell well short of the Provincial Exhibition in Prague in 1891, while the newspaper *Lidové noviny* considered it proof of the 'all-round cultural impotence' of Bohemian Germans, and the satirical magazine Šípy ironically called it the 'world fair of the Liberec Teutons', whose exhibits are 'the sticks, canes and stones they use to beat Czechs about the head with'.[37] More balanced reporting on the Liberec exhibition was provided by the economic magazine *Obzor národohospodářský*, which, while it levelled sharp criticism against the political tone of the event, was nonetheless able to highlight some positive aspects.[38]

A very different point of view was presented by the German-language press, not just in Bohemia but across the Monarchy and in Germany. Lengthy articles on the exhibition were published, for instance, by *Breslauer Zeitung*, the Viennese *Neuigkeits-Welt-Blatt* or *Der Alpen-Bote* in Steyr.[39] The event was given detailed coverage in the leading Viennese daily *Neue Freie Presse*, which in this context emphasized how advanced Bohemian Germans were and described the host town as the German capital of Bohemia.[40] Of the German-language newspapers in Bohemia, the Liberec newspapers *Reichenberger Zeitung* and *Deutsche Volkszeitung* understandably focused extensively on the exhibition. They played it up as the achievement of all Germans in Bohemia, accomplished through their national unity and they described the town of Liberec as the natural centre of Bohemian Germans.[41]

The German-language press in other Bohemian towns also covered the exhibition and ascribed it considerable political significance. For example, one article in *Aussiger Tagblatt*, published in Ústí nad Labem, highlighted the political significance of the event and expressed the hope that it would help to cement German unity in Bohemia.[42] The exhibition sparked great interest in the German liberal newspapers in Prague. Even a month before the opening, *Prager Tagblatt* printed a long article on the event that concluded with an indirect call to attend the Liberec exhibition and to view it primarily in political terms.[43] *Bohemia*, another German daily published in Prague, devoted its first two pages to the exhibition on its opening day. The article referred to the opening of the exhibition which was described as a German national holiday and highlighted that Germans from all over Bohemia were taking part.[44]

These reactions in the press indicate that the organizers' aim of putting on a comprehensive showcase of the products of Bohemian-German labour at the Liberec exhibition in an effort to support Germans' political demands was broadly welcomed by Germans across Bohemia. But the other aim of the organizers, to use the exhibition to support Liberec's ambition to become the generally recognized capital of Germans in Bohemia, fell flat in most cases. Newspaper articles published outside Liberec that otherwise praised the event and advocated for the unity of Germans in Bohemia either systematically ignored any mention of a Bohemian-German capital or played down the idea of Liberec occupying such a position. In the abovementioned article in *Prager Tagblatt*, for example, the idea of Liberec's status as the German capital of Bohemia was dismissed with a sneer: 'Liberec, which likes to call itself the capital of German Bohemia …'.[45]

After the exhibition, the organizers nevertheless claimed that, thanks to the event, Liberec came to be recognized as the pre-eminent town of the Bohemian Germans.[46] But such a statement did not reflect reality. Although Liberec sought such a position and the town leaders continued to promote the idea in various ways, other German towns in the region took a more reserved view of the idea, at least up until the First World War.[47]

As regards interest from visitors, the Liberec exhibition saw attendance figures like no other pre-war German exhibition in Bohemia, but it still failed to meet the ambitious expectations of the organizers. While 1,265,423 visitors was an unprecedented figure in the history of exhibition activities organized by Bohemian Germans, it was nowhere close to the number of visitors that attended the Jubilee Provincial Exhibition in Prague in 1891, which was almost twice as high. The organizers chalked the lower interest up to bad weather and a lack of adequate cooperation from the railroads. These facts certainly left a mark on attendance – it rained on more than half of the 137 days of the exhibition and the railroads truly did not offer any better deals for visitors.[48] But an important role was also played by the fact that the exhibition did not excite Bohemian Germans with the kind of fervour that its organizers envisioned in its overly ambitious plans.

The lower attendance rate was one of the reasons for the considerable budget deficit that followed the event. Although the exhibition took in 1,563,785 crowns in revenue, that figure was around a quarter of a million short of the original proposed budget. The expenses, amounting to more than two million crowns, exceeded the original estimate by roughly a quarter of a million crowns, leaving a deficit of approximately half a million crowns.[49] While the debt was settled relatively quickly with the help of the town and other underwriters of the exhibition, the negative financial balance cast a pall on the overall image of the event.

Conclusion

Before the First World War, exhibitions in German-speaking regions began to assume a strong nationalist undertone and increasingly, all the large German exhibitions from this time profiled themselves clearly as events with a distinctly German focus. They differed by the intensity of their confrontational rhetoric and above all they were attached to political demands. Some exhibitions were moderate in this respect, like the one in Ústí nad Labem in 1903. By contrast, the organizers of the exhibitions in Chomutov and particularly in Liberec proclaimed support for political arguments as the main objective of the entire project. The Liberec exhibition sought to demonstrate the unity of Bohemian Germans, provide a graphic illustration of their outstanding achievement in the economic and cultural spheres, and thereby support the political demands of Germans in Bohemia. By presenting the German-speaking regions of Bohemia as a consolidated entity they hoped to aid efforts to obtain recognition for a separate German-speaking territory.

The Liberec organizers succeeded in putting together a grand exhibition that far surpassed all previous German exhibitions in Bohemia. The size and grandeur of the exhibition grounds, the number of exhibitors and visitors, the variety of exhibits, and the strong response in the press, not just in the Bohemian lands and the Habsburg Empire, but also in Germany, were unprecedented. But in some regards their grand ambitions were left unfulfilled. Although the exhibition had a record number of visitors, there were not as many as the organizers had imagined, resulting in a substantial budget deficit. Even the willingness of other towns to take part in the collective Exhibition of Towns disappointed expectations. And Liberec's effort to be recognized as the capital of Bohemian Germans thanks to the exhibition fell flat. The long-term political goal of Bohemian Germans to create an administratively separate German-speaking territory in Bohemia ultimately failed too.

Considering the Liberec exhibition from a different angle, it was, in many respects, successful. Like other exhibitions in this period, it fulfilled the function of mediating contacts between producers and consumers and providing information about supply and demand in the market and generally about the industrial, craft and agricultural production in the region.

It provided visitors with a unique opportunity to learn about the work of Bohemian-German artists. It also offered information on history and other areas. It was instrumental in heightening the visibility of the town of Liberec and increasing local tourism. And it was in this light that many of its visitors and exhibitors viewed it. Its engaging appearance, the quality of the exhibits on show, and the variety of accompanying events in all likelihood left a stronger impression on them than the nationalist rhetoric and proclaimed political aims.

Nevertheless, it was because of these political aims that the Liberec exhibition was ever organized as such a large and grandiose event. Like other large exhibitions held in Bohemia between the 1890s and the First World War, in both the German- and Czech-speaking regions, it formed part of the nationalist conflict that reached with ever-growing intensity into various spheres of the life of society, including the economy, the arts and everyday private relations. Yet these exhibitions cannot be regarded simply as 'mirrors' of the contemporary situation because they did not just play a passive role in the Czech-German nationalist conflict. Given their economic and cultural significance and the enormous attention they garnered from society at that time, they became an active factor in and influenced the development of the Czech-German relations.

Notes

1 On the national question in Bohemia in the second half of the nineteenth century and the beginning of the twentieth century, see: Catherine Albrecht, 'The Bohemian Question', in *The Last Years of Austria-Hungary. A Multi-National Experiment in Early Twentieth-Century Europe*, ed. Mark Cornwall (Exeter, 2002), 75–96; Gary B. Cohen, *The Politics of Ethnic Survival: Germans in Prague, 1861–1914* (Princeton, 1981); Pieter M. Judson, *Guardians of the Nation. Activists on the Language Frontiers of Imperial Austria* (Cambridge, MA and London, 2006); Nancy M. Wingfield, *Flag Wars and Stone Saints. How the Bohemian Lands Became Czech* (Cambridge, MA and London, 2007); Jeremy King, *Budweisers into Czechs and Germans. A Local History of Bohemian Politics, 1848–1948* (Princeton and Oxford, 2002); Derek Sayer, *The Coasts of Bohemia. A Czech History* (Princeton, 1998); Jiří Kořalka, *Tschechen im Habsburgerreich und in Europa 1815–1914. Sozialgeschichtliche Zusammenhänge der neuzeitlichen Nationsbildung und der Nationalitätenfrage in den böhmischen Ländern* (Vienna and Munich, 1991); Otto Urban, *Die tschechische Gesellschaft 1848 bis 1918* (Vienna, 1994).

2 On the national aims, imperial politics and the issue of national identity at international exhibitions, see Paul Greenhalgh, *Ephemeral Vistas. The Expositions Universelles, Great Exhibitions and World's Fairs, 1851–1939* (Manchester, 1988); Maurice Roche, *Mega-Events and Modernity. Olympics and Expos in the Growth of Global Culture* (London and New York, 2000), 33–64. On the political motivations behind the participation and non-participation of Germany at World Fairs, see Christoph Cornelißen, 'Die politische und kulturelle Repräsentation des Deutschen Reiches auf den Weltausstellungen des 19. Jahrhunderts', *Geschichte in Wissenschaft und Unterricht* 52 (2001): 148–161;

Eckhardt Fuchs, 'Das Deutsche Reich auf den Weltausstellungen vor dem Ersten Weltkrieg', *Comparativ* 9, no. 5–6 (1999): 61–88; Heinz-Alfred Pohl, 'Die Weltausstellungen im 19. Jahrhundert und die Nichtbeteiligung Deutschlands in den Jahren 1878 und 1889', *Mitteilungen des Instituts für Österreichische Geschichtsforschung* 97 (1989): 381–425. On nationalism at provincial and regional exhibitions in Germany, see Thomas Großbölting, *'Im Reich der Arbeit.' Die Repräsentation gesellschaftlicher Ordnung in den deutschen Industrie- und Gewerbeausstellungen 1790–1914* (Munich, 2008), 385–394. There was also a strong nationalist tenor, for instance, to the Hungarian Provincial Exhibition in Budapest in 1885 – see Wilhelm Balogh, Franz Toldy and Moriz Gelleri, *Führer durch die Ausstellung* (Budapest, 1885).

3 A report on the industrial exhibition in 1841 was published in *Bohemia, ein Unterhaltungsblatt*, 17 October 1841, 3–4; it is referred to in Johann Slokar, *Geschichte der österreichischen Industrie und ihrer Förderung unter Kaiser Franz I.* (Vienna, 1914), 236; but K. E. Carpenter in his survey of the subject makes no mention of it: Kenneth E. Carpenter, 'European Industrial Exhibitions before 1851 and Their Publications', *Technology and Culture* 13 (1972): 465–486; a report on the exhibition in 1852 was presented in instalments in *Reichenberger Wochen-Blatt* 1 (1852): 225, 231, 238, 244, 263–264, 271–272 and 278–279.

4 Václav Houfek and Tomáš Okurka, 'Německé muzejnictví a výstavnictví v Čechách', in *Hledání centra. Vědecké a vzdělávací instituce Němců v Čechách v 19. a první polovině 20. století*, ed. Kristina Kaiserová and Miroslav Kunštát (Ústí nad Labem, 2011), 195–234, esp. 209–211.

5 The circular distributed to industrialists by the organization committee for the exhibition in July 1889 was published in an official publication on the exhibition: *Sto let práce. Zpráva o všeobecné zemské výstavě v Praze 1891* (3 vols, Prague, 1893–95), 1–2: 44.

6 On the Jubilee Provincial Exhibition, see: Catharine Albrecht, 'Pride in Production: The Jubilee Exhibition of 1891 and Economic Competition between Czechs and Germans in Bohemia', *Austrian History Yearbook* 24 (1993): 101–118; Milan Hlavačka, *Jubilejní výstava 1891* (Prague, 1991).

7 On the Czechoslavic Ethnographic Exhibition, see Marta Filipová, 'Peasants on Display. The Czechoslavic Ethnographic Exhibition of 1895', *Journal of Design History* 24, no. 1 (2011): 15–36; Sayer, esp. 95–96 and 124–127; Martin Wörner, *Vergnügung und Belehrung. Volkskultur auf den Weltausstellungen 1851–1900* (Münster, 1999), 98–100.

8 *Catalog der Gewerbe & landwirtschaftlichen Ausstellung in Eger 1892* (Cheb, 1892), n.p.; *Katalog zur Gewerbe-, Industrie- und landw. Ausstellung in Aussig 1893* (Ústí nad Labem, 1893), 3–4; *Offizieller Katalog der Spezial-Ausstellung technischer Arbeitsbehelfe für das Kleingewerbe auf der Nordböhmischen Industrie- und Gewerbeausstellung in Teplitz 1895* (Teplice, 1895), 19–21; *Führer durch die Nordostböhmische Gewerbe-Ausstellung Hohenelbe 1896* (Vrchlabí, 1896), 3–5.

9 On the Badeni crisis, see (for example) Albrecht, 'Bohemian Question', 79–83; Robert A. Kann, *A History of the Habsburg Empire 1526–1918* (Berkeley, 1974), 439–442; for more details, see Berthold Sutter, *Die Badenischen Sprachverordnungen von 1897* (2 vols, Graz, 1960–65).

10 *Officieller Führer in die Nordwestböhm. Ausstelung für deutsche Industrie, Gewerbe und Landwirtschaft in Brüx 1898* (Most, 1898), III–XV.

11 *Officieller Führer*, III–XV.

12 *Deutsche Gewerbe-, Industrie & landw. Ausstellung Tetschen 1902* (Děčín, 1902), 1.

13 The Association of Industries, Trade and Business, Allgemeine Deutsche Ausstellung für Gewerbe, Industrie und Landwirtschaft, Teplice Associations Fund, inv. no. 65, fol. 14, State Regional Archives Litoměřice – State District Archive Teplice.

14 Tomáš Okurka, *Všeobecná německá výstava v Ústí nad Labem 1903* (Ústí nad Labem, 2005).

15 Tomáš Okurka, 'Českoněmecká zemská výstava v Chomutově 1913', in *Comotovia 2007. Sborník příspěvků z konference věnované výročí 550 let udělení znaku města Chomutova (1457–2007)*, ed. Petr Rak (Chomutov, 2008), 160–185.

16 Reprinted in *Katalog der Deutschböhmischen Ausstellung Reichenberg 1906* (Liberec, 1906), 44–45.

17 *Katalog der Deutschböhmischen Ausstellung*, 44–45.

18 *Reichenberg und die Deutschböhmische Ausstellung 1906* (Liberec, 1906), 15.

19 Miloslava Melanová, *Liberecká výstava 1906* (Liberec, 1996), 11.

20 Deutschböhmische Ausstellung, The Archive of the Town of Liberec Fund, inv. no. 159, fol. 137, State Regional Archives Litoměřice – State District Archives Liberec.

21 See *Katalog der Deutschböhmischen Ausstellung*, 62–70.

22 Cohen, 245–246, 254.

23 Miloslava Melanová, 'Hlavní město Německých Čech? Liberec 1848–1918', in *Hledání centra*, ed. Kaiserová and Kunštát (Ústí nad Labem, 2011), 285–307.

24 *Reichenberg und die Deutschböhmische Ausstellung*, 18.

25 See the overview of basic facts on world and international exhibitions in Winfried Kretschmer, *Geschichte der Weltausstellungen* (Frankfurt am Main and New York, 1999), 288–303.

26 Marco Pozetto, *Max Fabiani. Ein Architekt der Monarchie* (Vienna, 1983).

27 *Die Deutschböhmische Ausstellung Reichenberg 1906. Industrie-, Gewerbe-, Kunst- und Landwirtschafts-Ausstellung der Deutschen Böhmens* (Liberec, 1909), 1: 136.

28 *Katalog der Deutschböhmischen Ausstellung*, 72–281.

29 *Die Deutschböhmische Ausstellung*, 2: 181–200.

30 *Deutschböhmen als Wirtschaftsgroßmacht* (Liberec, 1903).

31 *Die Deutschböhmische Ausstellung*, 2: 337–338 and 2: 348–349.

32 See, for instance, Friedrich Wieser, *Die deutsche Steuerleistung und der öffentliche Haushalt in Böhmen* (Leipzig, 1904), and the response of the Czech National Council in *Die nationale Steuerleistung und der Landeshaushalt im Königreiche Böhmen. Antwort auf die Erwägungen des Prof. Dr. Freih. Wieser* (Prague, 1905).

33 See Kurt Wilde, 'Die erste Deutsche Städte-Ausstellung 1903. Zeitpunkt und geschichtlicher Hintergrund', *Dresdner Hefte. Beiträge zur Kulturgeschichte* 18, no. 63 (2000): 29–35.

34 Städte-Ausstellung, Deutschböhmische Ausstellung, inv. no. 159, fol. 136, p. 1, State District Archives Liberec, The Archive of the Town of Liberec Fund.

35 See the letters of the exhibition committee in Deutschböhmische Ausstellung, inv. no. 159, fol. 136, State District Archives Liberec, fond The Archive of the Town of Liberec Fund.

36 *Katalog severočeské hospodářské a živnostensko-průmyslové výstavy severních a severovýchodních Čech v Mladé Boleslavi roku 1912* (Mladá Boleslav, 1912), 4.

37 *Lidové noviny*, 17 May 1906, 2; *Národní listy*, 21 June 1906, 1; *Šípy*, 19 May 1906, 2.

38 František Rössler, 'Tři německé výstavy roku 1906 (Drážďany – Liberec – Norimberk)', *Obzor národohospodářský* 11 (1906): 353–362, 401–411, 460–474.

39 The organizers followed and filled several notebooks with cut-outs of responses to the event in the papers: see Deutschböhmische Ausstellung, inv. no. 159, fol. 139, State District Archives Liberec, fond The Archive of the Town of Liberec Fund.

40 *Neue Freie Presse*, 22 April 1906, 2–3.

41 See, for instance, *Reichenberger Zeitung*, 17 May 1906; *Deutsche Volkszeitung*, 17 May 1906.

42 *Aussiger Tagblatt*, 19 May 1906, 1–2; a similar assessment is given in *Teplitz-Schönauer Anzeiger*, 3 October 1906, 1–2.

43 *Prager Tagblatt*, 15 April 1906, 49–52.

44 *Bohemia*, 17 May 1906, 1.

45 *Prager Tagblatt*, 15 April 1906, 49–52.

46 Letter to the Executive Committee of the Exhibition to the Town Council of Liberec, 9 October 1906, Deutschböhmische Ausstellung, inv. no. 159, fol. 137, State District Archives Liberec, fond The Archive of the Town of Liberec Fund.

47 Cf. Kaiserová and Kunštát, 403–409.

48 See Melanová, *Liberecká výstava*, 51.

49 The closing balance is printed in *Die Deutschböhmische Ausstellung* 1: 161–162.

Bibliography

Manuscript Sources

Deutschböhmische Ausstellung. The Archive of the Town of Liberec Fund. State Regional Archives Litoměřice – State District Archives Liberec, inv. no. 159, fol. 136–139.

The Association of Industries, Trade and Business. Teplice Associations Fund. State Regional Archives Litoměřice – State District Archive Teplice, inv. no. 65, fol. 14.

Printed Primary Sources

'Bericht über die Reichenberger Gewerbsprodukten-Ausstellung'. *Reichenberger Wochen-Blatt*, 1852, 225, 231, 238, 244, 263–264, 271–272 and 278–279.

'Correspondenz aus Böhmen'. *Bohemia, ein Unterhaltungsblatt*, 17 October 1841, 3–4.

'Deutschböhmische Ausstellung 1906'. *Teplitz-Schönauer Anzeiger*, 3 October 1906, 1–2.

'Die deutschböhmische Ausstellung in Reichenberg'. *Neue Freie Presse*, 22 April 1906, 2–3.

'Die Deutschböhmische Ausstellung in Reichenberg'. *Reichenberger Zeitung*, 17 May 1906, 1.

'Die Reichenberger Ausstellung und die Kaiserreise'. *Aussiger Tagblatt*, 19 May 1906, 1–2.

'Ein deutscher Ehren- und Festtag!'. *Deutsche Volkszeitung*, 17 May 1906, 1.

'Ein Fest deutscher Arbeit'. *Bohemia*, 17 May 1906, 1.

'Výstava Němců v Čechách'. *Lidové noviny*, 17 May 1906, 2.

Balogh, Wilhelm, Franz Toldy, and Moriz Gelleri. *Führer durch die Ausstellung*. Budapest, 1885.

Catalog der Gewerbe & landwirtschaftlichen Ausstellung in Eger 1892. Cheb, 1892.

Deutschböhmen als Wirtschaftsgroßmacht. Liberec, 1903.

Deutsche Gewerbe-, Industrie & landw. Ausstellung Tetschen 1902. Děčín, 1902.

Die Deutschböhmische Ausstellung Reichenberg 1906. Industrie-, Gewerbe-, Kunst- und Landwirtschafts-Ausstellung der Deutschen Böhmens. Liberec: Gebrüder Stiepel, 1909.

Die nationale Steuerleistung und der Landeshaushalt im Königreiche Böhmen. Antwort auf die Erwägungen des Prof. Dr. Freih. Wieser. Prague: Národní rada česká, 1905.

Führer durch die Nordostböhmische Gewerbe-Ausstellung Hohenelbe 1896. Vrchlabí, 1896.

Katalog der Deutschböhmischen Ausstellung Reichenberg 1906. Liberec, 1906.

Katalog severočeské hospodářské a živnostensko-průmyslové výstavy severních a severovýchodních Čech v Mladé Boleslavi roku 1912. Mladá Boleslav: Hejda & Zbroj, 1912.

Katalog zur Gewerbe-, Industrie- und landw. Ausstellung in Aussig 1893. Ústí nad Labem: Verlag des Ausstellungs-Comités, 1893.

Národní listy, 21 June 1906: 1.

Officieller Führer in die Nordwestböhm. Ausstelung für deutsche Industrie, Gewerbe und Landwirtschaft in Brüx 1898. Most, 1898.

Offizieller Katalog der Spezial-Ausstellung technischer Arbeitsbehelfe für das Kleingewerbe auf der Nordböhmischen Industrie- und Gewerbeausstellung in Teplitz 1895. Teplice, 1895.

Reichenberg und die Deutschböhmische Ausstellung 1906. Liberec, 1906.

Rössler, František. 'Tři německé výstavy roku 1906 (Drážďany – Liberec – Norimberk)'. *Obzor národohospodářský* 11 (1906): 353–362, 401–411, 460–474.

Šípy, 19 May 1906, 2.

Sto let práce. Zpráva o všeobecné zemské výstavě v Praze 1891. Prague: Nákladem Výkonného výboru Všeobecné zemské jubilejní výstavy, 1893–95.

Urban, W. 'Deutschböhmische Ausstellung in Reichenberg'. *Prager Tagblatt*, 15 April 1906, 49–52.

Wieser, Friedrich. *Die deutsche Steuerleistung und der öffentliche Haushalt in Böhmen*. Leipzig: Duncker & Humblot, 1904.

Secondary Sources

Albrecht, Catharine. 'Pride in Production: The Jubilee Exhibition of 1891 and Economic Competition between Czechs and Germans in Bohemia'. *Austrian History Yearbook* 24 (1993): 101–118.

Albrecht, Catherine. 'The Bohemian Question'. In *The Last Years of Austria-Hungary. A Multi-National Experiment in Early Twentieth-Century Europe*, edited by Mark Cornwall, 75–96. Exeter: University of Exeter Press, 2002.

Carpenter, Kenneth E. 'European Industrial Exhibitions before 1851 and Their Publications'. *Technology and Culture* 13 (1972): 465–486.

Cohen, Gary B. *The Politics of Ethnic Survival: Germans in Prague, 1861–1914*. Princeton: Princeton University Press, 1981.

Corneließen, Christoph. 'Die politische und kulturelle Repräsentation des Deutschen Reiches auf den Weltausstellungen des 19. Jahrhunderts'. *Geschichte in Wissenschaft und Unterricht* 52 (2001): 148–161.

Filipová, Marta. 'Peasants on Display. The Czechoslavic Ethnographic Exhibition of 1895'. *Journal of Design History* 24, no. 1 (2011): 15–36.

Fuchs, Eckhardt. 'Das Deutsche Reich auf den Weltausstellungen vor dem Ersten Weltkrieg'. *Comparativ* 9, no. 5–6 (1999): 61–88.

Greenhalgh, Paul. *Ephemeral Vistas. The Expositions Universelles, Great Exhibitions and World´s Fairs, 1851–1939*. Manchester: Manchester University Press, 1988.

Großbölting, Thomas. *'Im Reich der Arbeit.' Die Repräsentation gesellschaftlicher Ordnung in den deutschen Industrie- und Gewerbeausstellungen 1790–1914*. Munich: Oldenbourg, 2008.

Hlavačka, Milan. *Jubilejní výstava 1891*. Prague: Techkom, 1991.

Houfek, Václav, and Tomáš Okurka. 'Německé muzejnictví a výstavnictví v Čechách'. In *Hledání centra. Vědecké a vzdělávací instituce Němců v Čechách v 19. a první polovině 20. století*, edited by Kristina Kaiserová and Miroslav Kunštát, 195–234. Ústí nad Labem: Albis International and Univerzita J. E. Purkyně, 2011.

Judson, Pieter M. *Guardians of the Nation. Activists on the Language Frontiers of Imperial Austria*. Cambridge, MA and London: Harvard University Press, 2006.

Kaiserová, Kristina, and Miroslav Kunštát, eds. *Hledání centra. Vědecké a vzdělávací instituce Němců v Čechách v 19. a první polovině 20. století* (Ústí nad Labem, 2011).

Kann, Robert A. *A History of the Habsburg Empire 1526–1918*. Berkeley: University of California Press, 1974.

King, Jeremy. *Budweisers into Czechs and Germans. A Local History of Bohemian Politics, 1848–1948*. Princeton and Oxford: Princeton University Press, 2002.

Kořalka, Jiří. *Tschechen im Habsburgerreich und in Europa 1815–1914. Sozialgeschichtliche Zusammenhänge der neuzeitlichen Nationsbildung und der Nationalitätenfrage in den böhmischen Ländern*. Vienna: Verlag für Geschichte und Politik / Munich: Oldenbourg, 1991.

Kretschmer, Winfried. *Geschichte der Weltausstellungen*. Frankfurt am Main and New York: Campus, 1999.

Melanová, Miloslava. 'Hlavní město Německých Čech? Liberec 1848–1918'. In *Hledání centra. Vědecké a vzdělávací instituce Němců v Čechách v 19. a první polovině 20. století,* edited by Kristina Kaiserová and Miroslav Kunštát, 285–307. Ústí nad Labem: Albis International and Univerzita J. E. Purkyně, 2011.

Melanová, Miloslava. *Liberecká výstava 1906*. Liberec: Kalendář Liberecka, 1996.

Okurka, Tomáš. 'Českoněmecká zemská výstava v Chomutově 1913'. In *Comotovia 2007. Sborník příspěvků z konference věnované výročí 550 let udělení znaku města Chomutova (1457–2007)*, edited by Petr Rak. Chomutov: Město Chomutov, 2008.

Okurka, Tomáš. *Všeobecná německá výstava v Ústí nad Labem 1903*. Ústí nad Labem: Město Ústí nad Labem, 2005.

Pohl, Heinz-Alfred. 'Die Weltausstellungen im 19. Jahrhundert und die Nichtbeteiligung Deutschlands in den Jahren 1878 und 1889'. *Mitteilungen des Instituts für Österreichische Geschichtsforschung* 97 (1989): 381–425.

Pozetto, Marco. *Max Fabiani. Ein Architekt der Monarchie*. Vienna: Edition Tusch, 1983.

Roche, Maurice. *Mega-Events and Modernity. Olympics and Expos in the Growth of Global Culture*. London and New York: Routledge, 2000.

Sayer, Derek. *The Coasts of Bohemia. A Czech History*. Princeton: Princeton University Press, 1998.

Slokar, Johann. *Geschichte der österreichischen Industrie und ihrer Förderung unter Kaiser Franz I*. Vienna: Tempsky, 1914.

Sutter, Berthold. *Die Badenischen Sprachverordnungen von 1897*. Graz: Böhlau, 1960–65.

Urban, Otto. *Die tschechische Gesellschaft 1848 bis 1918*. Vienna: Böhlau, 1994.

Wilde, Kurt. 'Die erste Deutsche Städte-Ausstellung 1903. Zeitpunkt und geschichtlicher Hintergrund'. *Dresdner Hefte. Beiträge zur Kulturgeschichte* 18, no. 63 (2000): 29–35.

Wingfield, Nancy M. *Flag Wars and Stone Saints. How the Bohemian Lands Became Czech*. Cambridge, MA and London: Harvard University Press, 2007.

Wörner, Martin. *Vergnügung und Belehrung. Volkskultur auf den Weltausstellungen 1851–1900*. Münster: Waxmann, 199.

5

The Nation for Itself: The 1896 Hungarian Millennium and the 1906 Romanian National General Exhibition

Samuel D. Albert

Though two languages, ten years, and 390 miles stand between Budapest's 1896 Millennium Exhibition (*1896 évi Ezredéves Országos Kiállítás)* and Bucharest's 1906 National General Exposition (*Expoziţia Generală Română*), they share a great deal in common. Both events clearly and exemplarily illustrate the essential – albeit variable – role that national self-identification plays in smaller exhibitions. Occurring within the era of great exhibitions – which commenced with the 1851 London Great Exhibition and concluded with the 1939 New York World's Fair – the two exhibitions generally follow the model those larger exhibitions provided but these two exhibitions, and a whole range of lesser-known and lesser-considered exhibitions, are located at the margins.[1] Their marginality (the word is not used as a term of opprobrium, simply as a descriptor) does not diminish either their importance or their significance. Indeed, because they do occur on the margins; because they differ so greatly in focus and purpose from the exhibitions seen as characteristic and defining for this period – because they are major events designed for a smaller group of people, they are all that more important.

The margins on which these – and the numerous similar exhibitions of the time – can be placed are founded neither on geography nor size. The exhibitions took place in cities which are today perhaps thought of as exotic and remote but at the time of its exhibition, Budapest, a capital of the Hungarian part of the Austro-Hungarian Empire, an empire whose borders ranged from Switzerland to Russia and from Poland to the Balkans, was one of the fastest-growing European cities. Bucharest, while at the eastern edge of Europe and perceived by most Europeans as part of the distant and exotic Balkans, enjoyed close political and cultural ties with France and Italy.[2] Nor is size the measure; the two exhibitions under consideration were neither as grand nor as large as many other exhibitions of the period, but they were not insignificant in size or in scope. Their celebratory impetuses are not the basis of their marginalization either; the historical events they commemorate, though in the distant past and of particular significance more for the countries celebrating them, were still meaningful for Europe as a whole: the early migration of nomadic tribes into the European continent or the vast civilizing reach of the Roman Empire were key moments in the shaping of general European history. The marginality of these two exhibitions, which cannot be disputed, is twofold: a period of self-imposed ideological marginalization and a later historical marginalization imposed by academics.

The Hungarian and the Romanian organizers had very clear visions of their respective exhibitions: what material would be presented, to whom this carefully chosen material would be shown, and for what purpose the exhibition itself was ultimately organized. The consciously limited and focused audience is one of the key factors in locating these two particular exhibitions on the margins. This deliberately and knowingly imposed limitation on the exhibitions by their organizers, which initially marginalized them, has been subsequently reinforced by academic writing and research on exhibitions, especially by the intellectual framework developed. World's Fair research and literature have long focused on the most obvious 'great exhibitions': the 1851 London Exhibition, the 1889 and 1900 Paris World's Fairs, or the 1893 World's Columbian Exhibition, to name some obvious examples.[3] These large and prominent exhibitions enjoy an oversized influence, shaping and affecting the forms of later expositions as well as their interpretations, but they should be thought of as the exception, not the rule. Their structures, whether they survive or not; their displayed artefacts, even if now dispersed; their publications, pictures, and postcards; their entertainments and commemorations, while all interesting and the most visible and lasting remnants of the exhibitions, do not tell the complete story of exhibitions in this period. Numerous smaller and regional exhibitions of this period have long been ignored and subsequently excluded from the theoretical framework which has developed. Since this framework is based on the larger exhibitions, the developed theory can neither completely nor easily reconcile the numerous smaller exhibitions and the vastly different purposes which spawned them, resulting in an analysis which only further normalizes the already existing marginalization. Only recently has this trend been recognized and partly rectified by scholars such as Alexander C. T. Geppert, whose recent work, *Fleeting Cities: Imperial Expositions in Fin-de-Siècle Europe,* specifically focuses on a select but illuminating collection of five exhibitions which he considers 'nodes' of an international web. Their origins and achievements vary, but they illustrate the processes by which exhibitions come into being, gain meaning, and are subsequently interpreted.[4]

In their organization and design: with all manner of material displayed and the creation of a monumental structure, a 'clou', around which the exhibition is organized, almost every exhibition of this era, even significantly smaller ones, follows the persistent model presented by the great exhibitions. But the financial outlay necessary for a single large building, whether permanent or not, is neither feasible nor practical for smaller exhibitions with limited agendas and equally limited budgets. While exhibitions as a whole are generally ephemeral, smaller exhibitions tend to be even more ephemeral, few physical remnants survive, further contributing to their marginality. Yet towards the end of the nineteenth century, these smaller, marginal exhibitions occurred with greater frequency and exerted an outsized influence. They were usually geared towards national or regional audiences rather than international ones and often commemorated nationally significant events, further strengthening their limited local meaning and appeal. It is within this self-limiting framework that the Budapest and Bucharest exhibitions must be understood.

Among these many marginal exhibitions numerous commonalities can be found in their origins, in their organizations, and in the realizations. But two very different ways of understanding them exist. These two interpretive modes are bound, in part, to regional history. Most marginal exhibitions can be understood as elements reinforcing a larger national identity; by acknowledging and celebrating an event of regional importance, the centre binds the national periphery more tightly to itself.[5] But this interpretation is predicated upon ideological and political unity within the nation: the desire to assert and reinforce the positive power of the centre/periphery connection rather than underscore its discontents. In the Habsburg Empire, composed as it was of a variety of ethnic and national groups, bound together by an externally imposed imperial framework, marginal exhibitions, though recognizing and acknowledging the hegemony of the Empire, at the same time subversively emphasize the ethnic and cultural identities of the constituent polities, nurturing those identities which would eventually tear the Dual Monarchy asunder.[6] While the multi-ethnic problem is evident at the Budapest exhibition – the Cisleithanian portion of the Empire was home to at least half a dozen different nationalities only one of which actually descended from the Magyar tribes celebrated by the Millennium Exhibition – it was dramatically downplayed, almost ignored, by the exhibition's organizers. And while the exhibition celebrated a very narrow Hungarian/Magyar identity, it did not ignore the existence of the greater, Austro-Hungarian imperial identity, in which the Hungarian nation, a condominium partner in the Empire, was invested.[7]

The nationality problem was also present at the Bucharest exhibition though it is more subtle. Romania, unlike Hungary, was not greatly ethnically diverse; the regions of mixed Romanian and foreign ethnicity were located outside of the Romanian Kingdom. The nationality question in Bucharest then was not so much internal ethnic discord as national unease. The exhibition, in its presentations and participants, was used by its organizers to lay claim to regions outside of then-Romania by documenting and celebrating the unredeemed Romania populations.

The Exhibitions: Budapest

The Hungarian Exhibition celebrated the putative 1,000th anniversary of the arrival, under the leadership of the chieftain Arpad, of the Magyar tribes in the Carpathian Basin, with seven of the tribes settling in what is now Hungary and three in Transylvania. The precise date of arrival is not known. A committee of experts of the Hungarian Academy of Science, commissioned by the government, determined that a 12-year window – 888–900 – framed the date of the Magyar conquest and settlement; a more precise and exact date could not be determined. In 1893, for a variety of reasons, some scholarly, some organizational, the Hungarian government settled upon the year 895 as the date which would be celebrated. Thus the millennium year was originally to be 1895, but government inaction prompted a one-year postponement, in the same way the 400th anniversary of Columbus' 1492 discovery of America took place in 1893 instead of the more accurate 1892. The Millennium Exhibition in Budapest was only one element of a year-long series of national and nationalist events.[8] In almost every city with a Magyar population, the Millennium was marked by a public ceremony, often featuring the dedication of a commemorative statue or monument in addition to public speeches. Within greater Hungary, seven major millennium monuments were erected: Ópusztaszer and Pannonhalma, both still located within present-day Hungary; in Munkachevo in Ukraine; in Slovakia on the Zobor hill near Nyitra (Nitra) and the Dévény hill in Poszony (Bratislava); in Brassó (Brașov), now in Romania; and on the Zimony hill in Beograd (Belgrade).

On 2 May 1896, the Millennium Exhibition was officially opened by the head of the Austro-Hungarian Empire, Emperor Franz Josef I, though in his Hungarian guise as I. Ferenc Jozsef, King of Hungary. A peculiarity of Austria-Hungary is that there were two governments, a Hungarian and an Austrian one, headed respectively by a king and an emperor, but the two were the same person, Franz Josef, a situation described as the 'personal union'. These multiple and overlapping Habsburg identities can be confusing; Franz Josef embodied numerous titles – reflections of Austria's happy territorial expansion through marital rather than martial means – but it is clear that this festive opening in Budapest was to be seen not as an Imperial – that is, an Austro-Hungarian event – but solely as a Royal Hungarian event.[9] After presenting a short speech in front of an imaginative reconstruction of Attila's tent, Franz Josef spent two hours touring the Exhibition. During the following week, members of the Royal Family – the emperor, his wife Elizabeth, and their daughters Gizella and Maria Valeria accompanied by their husbands – attended a commemorative mass in the Matyas Church on the Buda castle hill and observed a parade by the local policemen, firemen and national guardsmen. During their two-week stay in Budapest, different members of the family, guided by its organizers and its architects, repeatedly visited the Exhibition.

The Millennium Exhibition was held in Budapest's City Park, a long-favoured outing destination situated on the eastern edge of the city. In 1885 it had been the site of the National General Exhibition from which, 11 years later, only a few buildings still remained: Miklós Ybl's Royal Pavilion, a small three-room building constructed as reception and refuge space for Franz Josef I when visiting the 1885 Exhibition, and subsequently converted into a branch of the famed restaurant Gerbeaud, as well as the Industry Hall, the 1885 exhibition's 'clou', a large glass and steel building modelled on the Exhibition Hall of the Vienna's 1873 World's Fair. The park was connected to the urban heart of Budapest by the Andrássy Avenue, a long, Parisian-style boulevard which had been a major element of the urban rationalization Budapest had begun in the 1870s.[10] Unlike the typical Parisian boulevards on which it was modelled though, the Andrássy Boulevard comprised three distinct and differing sections; as the thoroughfare flowed from the centre of Budapest outward, building height requirements lowered and the street widened. Just before its debouchment at the Heroes Square on the edge of the City Park, the street was the major artery through a villa district of free-standing houses, a marked contrast to the five-floor apartment buildings which lined its start. To facilitate public access to the park from the city's centre and prevent traffic jams on the new Andrássy Boulevard, a subway line, the *Millenniumi Földalatti Vasút*/Millennium Underground Line – the first subway on the European continent and a simple cut-and-cover line – was constructed.[11]

To commemorate the Millennium, to fittingly end the Andrassy Boulevard as a major modern street, and to demarcate the main entrance to the exhibition, three enduring structures were built to the plans of the architect Albert Schickedanz: the Fine Arts Museum, the Art Pavilion, and the Millennium Monument.[12] The construction of these monuments illustrates the larger national goals of the Millennium celebrations; they would remain as permanent monuments to Hungary's thousand-year existence, even long after the numerous structures of the exhibition itself had been removed from the park.[13] It should be noted as well that these three structures were not actually in the park itself: they were all located in the Heroes Square.

The Millennium Exhibition itself comprised two major sections: the Historical Exhibition and the Modern Exhibition. The purpose of the Historical Exhibition was stated in the accompanying catalogue:

> On the occasion of the thousandth anniversary of the arrival of the Magyars, at the exhibition to be held in Budapest, the outstanding portion must be the historical exhibition which will illuminate the main phases of our cultural development: historical, architectural, and artistic.
>
> We must show what the cultural stand of our present-day country was in the ninth century, when our ancestors arrived, and how it has developed in the ensuing years up to the present day.
>
> The historical exhibition must be retrospective, grouping available historical materials into illustrative periods, with a focus on ecclesiastical life, cultural development, private life, and the conduct of warfare.[14]

The Historical Exhibition, located on a small island in the lake of the City Park, consisted of four temporary buildings constructed of staff (a building material consisting of plaster mixed with fibre), each in a style coeval with the artefacts contained and displayed within. On its own small peninsula stood the re-creation of a primitive fishing village; in the village's reed hut were dioramas and displays of aboriginal occupations: shepherding, hunting and fishing.[15] The three other buildings on the island, of more traditional forms and designed by the architect Ignác Alpár, featured architectural elements copied from historic structures throughout greater Hungary. These recreated features, as did the materials and artefacts displayed within, dated from the periods the buildings themselves represented.

The four major exhibition groups called for in the enabling legislation, had been expanded to encompass 12 distinct areas of Hungarian history: architecture, applied arts, theatre and music, history of education and science, government and politics, public festivities, and primitive occupations. The thousand-year history of Hungary was divided into eight discrete periods, and more than one period was displayed in each structure.[16] The building housing artefacts from the earliest period of Magyar history, 896 until 1300 – from the arrival of the tribes under Arpád until the accession of Matyas Corvinus – was characterized as Romanesque; the structure housing them consisted of a reduced copy of the Romanesque church at Ják, with an additional, completely invented cloister. The artefacts of the Matyas Corvinus (Hunyadi) epoch, the Golden Age of Hungary, which dated from 1301 until 1699, were housed in a conglomeration of Gothic and Renaissance buildings. While this date range far exceeds the actual period of Matyas' or even the House of Hunyadi's rule, it includes within it the period of the Ottoman Turkish occupation (from 1562, the Battle of Mohács, which signalled the end of an independent Hungary, until 1699, the Treaty of Karlowitz, which, with the cession of Hungarian territory to the Habsburgs, marked the end of Ottoman occupation though not the re-emergence of a sovereign and independent Hungary). Among the structures cited in this amalgamation were the Csütörtökhely church, the tower of the Schässburg town hall and the castle at Vajdahunyad, Matyas' recently restored ancestral home.[17] The latter two, drawn from Transylvania, constituted an acknowledgement of that region's cultural importance. The period of 1700 to 1867, commencing with the era of Empress Maria Theresa and ending with the emergence of an autonomous, nationally self-aware Hungary, was characterized as Baroque. Its structure derived from the buildings of the Vienna Hofburg which had only recently been expanded as part of Vienna's modernization and from Fischer von Erlach's Belvedere Palace, as well as the Hungarian Imperial Summer Residence at Gödöllő, one the most significant Baroque structures in Hungary. The peculiar periodization of the exhibition with the overly expansive dates associated with King Matyas and Maria Theresa are reflections of the exhibition's organizers' desire to mould the historical past to serve the present; the periodization downplays tragic moments in Hungarian history, subsuming them to more glorious occasions.

The artefacts within the Historical Display did not span the full thousand years which the Millennium celebrated. That display concluded with 1867, the year of the eponymous Compromise, which had inaugurated Hungary's transformation into a modern, capitalist, industrialized European state. Originally, a presentation on the 25 years, 1867–92, of Franz Joseph's reign was to have been the concluding section of the Historical Exhibition, but it became an independent display within the Modern Exhibition housed in its own pavilion, physically and mentally separating contemporary Hungary from the historical Magyars.

The Historical Exhibition, initially conceived as an afterthought and a supplement to the materials of the Modern Exhibition, grew in importance to become the focus of the exhibition and was ultimately seen by the organizers as the Exhibition's primary purpose. The Millennium Exhibition itself was intended to be a statement to the world by the Magyars about themselves; as Gábor Baross, the trade minister who sponsored the original enabling legislation, said: 'The entire cultured world will be witness, and what we can and will show there, must be of us, from our own strength.'[18] The Historical Exhibition was a life-sized, verisimilar record and textbook of Hungarian historical and Hungarian cultural achievements; it showed 'the history of centuries and highlighted moments of the architectural history of the Fatherland, from St. Stephen to the present'.[19] The exhibition organizers felt that the material they had brought together was so significant, that

> the gathered material of our cultural history [the Historical Exhibition] should be collected for posterity in a large illustrated work, which could not lack in any middle- or upper-school. Indeed, individual images from that work should be found on the walls in all primary schools and kindergartens, in order to educate the small ones about the outstanding incidents in our thousand year political history, that the feelings of the Millennium festivities should be imbued into the children at as young an age as possible, that they should learn from our history to honour the strength of our heroes and the natural virtue of our people in order to be prepared for the next thousand years.[20]

The copies and reproductions of history had become more valued than the originals.

Arrayed adjacent to the Historical Exhibition was the Modern Exhibition. Markedly larger than the Historical Exhibition and consisting of more than 100 structures, this portion of the Exhibition was devoted to presenting diverse aspects of modern Hungarian industrial and technological development and production. Mostly housed in their own individual and representational buildings, these exhibitions ranged from the Törley Champagne Pavilion and Restaurant to a historical Cyclorama to the Nobel Dynamite Pavilion. Several regional pavilions representing non-Magyar ethnic areas administered by Hungary – Croatia and Bosnia-Hercegovina – were constructed as well, presenting industrial and cultural achievements of the regions.[21]

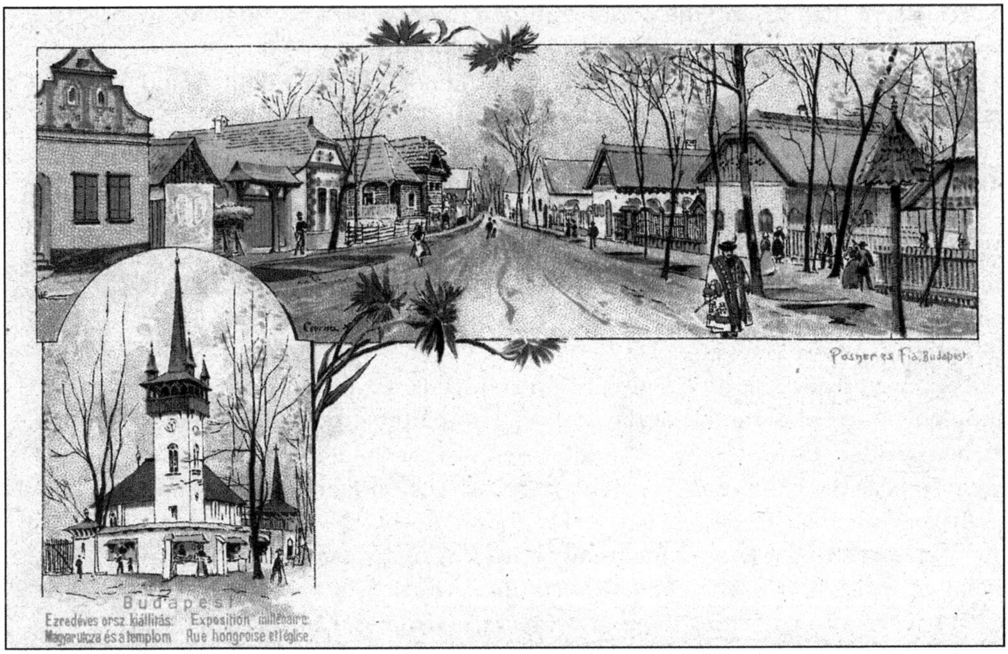

5.1 The 1896 Millennium Exhibition, Budapest,
The Ethnographic Village, postcard. Author's collection

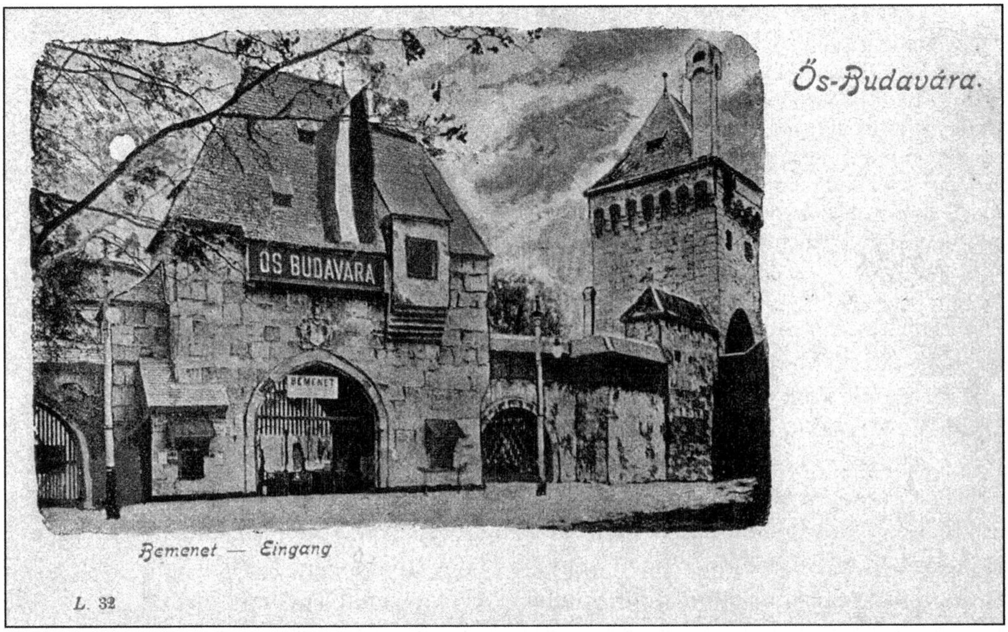

5.2 The 1896 Millennium Exhibition, Budapest,
Ős-Budavára, entrance, postcard. Author's collection

At the heart of the Modern Exhibition stood the Millennium, or Ethnographic Village (Fig. 5.1). Consisting of 24 house replicas stemming from all over greater Hungary, the village was a living display of rural Hungarian culture. Twelve of the constituent houses were Magyar; twelve were of 'nationalities', non-Magyars residing in greater Hungary: Bulgarians, Slovaks, Rumanians and Danube Germans. Besides the 24 display houses, which were based on the ethnographer János Jánko's well-researched and well-documented plans and financed and constructed by various counties, there were several other structures. Three Great Plain's shepherd huts and a gypsy hut – financed by the Archduke Joseph, son-in-law of Franz Josef's daughter Gisela – and several public building, including a model hospital, designed and financed by the Public Health section of the Hungarian Interior Ministry, filled out the display. Also located in the village were a school, built by the Ministry of Education and Religion, and a community hall, donated by the organizers of the Exhibition.[22] A church modelled on one in Kalotaszeg, a Transylvanian region renowned for its embroidery and site of Jánko's first ethnographic study, stood in the centre of the village.[23] The building, despite its form, was not used for sacred purposes but rather for the display of artefacts from Count Jenő Zichy's Caucasian expeditions,[24] which had sought to discover the ancient eastern origins of the Magyars.[25]

In and of itself, the Millennium Exhibition's ethnographic display was neither novel nor particularly noteworthy.[26] By the end of the nineteenth century, such displays were common at fair and exhibitions; regionally, both the L'viv 1894 Land Exhibition and Prague's 1895 Czechoslavic Exhibition featured ethnographic displays.[27] In all these cases, seemingly scientific ethnographic representation was employed as a tool to coalesce and more sharply focus national identity. What differentiates Budapest's village is the material chosen; not only are traditional ethnic Hungarian forms displayed, but also those of other nationalities resident under Hungarian political control. Rather than a simple ethnically pure national identity, a complex multi-ethnic one is presented: not a singular national identity, but a complex state identity, one subsumed to the Hungarian state, even if the constituent members themselves are not ethnic Hungarians. And one of the unstated but clearly evident goals of the exhibition overall was to support this Hungarian-led supra-national identity.

To the west of the Modern Exhibition was *Ős-Budavára*/Buda under the Turks, one of the two entertainment areas constructed for the Exhibition (Fig. 5.2). Designed by the well-known Austrian amusement park architect Oskar Marmorek, who had constructed *Alt-Wien* at the 1892 Vienna Music and Theatre Exhibition as well as the Vienna Prater's *Venedig in Wien*, *Ős-Budavára* was a verisimilar re-creation, based on period documentation, of the Buda castle hill at the time of the Habsburg victory over the occupying Ottoman Turks; but the various buildings which it recreated were dedicated to amusement: theatres, restaurants, cafes and concert stages.[28] In the centre of the complex stood a mosque modelled on one in Sarajevo. Daily, after purchasing tickets, exhibition-goers could observe the prayers of Muslims brought from the recently annexed Bosnia-Herzegovina.[29] A second amusement area, *Konsztantinápoly Budapesten*/Constantinople in Budapest, was located in the Lágymányos district of Budapest, adjoining the Danube, some distance from the actual exhibition, near the site of the present-day technical university, a placement necessary because of the extensive use it made of water.[30]

By the date of the exhibition, the presence of an amusement area, particularly one employing Middle Eastern exoticism for entertainment purposes was a regular and predictable occurrence, such as 1893 Chicago's 'Cairo Street'.[31] Yet *Ős-Budavára* illuminates one of the fundamental paradoxes underlying the exhibition: the difficulty, within a triumphant story of the thousand-year history of the Magyars, of presenting the roughly three centuries (1562–1867) when Hungary was occupied by the Ottoman Turks and subsequently ruled by the interloper Habsburgs. *Ős-Budavára* uses the exotic and the alien as entertainment, a minimization of national trauma. But the alien exploited is not distant but local: the amusement area references the very city where it is located. The Buda castle hill, the site on which it was 'modelled', was only a short underground ride away. On the edge of an exhibition which celebrated the triumphs of the Hungarian nation, the trauma of the 150-year Ottoman Turk occupation and the subsequent installation of a foreign, Habsburg, authority is lessened by being converted into amusements.

The Exhibitions: Bucharest

Almost ten years later to the day, on 6 June 1906, Carol I, King of Romania, opened the National General Exhibition in Bucharest (Fig. 5.3).[32] This exhibition, variously called *Expoziţia Generală Română*/Romanian National Exposition, *Exhibitio daco-romana*/Daco-Romanian Exhibition, or *Expoziţia Naţională*/ National Exposition, was a triple commemoration: it celebrated 25 years since Carol's coronation, 40 years since his accession to the Princeship of the United Principalities of Romania, and 1,800 years since Trajan's Dacian campaign.[33]

Unlike the Budapest exhibition, held on an already developed site and reusing several previously erected exhibition buildings, the Bucharest exhibition was built on virgin terrain. While this meant outlays for development and construction would be significantly larger, it also freed the exhibition from the necessity of conforming to any earlier layout or accommodating any existing structure. The site of the exhibition was the Filaret park, a rhomboid-shaped piece of land in the south of Bucharest which had been previously known as 'Field of Liberty'.[34] In the revolution of 1848, the park was where Wallachian revolutionaries had sworn allegiance to flag and government.[35] Preparation of the unencumbered 41-hectare site as suitable grounds for a national exhibition required extensive development: a total of 575,000 cubic metres of earth were moved, 4,000 trees and over 90,000 bushes and shrubs were planted, and 35,000 cubic metres of exhibition space were constructed

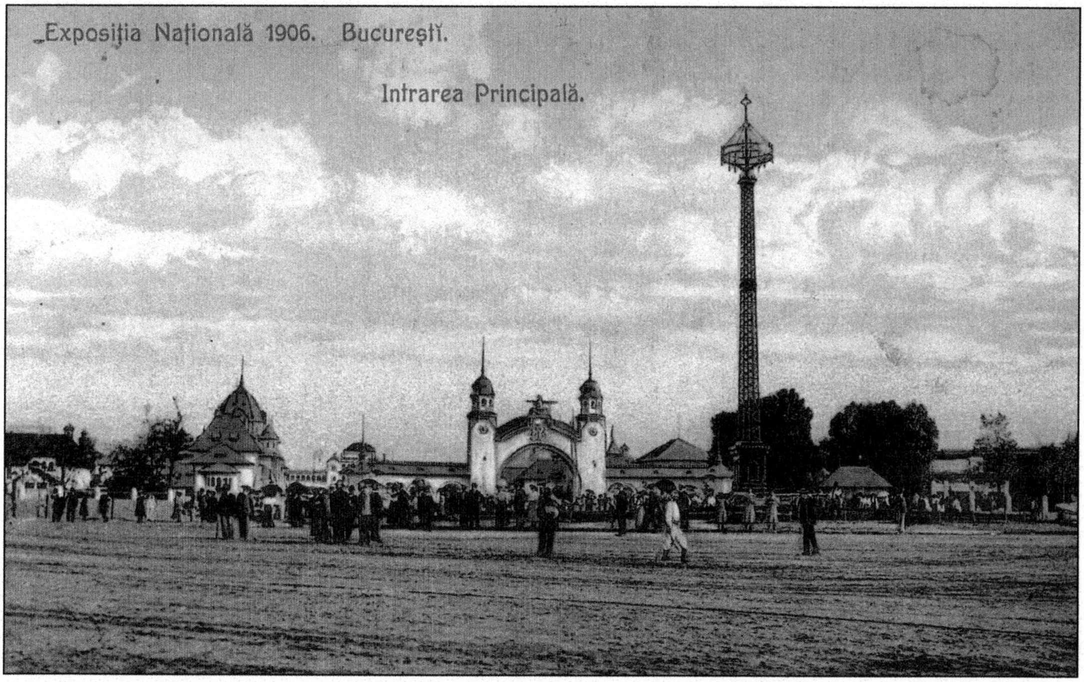

5.3 The 1906 National General Exhibition, Bucharest, main entrance, postcard. Author's collection

in the year leading up to the exhibition.[36] The exhibition's landscape designer, the French-born Eduard Redont, then practising in Bucharest – an example of the close cultural ties between the two countries – was unconstrained in his planning and could produce a design which reflected Romania's vision of itself. While the plantings were rather naturalistic, the buildings' disposition was formal, a beaux-arts arrangement with grand allées, symmetry and clear sight lines. The formality of the design and its expression of Romanian history were reinforced by naming of the allées, the squares and their intersections after great figures and principles from Romanian history.[37] The main entrance of the exhibition, at the north end of the north–south oriented park, was demarcated by a large triumphal arch. Southward from the arch ran a grand allée, the *Alea Independenței*/Independence Allée, which was flanked by major exhibition buildings, such as the Industrial and Agricultural pavilions. The allée passed through the King Carol I Esplanade, flanked on its west side by the Forestry and Mining pavilions and on its east side by name-giving Royal Pavilion, and then continued into Trajan Square. There, the allée – though it physically terminated at a viewing pavilion overlooking the lake in the middle of the park – visually continued on to the other side of the lake at *Muzeul trecutului nostru*/Museum of Our Past (now an art museum) on the same axis.

Dispersed throughout the park were a total of 165 buildings devoted to displaying aspects of industrial, economic, social and political development in the four decades of Carol I's reign, both within Romania proper and among the unredeemed populations. A number of institutions and industries had their own dedicated pavilions such as the Chamber of Commerce, the Ministry of

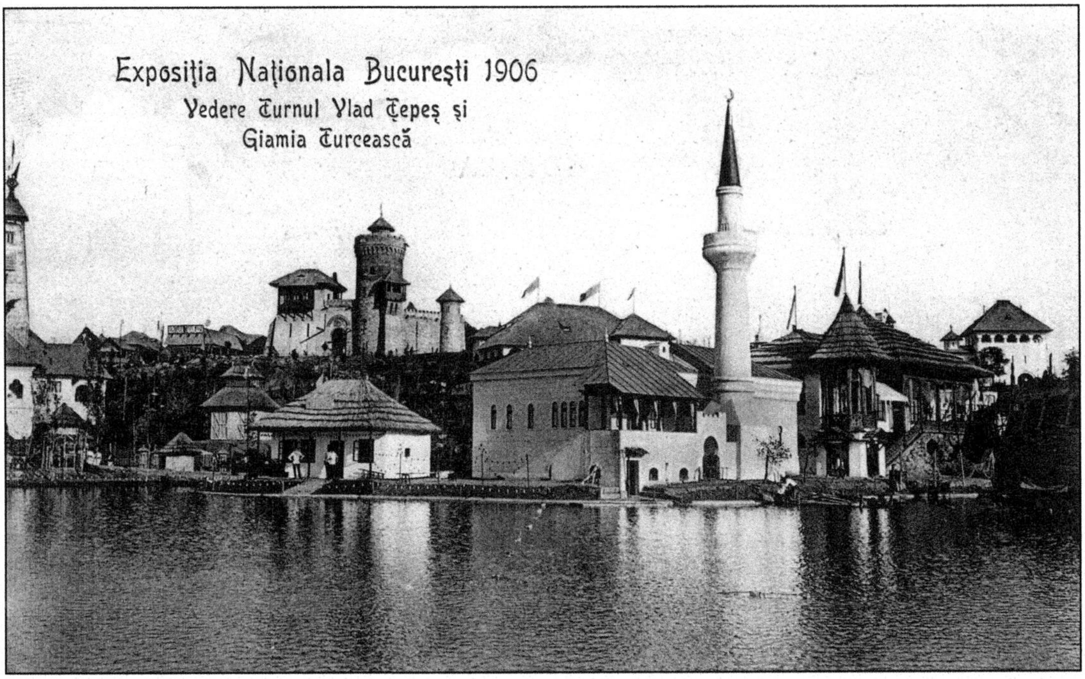

5.4 The 1906 National General Exhibition, Bucharest, view with Vlad Ţepeş' tower and Turkish mosque, postcard. Author's collection

Public Works, the Post Office and the Oil Industry while a special pavilion was devoted to domestic handicraft production. Informally interspersed among the larger pavilions were smaller structures. Some were regional pavilions, dedicated to or sponsored by ethnic Romanian populations residing outside of Romania proper. Because of Czarist restrictions, the Bessarabian pavilion had to be domestically financed, but the pavilions from Bucovina, the Banat and Macedonia were financed by the ethnic Romanian populations of those regions. A Transylvanian pavilion, organized and financed by ASTRA (*Asociaţiunea Transilvană pentru Literatura Română şi Cultura Poporului Român*/ The Transylvanian Association for Romanian Literature and Romanian Popular Culture), the major Romanian cultural organization of Hungarian-controlled Transylvania, was one of two displays devoted to the region; the other was a special section of the Hungarian display, housed within Hungary's own national pavilion.[38]

Replicas of several traditional Romanian building types (a Wallachian inn and a *Cula*, a traditional country house, as well as a reproduction of the Church of St Nicholas and a mosque) were to be found at the exhibition. While these buildings – some of which were initially temporary structures later reconstructed in more permanent materials – were used for displaying artefacts, a number of replicas at the exhibition were simply coverings for more conventional buildings: the park's water tower, which still stands today, is loosely based on the tower of the Bran castle of Vlad Ţepes, more commonly known as Dracula (Fig. 5.4). At the centre of the exhibition and of the park stood a lake which was used for aquatic activities: gondolas could be rented and a 'wasser chute', a flume ride, was sited on the shore. On the edge of the lake was a reproduction of Port Arthur, where the 1904 Battle of Port Arthur, the signal event of the Russo-Japanese War, was regularly re-enacted.

In addition to the regional and the national pavilions, several foreign countries had pavilions in Bucharest; Italy, France, Switzerland, Austria, Germany and Hungary all erected their own representational structures. The Danube River Commission had a pavilion located directly on the lake, a site of obvious importance. Established in 1856, at the close of the Crimean War and thus 50 years old at the time of the Bucharest exhibition, the international commission regulated shipping and the maintenance of the mouth of the Danube river, the economic lifeline which runs through Central Europe. The presence of these foreign pavilions, in marked contrast to the Budapest exhibition, which had almost exclusively (greater) Hungarian exhibitors, illustrates the differing focus of each exhibition.

One of the more peculiar, but not particularly unusual, features were the separate Austrian and Hungarian national pavilions. It was not the first occasion on which they had exhibited separately, even while still politically joined within the framework of the Compromise of 1867 as the Austro-Hungarian Empire. At the 1904 St Louis World's Fair, the same Empire exhibited in two different pavilions with two different organizing committees.[39] Often interpreted as manifestation of increasing dissonance within the Empire, at the Bucharest exhibition that explanation can only partially account for the separation. Of all the foreign countries represented, Hungary had the closest – and most tumultuous – relationship with Romania, sharing, in the region of Transylvania, a point of unity as well as contention. Though considered by Hungarians as the repository of the true Hungarian spirit, the rural antidote to the emerging urban illness of modernization, and the well-spring of true Hungarian design, Transylvania's population was not solely Magyar; significant German and Romanian populations with their own popular traditions as well as smaller pockets of other ethnicities were to be found there.

Invitations to the exhibition and the participation of foreign countries can be explained culturally and geographically, though the two are not mutually exclusive. Austria-Hungary's participation is clearly understandable as a geographic necessity: the Empire was physically and economically one of the largest neighbours of the Kingdom of Romania; Hungary additionally shared cultural ties. France, Italy and Switzerland were cultural relations; they were all Romance countries whose participation reinforces and underscores the importance of Trajan's conquest and the subsequent Romanization of Dacian culture, which, as the exhibition wished to present, lead directly to present-day Romanian culture. The German Empire was a special case, having a double geographic importance; it was a counter-weight to Austria-Hungary in vying for cultural superiority in Central Europe, but it was also culturally important: the King of Romania, Carol I, was of German origin.

While there were entertainment opportunities at the Bucharest Exhibition, too, such as the re-enactment of naval battles on the lake and concerts and dramatic performances in the Arena, there was no single amusement area comparable to either *Ős-Budavára* or *Konsztantinápoly Budapesten*. The exhibition's festive opening, which took place in the *Arenele Romana*/Roman arena, one of the originally permanent structures of the exhibition, was attended by the members of the Romanian government, the royal family, foreign guests and the accredited diplomats in Bucharest with the exception of the Imperial Ottoman Legation, which boycotted the event because the honour guard was the same regiment which had defeated the Turks in the Russo-Turkish War's siege of Plevna.[40]

The Exhibitions: Overlaps and Differences

In physical scope and historic range, the Bucharest exhibition was significantly smaller than the Budapest exhibition, but not in area or number of visitors. This was not accidental; the organizers of each exhibition had very clear but very different visions of their respective exhibitions: what they would present and to which audience they would present it. The Budapest exhibition was designed to highlight and celebrate the particularly Magyar aspects of Hungarian history. With its diminishment of the 350 or so years of occupation by the Ottoman Turks followed by the Habsburg dominance, with its emphasis on the early years of independent Hungary and the re-emergence of autonomous Hungary – populated not just by Magyars but other nationalities whose own histories were to be subordinated to that of the Magyars – the exhibition was an internally focused celebration. It minimized constituent nationalities' yearnings for independence and national acknowledgement within the Hungarian portion of the Empire to inferior positions. Almost all the products of the exhibition, from the poster advertising the exhibition, with its raucous illustration of the original Magyar tribes, to the Millennium Village with its clear subordination of the 'nationalities' – non-Magyar groups residing in Hungary – to the presence of *Ős-Budavára*, which transformed national tragedy into light-hearted entertainment, the Millennium Exhibition underscored and lionized the Magyar elements of Hungarian history. The Bucharest exhibition, in contrast, though it proclaimed Trajan's victory to be one of the moments celebrated, centred itself on the reign of Carol I, focusing on the present and on the national. In its documentation of the economic, political and cultural achievements of Romania during Carol's rule, it clearly presented him as a successor to Emperor Trajan who was responsible for the introduction of Roman culture to Romania.[41] This hoary romance heritage, the mystic chords of memory and culture which tied Romania to its Latin brethren, geographically removed as they might be, continued on after the exhibition. Throughout the 'teens and 'twenties of the following century, the ideal of the Romanians as descendants of the Romans and a cultural influence for good in the region continued.[42]

The focuses of the two exhibitions can be even more clearly discerned from commemorative materials produced. As with any exhibition of the period, marketing of souvenirs and keepsakes was vigorous. The subjects of these souvenir materials often illuminate the goals of the exhibition and its organizers better than any written explanation or manifesto; such objects are the exhibition writ small. But using such materials is not without pitfalls. Much of these ephemera are objects whose production quantities cannot be easily measured. Postcards, souvenir programmes and flyers, or commercial objects such as glasses, plates or mugs imprinted with images or dates record the popularity and the popular perception of exhibitions. But the ease with which these objects can be produced – while making them more fascinating – also makes them less reliable. More specialized objects, commercially bound books or medals and medallions are more reliable indicators because of the planning, investment and effort required for their production. While there is always a certain amount of kitsch produced as a result of exhibitions, voluminous serious documentation also emerges.

Publications produced for the 1896 Millennium show that it was clearly conceived as a celebration of the nation but not a celebration of the constituent ethnicities. All over greater Hungary, even in areas which had not originally been conquered by the Magyars, parades, speeches and the dedication of monuments took place, asserting Hungarian supremacy. The sheer volume of commemorative materials, guide books and other printed publications illustrates the seriousness with which this seminal event was promoted, as does the fact that the Hungarian government donated numerous copies of these to foreign libraries. Under the rubric 'Millennium' the standard reference for Hungarian books in print covering the millennium years lists almost three pages – close to 300 publications – including more than a page under the sub-heading 'Millennium Speeches and Celebrations', which consists of close to 125 documents.[43] Obviously some of these are only a few pages at best, but the fact that they were printed, that a record for posterity was created, renders their small size unimportant. The national significance and the national celebration, not just in Budapest – but throughout greater Hungary – of the Millennium, shows its importance and the desire of the organizers to clearly link the Magyar past with the Hungarian present. The imposition of a Magyar vision of the past over an area which was not actually part of that past was a type of internal colonization, the assertion of political and social power by a group which, though in political control, was only barely the numerical majority.

The Bucharest exhibition did not share that same goal; its purpose was the representation of the royal achievement, the lionization of Bucharest, the royal family, and the role that Romania played in the region under their leadership. While there was a clear historical element (the emphasis of the many connections between the ancient Roman past and the modern Romanian present), the purpose of the exhibition was not to celebrate the nation, but the nationality. While similar numbers of postcards with exhibition views seem to have been produced, the quantity of more serious printed material was only slight. Few speeches or commemorative addresses seem to have been printed, but numerous catalogues, documenting aspects of the exhibition, as well as several German-language guidebooks were produced.[44] The clearest indication of the organizers' goals can be seen in the production of prize and souvenir medals. Every medal bears the image of Carol I and most of them appose him with Trajan; the ideal behind the physical layout of the exhibition, the perambulatory connection between Carol and Trajan, is made visible and tangible here.

The Magyars, in their exhibition, looked back at their long historical presence in the Carpathian basin but ignored many of the emerging problems already foretelling of the downfall of the Dual Monarchy in which they were so invested. The vision they presented spoke of the achievements and successes of the Hungarian portion of Empire, not of the health and future of the Austro-Hungarian Empire as a whole. The Romanians, on the other hand, used their exhibition as a means of presenting the vision of a vibrant nation, hale and unified under the firm leadership of Carol I. And even as the achievements of his reign and kingdom were celebrated, the separation of Romanians outside of Romania, especially Transylvania, was underscored. But these very specialized, very tightly focused messages also mean that these exhibitions differ significantly from the great exhibitions of the era; this tight national focus places them on the margins.

The two exhibitions no doubt seem, as indeed many of the smaller exhibitions of this period might, quaint, in comparison to exhibitions such as the Paris *Exposition Universelle* of 1889 or 1900, the Chicago Fair of 1893 or the St Louis Fair of 1904, but only if one compares them without keeping in mind their respective purposes. These various exhibitions at the political or geographical margins provided, as they were designed and intended to, a regional or national point of accretion. They were never meant to compete with the grand international exhibitions in size, scope or purpose, and to use those larger exhibitions as a yardstick disregards the smaller fairs' essential purposes.

Both exhibitions, the Hungarian Millennium Exhibition of 1896 and the Romanian National General Exhibition of 1906, had lasting physical and cultural effects on their respective nations. The parks where the exhibitions were held still exist and can still be visited. Some of the structures fabricated in durable materials, either for the exhibition itself or afterward, remain as well, though often used for purposes far alien from their original ones.[45] But the effects of the exhibition, the cultural artefacts that we can see today, are more profound than the simple physical structures: each exhibition focused on a moment of national consciousness, one, the Hungarian exhibition, gazing inward and alone; the other, the Romanian, gazing outwards, presenting a national face to international partners. Despite these profoundly different orientations, in each case, the greatest beneficiary of the exhibition was the nation itself.

Notes

1 This conceptualization and periodization of exhibitions is drawn from: Paul Greenhalgh, *Ephemeral Vistas: The Expositions Universelles, Great Exhibitions, and World's Fairs, 1851–1939*, Studies in Imperialism (Manchester, 1988), esp. chs 1 and 5; as well as Tony Bennett, 'The Exhibitionary Complex', *new formations* 4 (1988): 73–102.

2 Vesna Goldsworthy has explored the literary construction of the 'distant' Balkans and many of the stereotypes assigned to it, which, it seems, has somewhat influenced understandings of not just the exhibitions, but cultural activities in the region. See Vesna Goldsworthy, *Inventing Ruritania: The Imperialism of the Imagination* (New Haven, 1998), esp. 73–86 and 101–111.

3 A good overview of these exhibitions (and numerous others) is provided in Paul Greenhalgh, *Fair World: A History of World's Fairs and Expositions, from London to Shanghai, 1851–2010* (Winterbourne, 2011). Also see Alexander C. T. Geppert, Jean Coffey and Tammy Lau, *International Exhibitions, Expositions Universelles and World's Fairs, 1851–2005: A Bibliography* (Berlin, 2006). Researchers also focus on more and more specialized aspects of these better-researched exhibitions. A recent example is Van Troi Tran, *Manger et boire aux Expositions universelles: Paris, 1889, 1900*, Tables des hommes (Rennes, 2012).

4 Alexander C. T. Geppert, *Fleeting Cities: Imperial Expositions in Fin-de-Siècle Europe* (Basingstoke, 2010). His discussion on fairs and exhibitions and the rise of 'exposition studies' at pp. 6–12 is particularly helpful for understanding the argument presented here.

5 See Wilson Smith, 'Old London, Old Edinburgh: Constructing Historic Cities', in this volume.

6 Numerous examples of these smaller exhibitions abound, but few have been analysed in any depth. One such analysis, however, is Marta Filipová, 'Peasants on Display: The Czechoslavic Ethnographic Exhibition of 1895', *Journal of Design History* 24, no. 1 (2011): 15–36.

7 It is important to distinguish between Magyar and Hungarian. Magyars are primarily those people descended from the original tribes which arrived with Arpad, though in the nineteenth century self-identification as Magyar was possible and indeed quite common. Within the trans-Leithanian portion of the Empire, commonly referred to as the Hungarian half, resided numerous ethnic groups: Slovaks, Bulgarians, Croats, Germans, Romanians as well as Magyars. Though they constituted only about half of the total regional population, ethnic Magyars enjoyed a privileged cultural and political position, a result of the Compromise of 1867. 'Hungarian', within this article, will refer to the Kingdom of Hungary, the trans-Leithanian portion of the Empire; 'Magyar' to the smaller, politically powerful ethnic group within Hungary.

8 These events are discussed in more detail in Bálint Varga, 'Contested Lieux de Mémoire in Central Europe: The Case of Devín and Brasov' (Master's

thesis, Central European University, 2007), esp. pp. 27–33; Bálint Varga-Kuna, *Celebration and Conflict: The Reception of the Hungarian Millennium Memorials in Theben and Kronstadt* (Vdm Verlag, 2008); and András Gerő, *Imagined History: Chapters from Nineteenth- and Twentieth-Century Hungarian Symbolic Politics*, East European Monographs (Boulder, CO and Wayne, NJ, 2006), esp. ch. 9, 'The Altar of the Nation: The Millennium Monument in Budapest'.

9 His official long title following the Compromise of 1867 read: 'His Imperial and Royal Apostolic Majesty, Francis Joseph I, by the grace of God Emperor of Austria; Apostolic King of Hungary, King of Bohemia, Dalmatia, Croatia, Slavonia, Galicia, Lodomeria, Illyria; King of Jerusalem, etc.; Archduke of Austria; Grand Duke of Tuscany, Cracow; Duke of Lorraine, Salzburg, Styria, Carinthia, Carniola, the Bukovina; Grand Prince of Transylvania; Margrave of Moravia; Duke of the Upper and Lower Silesia, Modena, Parma, Piacenza, Guastalla, Oswiecin, Zator, Cieszyn, Friuli, Ragusa, Zara; Princely Count of Habsburg, Tyrol, Kyburg, Gorizia, Gradisca; Prince of Trent, Brixen; Margrave of the Upper and Lower Lusatia, in Istria; Count of Hohenems, Feldkirch, Bregenz, Sonnenberg, etc.; Lord of Trieste, Kotor, the Wendish March; Grand Voivode of the Voivodship of Serbia, etc. etc.'.

10 Eszter Gábor, *Andrássy út*, A mi Budapestünk (Budapest, 2005) is a short but incisive guide to the history and construction of the boulevard.

11 Dorothy Barenscott considers the relationship between the modernization of the city's infrastructure and the development of national identity in some detail in 'Articulating Identity through the Technological Rearticulation of Space: The Hungarian Millennium Exhibition as World's Fair and the Disordering of Fin-de-Siècle Budapest', *Slavic Review: Interdisciplinary Quarterly of Russian, Eurasian, and East European Studies* 69, no. 3 (2010): 571–590, at 573–582.

12 A profoundly thorough documentation and analysis of Schickedanz's work can be found in Eszter Gábor and Mária Vero, *Schickedanz Albert (1846–1915): ezredévi emlékművek múltnak és jövőnek = Millennium Monuments for the Past and the Future: kiállítás a Szépművészeti Múzeumban, 1996. szeptember 19-tol december 31-ig* (Budapest, 1996).

13 Alice Freifeld thoroughly discusses the differences between the 1885 and 1896 events, with the shift from exhibition to commemoration. See Alice Freifeld, 'Marketing Industrialism and Dualism in Liberal Hungary: Expositions, 1842–1896', *Austrian History Yearbook* 29 (1998): 63–91.

14 'Az 1896-iki ezredéves országos kiállítás történelmi főcsoportjának programja.

'A honfoglalás ezredéves megünneplése alkalmából 1896-ban Budapesten rendezendő országos kiállítás kimagasló részét a történelmi főcsoportnak kell képeznie, mely Magyarország kulturális fejlődésének főmozzanatait: történeti, műemléki és művészi megvilágításában mutatja be.

'Szemlélhetővé kell tennünk, hogy minő kulturális állapotban' volt mai hazánk a IX. században, a mikor őseink azt elfoglalták s minő kulturális mozzanatok magaslanak ki a honfoglalást követő századokban egész a legújabb időkig.

'A történelmi főcsoport retrospektív jellege követeli meg, hogy a rendelkezésre álló anyagot hazánk történetének korszakai szerint csoportosítsuk, kiváló tekintettel az egyházi életre, a közszellem nyilvánulására, a magánélet viszonyaira és a hadviselésre.' Sándor Matlekovitcs, *Magyarország közgazdasági és közművelődési állapota. Az 1896. évi ezredévkiállítás eredménye* (Budapest, 1897), 4–5. Translation by the author.

15 Like all the displays, the Primitive Occupations had an extensive catalogue, showing the breadth of the display as well as the range of lenders called upon. See Géza Nagy and Otto Herman, *Amtlicher Katalog der historischen Hauptgruppe*, trans. Willibald Semayer (Budapest, 1896).

16 See *Amtlicher Katalog der historischen Hauptgruppe* (Budapest, 1896); Zoltán Bálint, *Die Architektur der Millenniums-Austellung: mit Unterstützung der Ausstellung-Direction des ungarischen Ingenieur- und Architekten-Vereines, sowie der Architekten, welche bei den Ausstellungsbauten mitwirkten* (Vienna, 1897).

17 The castle, while seen as a symbol of Magyar/Hungarian history, had been recently restored by Friedrich Schmidt and his Vienna-based students at the Akademie der bildenden Künste, an episode explored in greater detail in József Sisa, 'Neo-Gothic Architecture and Restoration of Historic Buildings in Central Europe: Friedrich Schmidt and His School', *Journal of the Society of Architectural Historians* 61, no. 2 (2002): 170–187.

18 Matlekovitcs, 2.

19 *Neue Freie Presse*, 3 June 1896, 2.

20 Béla Czobor and Emerich von Szalay, *Die historischen Denkmäler Ungarns in der 1896–er Millenniums-Landesausstellung* (Budapest and Vienna, 1897), 17.

21 Ausstellungsbureau der b. h. Landesregierung, *Bosnien und die Hercegovina auf der Millenniums-Ausstellung in Budapest im Jahre 1896* (Budapest, 1896). For a broader discussion of the influence of such pavilions, see Dragan Damjanović, 'Herman Bollé and Croatian Pavilions at the Exhibitions in Trieste (1882) and Budapest (1885 and 1896)', *Centropa* 10, no. 3 (2010): 231–243.

22 János Jankó, *Az ezredéves országos kiállítás néprajzi faluja* (Budapest, 1897), 8.

23 *Kalotaszeg magyar népe: néprajzi tanulmány* (Budapest, 1892).

24 Zichy organized and financed three expeditions whose purpose was to find and document the eastern origins of the Magyars. The first, in 1895, was to the Caucasus, the second, in 1896, was to Turkestan and the third and final one was the 1897–98 journey to China. For more see László Károly Marácz, 'Gábor Bálint de Szentkatolna (1844–1913) and the Study of Kabardian', in *Exploring the Caucasus in the Twenty-First Century: Essays on Culture, History and Politics in a Dynamic Context*, ed. Françoise Companjen, László Károly Marácz and Lia Versteegh (Amsterdam, 2010), esp. 35–39.

25 Jenő Zichy and János Jankó, *Kaukázusi és Közepázsiai utazásai* (Budapest, 1897).

26 A more detailed history of the village can be found in Rebecca Houze, 'Home as a Living Museum: Ethnographic Display and the 1896 Millennium Exhibition in Budapest', *Centropa* 12, no. 2 (2012): 131–151.

27 For Ľviv see Markian Prokopovych, *Habsburg Lemberg: Architecture, Public Space, and Politics in the Galician Capital, 1772–1914*, Central European Studies (Lafayette, IN, 2009), esp. 242–274; and for Prague see Filipová, esp. 22–27.

28 For more on the recreations of old towns within the exhibition space, see Wilson Smith, 'Old London, Old Edinburgh: Constructing Historic Cities', in this volume.

29 For more on Ős-Budavara see 'Die Millenniums-Ausstellung in Budapest (Rekonstruktion der Königsburg-Ofen)', *Neudeutsche Bauzeitung* 2 (1896), 33–36; Markus Kristan and Samuel D. Albert, *Oskar Marmorek: Architekt und Zionist,*

1863–1909, Veröffentlichungen der Albertina (Vienna, 1996). A more recent and profound consideration of the idea of *Alt-Wien* is Wolfgang Kos and Christian Rapp, *Alt-Wien: die Stadt, die niemals war*, Sonderausstellung des Wien Museums (Vienna, 2004).

30 For more on Konsztantinápoly Budapesten, and the Exhibition in general, see 'Konsztantinápoly Budapesten', *BudaPesti Negyed* 4, no. 1 (1996), 315–330.

31 For a recent analysis of 'Cairo Street', see István Ormos, 'The Cairo Street at the World's Columbian Exposition, Chicago, 1893', in *L'orientalisme architectural entre imaginaires et savoirs*, ed. Nabila Oulebsir and Mercedes Volait (Paris, 2009), 195–214.

32 Prince Karl of Hohenzollern-Sigmaringen had been elected Ruling Prince of Romania in 1866 after the overthrow of Alexandru Ioan Cuza. In1878, he declared Romania, which had been under the Ottoman Turkish Empire, an independent state. In 1881, he was proclaimed King of Romania.

33 Andrei Florin Sora, 'L'histoire à l'Exposition/L'histoire dans l'Exposition. L'Expostion Nationale Roumaine de 1906', in *In medias res: studii de istorie culturala*, ed. Andi Mihalache and Adrian Cioflânca (Iasi, 2007), 496.

34 A most complete history of the park, before, during and after the 1906 exhibition is Virgiliu Z. Teodorescu, *Un Parc Centenar – Parcul Carol I* (Bucharest, 2007).

35 Sora, 'L'histoire à l'Exposition', 496.

36 Sora, 'L'histoire à l'Exposition'.

37 Carmen Popescu, *Le style national roumain: construire une nation à travers l'architecture, 1881–1945*, Collection 'Art & société' (Rennes, 2004), 142.

38 This section produced one of the two catalogues devoted to Transylvania: *Catalogul secţiunei ungare a Exposiţiunei naţionale române din 1906* (Bucharest, 1906). The other was Expoziţunea generală română, *Catalogul etnografic al Românilor din Ungaria* (Bucharest, 1906).

39 See Gregory Knott, '"Felix Austria," or: Modernism and the Success of the Austrian Presence at the 1904 World's Fair', *Yearbook of German-American Studies* 41(2006): 77–86, for more about the politics of separate exhibitions. Some of the materials displayed at the Hungarian display, their meaning, and subsequent fate are also discussed in Juliet Kinchin, 'Modernity and Tradition in Hungarian Furniture, 1900–1938: Three Generations', *Journal of Decorative and Propaganda Arts* 24 (2002): 65–93.

40 An extensive discussion of the festivities can be found in Eleonora Cofas and Carmen Constantinescu, 'Expositiunea generala-romana din Bucureşti. Consideratii dupa noua decenii', *Muzeul national* IX (1997): 145–162.

41 Andrei Florin Sora, 'Imagini în imagini. Expoziţia Naţională Română (1906) în cărţile poştale ilustrate', in *Exercitii întru cunoastere: societate si mentalitati în noi abordari istoriografice*, ed. Mirela-Luminita Murgescu and Simion Câltia, Innovatio historiae (Iasi, 2003), 132–146.

42 The practical manifestations of this fascination are discussed in Robert Born, 'Römer und/oder Daker. Zur symbolischen Funktionalisierung der Antike in Rumänien von 1918 bis 1989', in *Neue Staaten, neue Bilder?: visuelle Kultur im Dienst staatlicher Selbstdarstellung in Zentral- und Osteuropa seit 1918*, ed. Arnold Bartetzky, Marina Dmitrieva and Stefan Troebst, *Visuelle Geschichtskultur* (Vienna, Weimar and Cologne: Böhlau, 2005), 257–271.

43 Géza Petrik, *Magyar könyvészet. Az 1886–1900. években megjelent magyar könyvek, hirlapok és folyóiratok, atlaszok és térképek összeállítása, tudományos, szak- és tárgymutatóval* (Budapest, 1913), 258–261.

44 Academia Româna Biblioteca, *Cresterile colectiunilor in anul 1906* (Bucharest, 1906), 47.

45 The structures of the historical exhibition in Budapest, which were reconstructed in more durable materials after the exhibition, now house the Agricultural Museum.

Bibliography

Academia Româna Biblioteca. *Cresterile colectiunilor in anul 1906*. Bucharest: Carol Göbl, 1906.

Amtlicher Katalog der historischen Hauptgruppe. Budapest: Kosmos, 1896.

Ausstellungsbureau der b. h. Landesregierung. *Bosnien und die Hercegovina auf der Millenniums-Ausstellung in Budapest im Jahre 1896*. Budapest: Pesti könyvnyomda, 1896.

Bálint, Zoltán. *Die Architektur der Millenniums-Austellung: mit Unterstützung der Ausstellung-Direction des ungarischen Ingenieur- und Architekten-Vereines, sowie der Architekten, welche bei den Ausstellungsbauten mitwirkten*. Vienna: A. Schroll, 1897.

Barenscott, Dorothy. 'Articulating Identity through the Technological Rearticulation of Space: The Hungarian Millennium Exhibition as World's Fair and the Disordering of Fin-de-Siècle Budapest'. *Slavic Review: Interdisciplinary Quarterly of Russian, Eurasian, and East European Studies* 69, no. 3 (Fall 2010): 571–590.

Bennett, Tony. 'The Exhibitionary Complex'. *new formations* 4 (Spring 1988): 73–102.

Born, Robert. 'Römer und/oder Daker. Zur symbolischen Funktionalisierung der Antike in Rumänien von 1918 bis 1989'. In *Neue Staaten, neue Bilder?: visuelle Kultur im Dienst staatlicher Selbstdarstellung in Zentral- und Osteuropa seit 1918*, edited by Arnold Bartetzky, Marina Dmitrieva and Stefan Troebst. Visuelle Geschichtskultur, 257–271. Vienna, Weimar and Cologne: Böhlau, 2005.

Catalogul secţiunei ungare a Exposiţiunei naţionale române din 1906. Bucharest: n.p., 1906.

Cofas, Eleonora, and Carmen Constantinescu. 'Expositiunea generala-romana din Bucureşti. Consideratii dupa noua decenii'. *Muzeul national* IX (1997): 145–162.

Czobor, Béla, and Emerich von Szalay. *Die historischen Denkmäler Ungarns in der 1896–er Millenniums-Landesausstellung*. Budapest and Vienna: M. Gerlach und Compagnie, 1897.

Damjanović, Dragan. 'Herman Bollé and Croatian Pavilions at the Exhibitions in Trieste (1882) and Budapest (1885 and 1896)'. *Centropa* 10, no. 3 (2010): 231–243.

'Die Millenniums-Ausstellung in Budapest (Rekonstruktion der Königsburg-Ofen)'. *Neudeutsche Bauzeitung* 2 (1896): 33–36.

Expoziţunea generală română. *Catalogul etnografic al Românilor din Ungaria*. Bucharest: n.p., 1906.

Filipová, Marta. 'Peasants on Display: The Czechoslavic Ethnographic Exhibition of 1895'. *Journal of Design History* 24, no. 1 (1 March 2011): 15–36.

Freifeld, Alice. 'Marketing Industrialism and Dualism in Liberal Hungary: Expositions, 1842–1896'. *Austrian History Yearbook* 29 (1998): 63–91.

Gábor, Eszter. *Andrássy út*. A mi Budapestünk. Budapest: Városháza, 2005.

Gábor, Eszter, and Mária Vero. *Schickedanz Albert (1846–1915): ezredévi emlékművek múltnak és jövönek = Millennium Monuments for the Past and the Future: kiállítás a Szépművészeti Múzeumban, 1996. szeptember 19-tol december 31-ig*. Budapest: Magyar Nemzeti Múzeum, 1996.

Geppert, Alexander C. T. *Fleeting Cities: Imperial Expositions in Fin-de-Siècle Europe*. Basingstoke and New York: Palgrave Macmillan, 2010.

Geppert, Alexander C. T., Jean Coffey and Tammy Lau. *International Exhibitions, Expositions Universelles and World's Fairs, 1851–2005: A Bibliography*. Berlin: Freie Universität Berlin, 2006.

Gerő, András. *Imagined History: Chapters from Nineteenth- and Twentieth-Century Hungarian Symbolic Politics*. East European Monographs. Boulder, CO and Wayne, NJ: Social Science Monographs, 2006.

Goldsworthy, Vesna. *Inventing Ruritania: The Imperialism of the Imagination*. New Haven: Yale University Press, 1998.

Greenhalgh, Paul. *Ephemeral Vistas: The Expositions Universelles, Great Exhibitions, and World's Fairs, 1851–1939*. Studies in Imperialism. Manchester: Manchester University Press, 1988.

Greenhalgh, Paul. *Fair World: A History of World's Fairs and Expositions, from London to Shanghai, 1851–2010*. Winterbourne, UK: Papadakis, 2011.

Houze, Rebecca. 'Home as a Living Museum: Ethnographic Display and the 1896 Millennium Exhibition in Budapest'. *Centropa* 12, no. 2 (2012): 131–151.

Jankó, János. *Az ezredéves országos kiállítás néprajzi faluja*. Budapest: Pesti Könyvnyomda, 1897.

Jankó, János. *Kalotaszeg magyar népe: néprajzi tanulmány*. Budapest: Athenaeum, 1892.

Kinchin, Juliet. 'Modernity and Tradition in Hungarian Furniture, 1900–1938: Three Generations'. *Journal of Decorative and Propaganda Arts* 24 (2002): 65–93.

Knott, Gregory. '"Felix Austria," or: Modernism and the Success of the Austrian Presence at the 1904 World's Fair'. *Yearbook of German-American Studies* 41 (2006): 77–86.

'Konsztantinápoly Budapesten'. *BudaPesti Negyed* 4, no. 1 (1996): 315–330.

Kos, Wolfgang, and Christian Rapp. *Alt-Wien: die Stadt, die niemals war*. Sonderausstellung des Wien Museums. Vienna: Czernin, 2004.

Kristan, Markus, and Samuel D. Albert. *Oskar Marmorek: Architekt und Zionist, 1863–1909*. Veröffentlichungen der Albertina. Vienna: Böhlau, 1996.

Marácz, László Károly. 'Gábor Bálint de Szentkatolna (1844–1913) and the Study of Kabardian'. In *Exploring the Caucasus in the Twenty-First Century: Essays on Culture, History and Politics in a Dynamic Context*, edited by Françoise Companjen, László Károly Marácz and Lia Versteegh, 27–46. Amsterdam: Pallas, 2010.

Matlekovitcs, Sándor. *Magyarország közgazdasági és közművelődési állapota. Az 1896. évi ezredévkiállítás eredménye*. Budapest: Pesti Könyvnyomda, 1897.

Nagy, Géza, and Otto Herman. *Amtlicher Katalog der historischen Hauptgruppe*. Translated by Willibald Semayer. Budapest: Kosmos, 1896.

Ormos, István. 'The Cairo Street at the World's Columbian Exposition, Chicago, 1893'. In *L'orientalisme architectural entre imaginaires et savoirs*, edited by Nabila Oulebsir and Mercedes Volait, 195–214. Paris: CNRS, 2009.

Petrik, Géza. *Magyar könyvészet. Az 1886–1900. években megjelent magyar könyvek, hirlapok és folyóiratok, atlaszok és térképek összeállítása, tudományos, szak- és tárgymutatóval*. Budapest: Az Eggenberger-féle könyvkereeskedés bizománya, 1913.

Popescu, Carmen. *Le style national roumain: construire une nation à travers l'architecture, 1881–1945*. Collection 'Art & société'. Rennes: Presses universitaires de Rennes [Bucharest]: 2004.

Prokopovych, Markian. *Habsburg Lemberg: Architecture, Public Space, and Politics in the Galician Capital, 1772–1914*. Central European Studies. Lafayette, IN: Purdue University Press, 2009.

Sisa, József. 'Neo-Gothic Architecture and Restoration of Historic Buildings in Central Europe: Friedrich Schmidt and His School'. *Journal of the Society of Architectural Historians* 61, no. 2 (2002): 170–187.

Sora, Andrei Florin. 'Imagini în imagini. Expoziţia Naţională Română (1906) în cărţile poştale ilustrate'. In *Exercitii întru cunoastere: societate si mentalitati în noi abordari istoriografice*, edited by Mirela-Luminita Murgescu and Simion Câltia, 132–146. Innovatio historiae. Iasi: Dominor, 2003.

Sora, Andrei Florin. 'L'histoire à l'Exposition/L'histoire dans l'Exposition. L'Exposition Nationale Roumaine de 1906'. In *In medias res: studii de istorie culturala*, edited by Andi Mihalache and Adrian Cioflânca, 491–504. Iasi: Editura Universitatii 'Alexandru Ioan Cuza', 2007.

Teodorescu, Virgiliu Z. *Un Parc Centenar – Parcul Carol I*. Bucharest: Muzeului Municipiului Bucureşti, 2007.

Tran, Van Troi. *Manger et boire aux Expositions universelles: Paris, 1889, 1900*. Tables des hommes. Rennes: Presses universitaires de Rennes, 2012.

Varga, Bálint. 'Contested Lieux de Mémoire in Central Europe: The Case of Devín and Brasov'. Master's thesis, Central European University, 2007.

Varga-Kuna, Bálint. *Celebration and Conflict: The Reception of the Hungarian Millennium Memorials in Theben and Kronstadt*. Vdm Verlag, 2008.

Zichy, Jenő and János Jankó. *Kaukázusi és Középázsiai utazásai*. Budapest: Ranschburg, 1897.

6

The Forefront of English Commercial Centres: Wolverhampton's Exhibitions of 1869 and 1902

Marta Filipová

By the stroke of a pen our pleasant [Hyde] Park – nearly the only spot where Londoners can get a breath of fresh air – is to be turned into something between Wolverhampton and Greenwich Fair.[1]

[Wolverhampton's] industries are many and varied, and its reputation as a commercial community and as the centre of social and political activity is world-wide.[2]

Wolverhampton's two attempts at joining the exhibitions network materialized in the South Staffordshire Exhibition of Arts and Industries of 1869 and the Wolverhampton Arts and Industrial Exhibition of 1902. As the first remark above (from *The Times*) indicates, as early as 1850, the image of Wolverhampton was not a positive one. The two ambitious projects could therefore be seen in part as a chance to rectify the reputation of the town and region as the antipode of nature and to turn it into an exemplar of vibrant economic, cultural and social life, 'the centre of social and political activity'.[3]

The international and universal exhibitions boom that followed the success of the Great Exhibition of 1851 affected many cities and towns in the UK that tried to copy and develop the exhibition model devised in London. Large cities like Edinburgh (1886), Manchester (1887) and Glasgow (1888) showed a similar will and ambition to showcase the best of their trade, manufacture and arts.[4] These spectacles, organized in order to promote local and national businesses and to educate citizens about the past, present and the future of regional industries and arts, also had other agendas. The most obvious one related to the conscious attempt to construct a local identity by using the displays of objects and related activities on the exhibition grounds.

Such motivations were visible even more in smaller exhibitions that took place around Britain in the second half of the nineteenth century and in a way recreated the universal exhibition model on a much more modest scale. Although such exhibitions could not claim to have an international reach or a long-lasting effect comparable to the legacy of, for example, the London exhibitions, they nevertheless demonstrate how the exhibition medium was used as an important tool for the creation of a cultural, geographical and political identity of the region. When considering 'exhibitions in the margins', the city of Wolverhampton in the West Midlands of England can be seen as a truly peripheral location to host an exhibition from the geographical as well as cultural point of view. Yet despite its provincial character in the nineteenth century, Wolverhampton was determined to put on two extensive events, in order to reinvent itself, if only temporarily. Its two showcases of regional industries, trade and culture, can be used as a pertinent case study of how the exhibition medium could be adapted to local needs.

Wolverhampton and the Black Country

The region in which the city of Wolverhampton is located is today the West Midlands, but historically it was part of South Staffordshire and popularly the area has been referred to as the Black Country. The Black Country is only an approximate geographical area in the English Midlands, and its borders are often disputed and contested, but it generally consists of the cities and towns of Dudley, Smethwick, West Bromwich, Oldbury and Wolverhampton.[5] The name refers to the historic existence of some of the most important industries in the UK in the nineteenth century – coal mining and ironworks. They also gave rise to the negative image of the region, because they were manifested in 'numerous collieries and furnaces, and the desolation of scenery'.[6]

The Black Country's landscape and environment had often been described by travellers, novelists or journalists in similarly unfavourable terms. Even before the middle of the nineteenth century James Mitchell, a Scottish antiquarian and scientist, noted that

> no one who has ever passed between Birmingham and Wolverhampton during the night can ever forget that scene. The blazing fires on every side, from the coal burning upon the ground, the flames proceeding from the chimneys of the great towers of the iron-furnaces present an impressive and even awful prospect.[7]

Charles Dickens likewise painted a grim picture of the Black Country, which he called a 'cheerless region' without much greenery.[8] The Wolverhampton of his book *The Old Curiosity Shop* had 'tall chimneys, crowding on each other, and presenting that endless repetition of the same dull, ugly form [... and pouring] out their plague of smoke, [which] obscured the light, and made foul the melancholy air.'[9]

Such images prevailed for a long time, but when the twentieth-century English critic and novelist Walter Allen quoted the comment from Betjeman's 1943 guide to *English Cities and Small Towns* to the effect that 'even a town like Wolverhampton looks splendid through Memory's telescope', he added, with a faint bit of optimism, that the author of the guide was in fact unfair: 'there are far worse places than Wolverhampton, even outside the Black Country.'[10]

In his architectural guide to Staffordshire, Nikolaus Pevsner was also quite uncompromising when he described the Black Country as consisting of six towns which 'could not be, urbanly speaking, much worse', but in an attempt to find a positive feature he added that in fact 'the unbiased traveller will find it less black than he had expected'.[11]

These perceptions were also based on the fact that throughout most of the nineteenth century the Black Country suffered from bad social and work conditions, high unemployment and overpopulation.[12] Alongside its desolated landscape, this affected the image and identity of not only the Black Country but also Wolverhampton too. Wolverhampton developed quickly, particularly at the beginning of the nineteenth century with the Industrial Revolution. Originally a small market town, it became a borough in 1848 and benefited from a network of canals and railways that linked it with other towns in the Black Country and with Birmingham.[13] In the second half of the nineteenth century, the main businesses included ironworks, japan ware, locks manufacture, galvanizing and, from the late 1870s, locomotive building.[14] Because of the number of jobs related to the natural resources, the population increased from 18,000 in 1821 to 50,000 in 1851, and to 68,000 at the time of the first exhibition in 1869.[15] Soon after, when coal mining and metalworking started declining, the influx of workers from Ireland, Wales and Scotland halted. Out-migration of miners to the North East or South Wales followed and by the second exhibition in 1902, the town's population growth started slowing down.[16] Nevertheless, by 1901 the town had reached almost 95,000 inhabitants.[17]

For most part of the century, coal was a powerful focus of local identity. When Queen Victoria visited Wolverhampton in 1866 to unveil the statue of the deceased Prince Consort during her first public trip after the Prince's demise, triumphal arches of coal were erected on the route of her procession.[18] One such arch consisted of coal, bars of iron, picks and shovels, that served as decoration, while another was composed of coal vases, baths and hardware goods 'arranged in an artistic design' with a 'pleasing and unique' effect.[19]

The industrial riches of the West Midlands were also proudly displayed in the Great Exhibition in London where the region made its mark successfully: the glass for the Crystal Palace came from the Chance factory in Smethwick, the iron was made at the Fox, Henderson & Co. Iron Works in Smethwick and Woodside, while the glass fountain located in the centre of the Palace, featured in many images of the interior, was created by the Ostler factory in Birmingham. There was also a display of a 2-metre high piece of coal from Tipton in the Black Country that, according to contemporary descriptions, weighed six tons.[20] However, until today, local historians point to the lack of acknowledgment in the history of the Great Exhibition of the fact that much of the material for the Crystal Palace was supplied by the industries of Staffordshire.[21] In this sense, recognition of the region's industrial or cultural contribution was already limited in the 1850s.

In the nineteenth century, the notion of the Black Country stemmed from the fact that the area had similar industries and manufactures, rather than from being a specific administrative regions with no fixed geographical borders. In this sense, its identity was based on an imagined sense of belonging together which consisted in, for instance, being related in one way or another with the local industries and trades or sharing the same dialect.[22] Importantly, regional identity can be created both from the inside by the people who inhabit the region and from the outside by those who act as observers of the region's typical traits, which, inevitably, often succumb to stereotypes. As the English historian Edward Royle argued, a 'region historically [speaking] is not a fixed concept but a feeling, a sentimental attachment to territory shared by like-minded people' and as much as national, or any other, identity, it can be understood as an imagined community.[23]

In the case of the Black Country, these like-minded people, who also stood behind the idea to organize the exhibitions, were the local enterprising classes. The middle classes in provinces increasingly identified with a particular location during the nineteenth century and this was often 'expressed through political and commercial interest groups ... involving new and assertive social classes who developed an image of themselves and their importance as a challenge to the dominance of London'.[24]

The identity they constructed was therefore in many ways dependent on its relationship to other actors in the same field. In the case of a regional identity, another urban unit which is often larger or more economically successful plays the crucial part of a model and/or competitor. London was the obvious role model of a metropolis which provinces and regions have used as a point of comparison. For Wolverhampton though, the key rivals were also more regional – other nearby towns, such as Stafford, but primarily Birmingham, which expanded throughout the nineteenth century from 70,000 inhabitants in 1801 to over half a million people by 1901.[25] For most of the nineteenth century, however, the region did not have a single regional capital and Birmingham was only slowly adopting this place. The city was a centre of highly skilled trades, including jewellery and gun making, while the towns in the Black Country, with Wolverhampton in the lead, could match this with their extractive industries and metal manufacture.[26] And while in the second half of the nineteenth century Birmingham was acknowledged to be more wealthy, expensive and with a more obviously identifiable middle class, Wolverhampton and the rest of the Black Country did have an active and 'progressive business elite'.[27] The predominant social composition was, admittedly, working-class, but the Black Country's so-called elites grew steadily out of the region's reasonable prosperity, demographic stability and close links between the towns that benefited from the ties to other urban unities, to the region and to the nation.[28] The middle and upper-middle classes consisted mainly of local industrialists who owned the largest companies in the region (such as Mander, Chance, Chubb), tradesmen, clerks and shopkeepers, who were relatively poor and small, yet they exercised substantial cultural, political and economic influence.[29] Although the few aristocrats of the Black Country, including Lord Granville and the Earls of Dudley, tended to stay outside of the region due to the increasing pollution and despoiling of the countryside, they also played an important role in the local economy and cultural and social life, investing – for example – in regional transport, such as canal and railway development. They were also involved in the two exhibitions.

The South Staffordshire Exhibition of Art and Industries, 1869

The staging of events that aspired to be comparable to the Great Exhibition could be understood as an attempt by regional administrations and town councils to showcase industrial, cultural and economic progress, to increase the competitiveness of local enterprises and industries and to put a specific, desirable image on display. The two exhibitions in Wolverhampton were thus the town's attempt to take a lead in the exhibitionary efforts of the region, especially because there was no immediate competition available. None of the other towns in the Black Country and not even Birmingham entered the arena of great and costly exhibitions on the scale as Wolverhampton, and as such the exhibitions of 1869 and 1902 were an attempt to gain advantage over its regional rivals.

While they copied some elements of the Great Exhibition in London and the 1867 *Exposition Universelle* in Paris, the organizers of the 1869 exhibition in Wolverhampton also took inspiration from smaller-scale events that had been held previously. These were often linked to the promotion of and instruction in specific regional manufacture and were most often organized by local mechanics' institutes in a hall or a single building, turned into an exhibition space just for this occasion. The earliest events in Wolverhampton were held in 1839 and 1840 and were associated with George Wallis (1811–91), an art educator and artist born in Wolverhampton, who became involved in the organization of the 1851 Exhibition and later worked as the keeper of the South Kensington Museum collection.[30] Both of these early events were rather limited in scale and contained a 'small but interesting' display of japan ware, locks, hardware, iron, steam engines and geological specimens from South Staffordshire as well as various paintings and other works of art.[31]

Similar small exhibitions were staged in other cities of the region too. Wolverhampton's largest competitor, Birmingham, entered the exhibitions field in 1839 with the first recorded exhibition of 'a few good pictures and sundry specimens of manufactures'.[32] A more comprehensive event in the same year was linked with the Birmingham Mechanics' Institute, an organization that was responsible for 'cheap instruction of the members in the principles of the arts they practise and in various braches of science'.[33] Displaying mechanical models, machinery, curiosities and so on, it was repeated next year, however, with a grave loss of profit and therefore it was discontinued.[34] Nevertheless, in 1849, the Institute also organized the Exhibition of the Manufactures of Birmingham and the Midlands Counties, which was briefly visited by Charles Dickens and Prince Albert.[35] The Industrial Exhibition led to the opening of a permanent exhibition building at Bingley Hall in 1865, which was attended by over 160,000 people.[36]

The first significant and substantial event, however, was the South Staffordshire Industrial and Fine Arts Exhibition of 1869 held at Molineux House, which was owned by a local business tobacco entrepreneur, Oliver McGregor, and which had grounds to the north-east of Wolverhampton's city centre also used for the event. Organized by a committee of local representatives from the Town Hall and businessmen under the presidency of the Earl of Lichfield, it was preceded by long preparatory and organizational works and a construction of an ephemeral pavilion to host a large array of exhibits.[37]

The Molineux House, built in 1720, was used as an entrance to the site with the temporary exhibition building of iron and glass located in the garden. In total, the building and the surrounding grounds contained approximately 130 industrial and over 700 artistic exhibits. The industrial displays consisted of products by local businesses mainly in iron and steel, such as nails and bolts, wrought iron frames, fences and chains, but also of crafted objects from the region, including displays of jewellery, crystal glass and locks and keys of the Chubb company.[38] Sciences were also represented by, for example, a nine metre long diagram of the solar system, and demonstrations of the use of electricity in communication as exemplified by various batteries, engines and the telegraph.[39] A large steam engine by the Cornwall works from Smethwick, a model of a mine with state-of-the-art equipment from Wigan Coal and Iron Company and various items of agricultural machinery by Henry Denton of Wolverhampton were amongst the larger objects on display.[40]

A common element in international exhibitions was the display of colonial objects and peoples. Regional exhibitions often attempted to include these 'exhibits' too, although they often faced the issue of limited resources and means. A few colonial objects were included in Wolverhampton to represent recent British exploits in Abyssinia, which included the 1868 armed expedition to free several missionaries held in the Ethiopian Empire. The display comprised a glass case with trophies, such as Ethiopian ceremonial robes, seals and weapons from the ownership of Queen Victoria and the Lords of the Admiralty, which served as a reminder of British overseas powers. Apart from such African ephemera, Indian silverware, ivory and wooden carvings, silks and swords were also borrowed from the India Museum in London.[41]

By and large, the colonial display was rather limited in size especially when compared to other exhibitions in the UK, such as the Great Exhibition or the 1862 International Exhibition in London. In this respect, the provinciality of the event was undeniable. Nevertheless, the organizers were conscious of the power of the visual spectacle that the exhibition represented. A display of flags and banners was put up on 11 May 1869 on the opening day of the

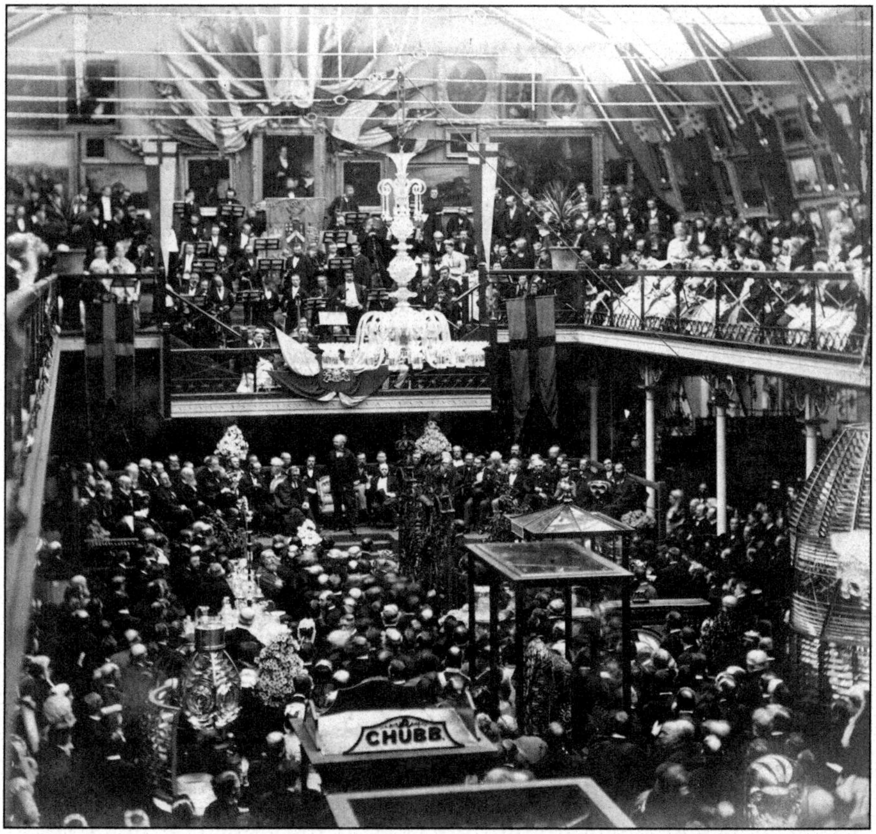

6.1 Opening of the South Staffordshire Exhibition, 1869, photograph. Wolverhampton Archives & Local Studies

South Staffordshire Exhibition in the town, and its decorations were similar to those accompanying the Queen's visit in 1866 (Fig. 6.1). Shops were closed so that people could attend or witness the procession of the local nobility, clergy, gentry 'and other influential personages' leading to the exhibition grounds.[42] The choreography of the parade was also reminiscent of the one that accompanied the Queen's visit to Wolverhampton. And, alluding to the Great Exhibition, the Mayor of Wolverhampton, Moses Ironmonger, claimed that 'the example then set had been followed by other nations and in different parts of the kingdom, until South Staffordshire now came to have its [own] Exhibition.'[43]

The event was officially opened by a range of aristocratic figures and influential businessmen of national as well as local importance. In addition to the Mayor, others present at the opening included Lord Granville, who had also been involved in the Great Exhibition as a commissioner in the Royal Commission for the Exhibition of 1851, the Earl of Lichfield and John Chubb, the locks and safes manufacturer. The opening ceremony, which took place in the main hall of the Molineux House, saw the ground floor and galleries filled up by standing and seated men and women. Open for five months until 29 October 1869, the Exhibition attracted visitors from all over Staffordshire as well as further afield. The final attendance figure was about 220,000, which included more than 10,000 children.[44]

Wolverhampton's 1869 contribution to the exhibitions network therefore combined the application of established features of its larger precedents in London (1851, 1862) and Paris (1867), seen in the content or the accompanying events, with the emphasis on displaying the local identity embedded in local products (such as coal, engines, locks). How successful and long-lasting this effort was will be discussed shortly.

The Wolverhampton Arts and Industrial Exhibition, 1902

Whereas the 1869 event referred to the South Staffordshire region in its title, the 1902 exhibition, called the Wolverhampton Arts and Industrial Exhibition, made the association with the city of Wolverhampton even closer (Fig. 6.2). In many ways inspired by the Glasgow Exhibition in the previous year, its aim was to become a cosmopolitan event with national and international exhibitors.[45] It was initiated by a local businessman, Thomas Graham, the owner of the *Express and Star* newspaper, and supported by other industrialists as well as the town council. The exhibition under the presidency of Earl of Dartmouth opened on 1 May in what is today the West Park located to the west of the city centre, with the fine art section housed in the Wolverhampton Art Gallery. The opening was attended by prominent guests who gave the event a special significance: their Royal Highnesses the Duke and Duchess of Connaught and Strathearn (the former being the brother of King Edward VII) were followed by a number of other people of importance, including the

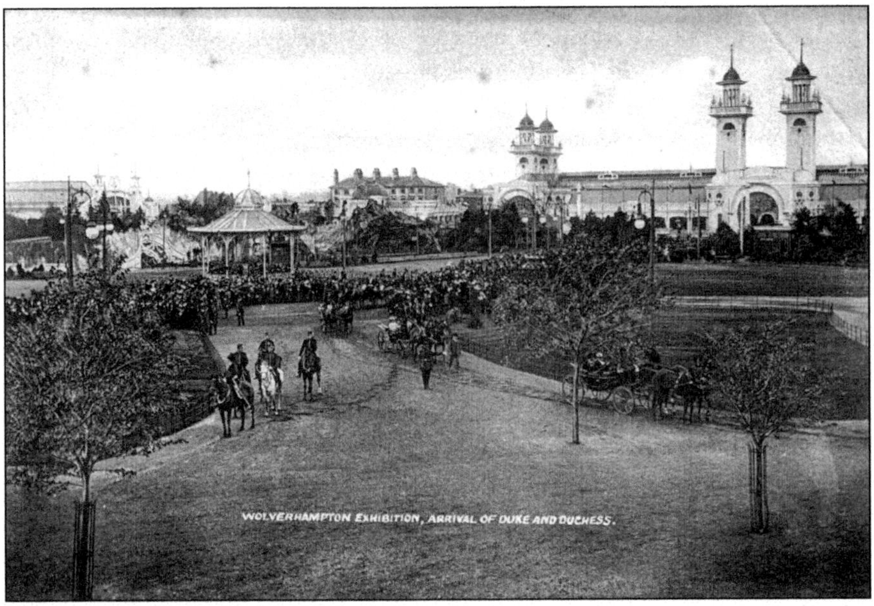

6.2 View of the Wolverhampton Arts and Industrial Exhibition, 1902, photograph. Author's collection

London Mayor with his sheriffs.[46] The visit was carefully staged and the party of guests, who arrived on a special train from London, was first taken to the local cultural institutions, the Art Gallery and the Concert Hall, after which they attended and officially opened the Exhibition.[47]

The exhibition was larger than that of 1869 and this reflected the ambitions of the organizers and exhibitors in many respects. As in the South Staffordshire exhibition, the grounds of the park were turned into a place of entertainment and instruction. The three main temporary buildings, erected here, were the Industrial Hall, Machinery Hall and the Canadian pavilion designed by Robert Walker and Thomas Ramsey of Glasgow, who placed them in a crescent. The two architects continued to work for national exhibition grounds at the Scottish National Exhibitions in Edinburgh (1908) and Glasgow (1911), while Walker collaborated with another Glasgow-based architect, James Miller (1860–1947), on a Canadian Pavilion for the Glasgow International Exhibition of 1901.

The inclusion of three temporary buildings made the exhibition grander in scale as well as in the designs. The pavilions were in part inspired by contemporary architecture, including Macintosh's features, with arches and emphasis of the central entrances, prompted in part by colonial architecture often seen on large exhibition grounds. In this respect, Wolverhampton Arts and Industrial Exhibition established concrete links with other international exhibitions and became a part of the exhibition network by creating more spectacular and elaborate buildings just for the purposes of the temporary displays than in 1869.

The organization of objects was also more thought-through and organized. The largest exhibition building of 107 × 52 metres – the Industrial Hall – was divided into three bays which contained displays by local and British manufactures, including silver work, safes, paints, upholstery, ironwork and glass, but it also included exhibits from abroad, such as Japanese arts and crafts and woodwork and carving from India. Some companies that debuted in the 1869 exhibition, such as Chubb's locks and safes or the Talbot bicycles, reappeared, while many made their first show with innovative displays. This included, for instance, a presentation of the Heinz company which offered tastings of soup, baked beans or tomato sauces served on a biscuit.[48]

The Machinery Hall was slightly smaller (107 × 40 metres) and was also split into three bays which presented 'instructive, interesting, and wonderful aggregation of machinery that claims the attention of the visitor'.[49] These comprised, for instance, the printing press, weaving machines and boilers and generators at work for the actual exhibition.[50] The Canadian pavilion, paid for by the Canadian government, housed the natural resources of the country, such as agricultural goods, fruit products, minerals, meat, whiskey and canned goods under the notion that 'Canada is indeed the land blessed with "milk and honey" – the land for emigrants'.[51] The most impressive exhibit, 'probably the best of the kind ever seen in this country', was over 500 specimens of grains, grasses and fodder plants in bottles.[52] More perishable foodstuffs, including butter, cheese and eggs, were placed in a large refrigerator. Samples of mineral resources, such as gold ores, iron and steel exhibits, clays, and especially timber, were accompanied by photographs of the different varieties of Canadian trees and paintings of the landscape.

Additional exhibits in special pavilions presented the latest inventions or improvements in building or engines or pneumatic tools that could be inspected while working. To facilitate entertainment, a concert hall was built offering two performances of the Continental Orchestra a day, recitals, 'biograph displays', or a ping-pong tournament.[53] On top of that, an illuminated spiral toboggan, the Canadian water chute (a frequent addition to world fairs and great exhibition but the first one in the Midlands), a fairy river, various restaurants and tea rooms as well as the Hall of Laughter with 25 mirrors were built on the exhibition grounds.

Yet despite the spectacular and entertaining character of the exhibition and the fact that only in the first month it was attended by more than 245,000 people, and 1.5 million of people entered the gates by the end of the event in November, the venture ended up with a loss of £30,000.[54] This failure is often ascribed to poor attendance, resulting from a lack of affordable means of transport, the unfinished feel of the site at its opening, as well as poor weather during the course of the event.[55]

Aims and Goals

Despite the attempt to connect with the international exhibitions around the UK, the main benefactors of the Wolverhampton exhibitions were the local community. Officially, the South Staffordshire Exhibition in 1869 had the clear financial goal of 'promoting the welfare of the Wolverhampton School of Practical Art and of the South Staffordshire Educational Association'.[56] The School was officially opened on 1 August 1854, inspired by the Great Exhibition; a number of individuals from Wolverhampton, including George Wallis and the local entrepreneur in paints and varnishes Charles Mander, had recognized the importance of education in arts and crafts, and the School was one of the first in Britain to have purpose-built premises.[57] The students of the School, financed by public subscription and a mortgage independent of government support, were provided with instruction in drawing, painting and modelling, but by the late 1850s the institution had fallen into financial difficulties and closed down temporarily in the early 1860s.[58]

Yet broader intellectual and educational benefit was also anticipated by the organizers to be brought to local artisans who 'might visit the treasures of art and the products of skilled industry, past and present, by suggesting ideas and planting seeds that should bear good fruit in the future, and instil to some extent ... a love and taste for the beautiful and the refined.'[59] The whole event was therefore meant to be 'agreeable and instructive to the people of Wolverhampton', as the organizers emphasized the need for better education in the town.[60] Instruction of children and the working classes was generally seen as one of the priorities when considering how to improve social conditions in the region in the second half of the nineteenth century. For instance, when the welfare of the so-called 'industrious classes' was discussed at a conference held in Wolverhampton in 1861, improvement in the offer of schools, colleges for working people and libraries but also music were identified as priorities.[61]

The 1869 exhibition took on the role of an educational event in many ways and, indeed, the number of municipal institutions offering other educational activities increased after it finished. While at the time of the exhibition Wolverhampton was 'assuming the character of a thoroughly up to date commercial centre', its cultural institutions were much less developed.[62] The Wolverhampton Literary Club was only founded in 1878, the local art gallery was not established until 1884, while the Grand Theatre opened in 1894 and the first moving pictures were brought to the town two years later. The first library opened in the year of the South Staffordshire Exhibition. More 'popular and rowdy' forms of recreation were common in the Black Country according to contemporary critics – these included wakes and fairs, prize-fights and animal sports, later to be joined by music halls.[63]

Although the Exhibition's real impact was predominantly on the local population, the organizers did aim at informing a wider audience about local industrial successes, too. In his opening speech, Earl Granville described the 1869 Exhibition's ambition: 'I think it was patriotic – if the word patriotism can be confined to a county – I think it was a good idea to show to the world what the manufactures of Staffordshire and its immediate neighbourhood are.'[64] *The Art Journal*, too, praised the organizers for their achievement and referred to Wolverhampton as a 'prosperous town'.[65] Still, it emphasized that it was primarily a regional event with the local population as the main recipients. The South Staffordshire Exhibition therefore ended up to 'excite widespread interest and benefit of all classes' locally, rather than permanently changing the image that the Black Country had outside of the region.[66]

Similarly, the primary purpose of the 1902 Exhibition was instruction and education of the local population but this time the organizers defined their national and international goal even more explicitly and wanted the exhibition 'to help the country in its great commercial battle'.[67] The catalogues and guidebooks communicated such ambitions too. What they called 'the modern wonders of the world in science, in art' were displayed 'for the inspection of the public'[68] and the event was 'wholly cosmopolitan in aim, and while it [reflected] Midland industries and Midland genius, it [was] in the fullest sense international in character'.[69] Wolverhampton and its many industries were presented in the exhibition guide as at 'the forefront of the great commercial centres of England'.[70]

Ultimately though, Wolverhampton's bold attempts at joining the national and international exhibitions network had limited impact and significance and failed to attract a wider attention. The main reasons for the lack of visitors from outside the region were mobility constraints, the lack of other entertainment and enticements the exhibition could offer and its location. For example in the case of the 1902 exhibition, despite the number of sites that provided special events, competitions or various adventures in the exhibition grounds, the town of Wolverhampton failed to attract visitors otherwise. Also attention of newspapers from outside the region was limited to the opening which was attended by members of the royal family and official guests from London.[71] No major national paper reported on the course of the exhibition or its results.

Art in a Regional Context

The programme of education promoted by both exhibitions included instruction in the visual arts, which was another important component of other exhibitions in the UK and abroad. The addition of artistic displays in the exhibitions that were held in predominantly industrial Wolverhampton could therefore be seen as a conscious attempt to present the town and the region as on a par with other major cultural centres that followed the rising popularity of displays of art and design in the second half of the nineteenth century. The art exhibits were also meant to attract further audiences.

In nineteenth-century Britain, displays of fine art at international exhibitions at this time were often marginalized at the expense of industrial and colonial exhibits, which more obviously showcased the nation's status as an industrial and colonial power.[72] Yet attention to fine art at British exhibitions underwent a significant change, and while at the 1851 Exhibition painting was scarce and sculpture was scattered around the building, in 1862 the South Kensington Exhibition gave fine art a dominant place and briefed the exhibiting countries to display what can be called 'national' art. By the late 1860s, presentations of national and international art collections had become indispensable components of large exhibitions, showing the rich artistic heritage of the organizing nation on one hand and the high level of art collecting by local aristocrats, members of the upper-middle classes and institutions on the other.

In the 1869 Wolverhampton exhibition, artistic objects were put on display together with the so-called industrial art, for the organizers aimed at presenting the region and the town as not only industrially advanced, but also cultured and artistic, as well as rich in the possession of artworks. Thus the distinction between fine art and industry that was typical for many other large exhibitions was completely absent in the case of the Wolverhampton exhibition.[73] Such joint display was partly prompted by the fact that the term 'industrial art', especially in the context of the 1869 Exhibition, referred to *all* objects of artistic value, even though they were – to a large extent – produced by machines. These works of art not only had the purpose 'to please the senses or intellect, but [... also to] perform a utilitarian function'.[74]

Thus when George Wallis described in his special report on the 1869 exhibition the contents of the art section, he listed what he called the 'art of the high class' as well as examples of the art industry of the past, with references to local manufactures.[75] These included objects as varied as steel toys, ornamental steel works, cabinet brass work, papier mâché Japanned tin, art metal work and fireclay and works in terracotta, mainly from local manufacturers, including, for example, Wedgwood medallions, Osler's chandeliers and crystal glass from Webb of Stourbridge, as well as 'a series of objects purchased from the Paris Exhibition of 1867 as illustrations of cheapness in art productions'.[76]

Also, the Vice-Chairman of the 1869 Exhibition, Nathaniel Neal Solly, had a similar view of the eclecticism of the art section, which he depicted as

> a survey of the varied and beautiful works of art …, the produce of the industry, taste and knowledge of past and present times, [which] may stimulate the rising generation in this neighbourhood, to rival what has been done before, and may call into activity, talent and genius which might have remained hidden, but for a spirit of healthy emulation, created by the sight and study of the art treasures of the South Staffordshire Exhibition.[77]

These 'beautiful works of art' of the present and the past, which contained about 700 oil paintings, watercolours, photographs, engravings, sculptures, art objects in metal or china, in cases or free-standing, were loaned from a number of owners: the local elite who wished to showcase its possessions and status, the South Kensington Museum and the India Collection in London.[78]

In general, though, the artistic and craft objects were scattered amongst other objects around the Molineux House and did not form a concise display (Fig. 6.3). Paintings alongside statues of various sizes and material were located on the balcony while the ground floor was dedicated to larger industrial objects, display cases and installations of trays, ceramics, ironware, jewellery or household items. Moreover, significant amount of room on the walls was dedicated to advertisement of local businesses, competing visually with the works of art.

The painting section featured, for instance, works by artists such as Turner, Gainsborough and David Cox but predominantly focused on local painters. French art, so common at displays in London, was almost entirely absent here and the few sculptures on show were predominantly neo-classical. Still, as Wallis remarked about the importance of the pieces, 'no such collection [had] ever been got together in Wolverhampton before and the effect upon the public taste of the district cannot fail to be of great values and importance in the future'.[79]

Fine art in Wolverhampton received more attention in the following years, at the Wolverhampton and Staffordshire Fine Art and Industrial Exhibition of 1884. This exhibition, organized by the Wolverhampton Corporation, was much smaller in size when compared to the 1869 and 1902 events because it

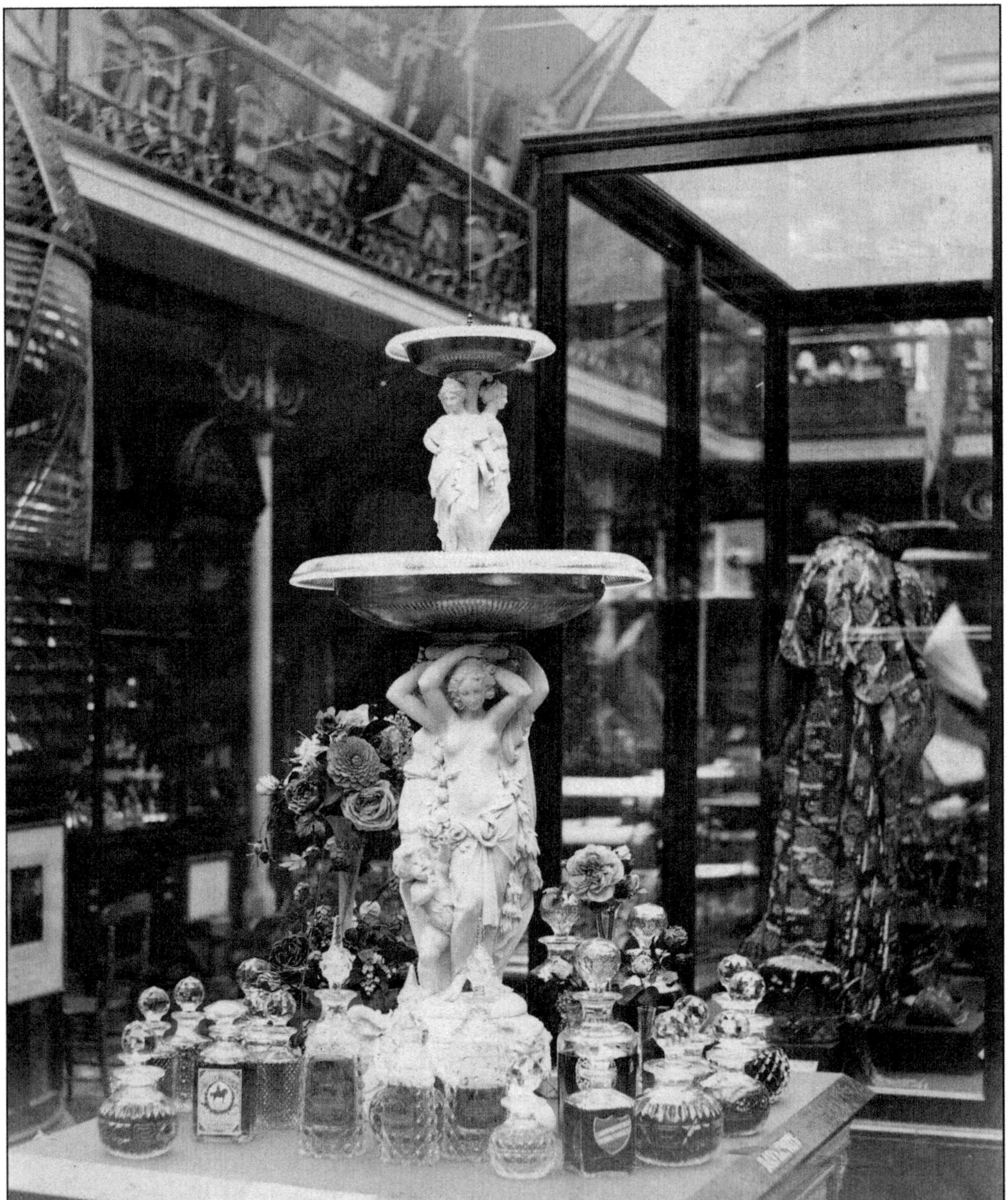

6.3 South Staffordshire Exhibition, 1869. View of the display in the Molineux House, photograph. Wolverhampton Archives & Local Studies

was confined to a few buildings in the city centre, including the newly built Art Gallery and the Industrial Hall.[80] Here card printers, carriages and a knitting machine were on display, while local businesses, most of which had also exhibited in 1869, presented their products for domestic and commercial use, such as stoves, ventilation systems and glass-blowing equipment.[81]

The 1884 Exhibition also extended to an Archaeological and Geological building which hosted a display of maps and photographs of places from all over the world, mining equipment, stuffed animals and fossils of the region. The archaeological section contained autographs of famous people, including royalty, American presidents and novelists, while the anthropological display presented objects from around the world, such as Japanese enamels, Chinese and Tibetan weapons, and garments from Zululand.

The fine arts in the Gallery, however, received more attention than previously. The Horseman Art Gallery, as it was originally known because of the monetary and intellectual involvement of the local contractor Phillip Horseman (1825–90), featured works loaned from other galleries, individual collectors from the Midlands and artists; some of these items were later turned into a permanent collection. In 1884, paintings were displayed in the rooms on the top floor and in the vestibule, while the ground floor was dedicated to needlework and works from the South Kensington museum, and engravings from Britain, Italy, Germany, Holland and France were shown in the basement. Over seven hundred pictures and 150 drawings ranged from British painters (for example, Cox, Hoppner, Turner or Reynolds) to old masters including Rubens, Rembrandt, Van Dyck and Frans Hals.[82]

By 1884 the separation of industrial objects, crafts and fine art was therefore a prominent feature and the segregation of art from industry was even more in evidence in the 1902 Wolverhampton Art and Industrial Exhibition, where the fine art section was given its own self-contained space in the Art Gallery. This approach was motivated both by the increasing emphasis put on artistic displays at exhibitions in London and elsewhere, which echoed in the regional exhibitions, but also by the prominence and attention that fine art received in Wolverhampton after the opening of the Art Gallery and its financial boost from the 1884 exhibition.

The generally larger 1902 exhibition could therefore display more clearly articulated and wide-ranging collection of artworks. Such emphasis on the cultural aspect of the exhibition was the organizers' attempt to turn the event into a more cosmopolitan and prestigious affair.[83] The fine art section in Wolverhampton was varied and comprised a large array of objects – original drawings, etchings, engravings, lithographs, mezzotints, woodcuts by Rembrandt, Dürer and Gainsborough, Japanese art (for example, by Hokusai), as well as manuscripts, early prints, recent prints, book binding and fans. Old masters and more contemporary painters including Thomas Gainsborough, George Romney, John Hoppner, Peter Lely, Alphonse Legros, William Turner and James Abbott McNeill Whistler were also displayed.

This time, French painting was represented through, for example, the work of Jean Baptiste Camille Corot (1796–1875) and the Barbizon School.

Alongside the historic pieces, the more contemporary works of art of 'The Modern School' were represented by, for example, the Scottish painter William Strang, the Impressionist Philip Wilson Steer (1860–1942) and a number of other artists linked to the New English Art Club, established in 1885 in London. The work of these artists, according to the commentator in the *Wolverhampton Exhibition Pictorial*, 'contains none of those conscious or unconscious archaisms which, while often charming, are brought in to veil defects in drawing'.[84] This was a reference to the Pre-Raphaelites and their followers who were, nevertheless, represented here too in one of the galleries through the work of Dante Gabriel Rossetti, John Everett Millais and Edward Burne-Jones.

The display, which appeared less provincial and more national and international than in 1869, was praised by the editor of the *Saturday Review* who left London and ventured to the provinces to visit the exhibition in the 'dingy enough quarter, the town of Wolverhampton'.[85] Despite his dismissive comment on the location, he praised the Exhibition and the organizers for retaining a defined taste in the selection of works where 'popular names have been of no account, youth and outsidership have been no hindrance to presence and good places; and the result is a limited, well arranged, and deeply interesting collection.'[86]

The attendance of about 1,000 visitors a day to the art section was also understood as a proof that the gallery was a 'happy indication of the interest that is taken in the high forms of culture'.[87] The displays of art, both fine and industrial, therefore became a key element of all Wolverhampton's exhibitions which aimed at educating the visitors as much as at showcasing the sophisticated culture in the region. And locating the arts in a separate section in the permanent Art Gallery also led to the creation of another public institution, both symbolic and material, which alongside the exhibition tried to create a positive image of the region on the outside.

Conclusion: Legacy and Aftermath

From very early on, Wolverhampton and the Black Country had been faced with negative comments about its people, landscape and deprivation made primarily by those from outside the region. The two exhibitions that took place in Wolverhampton in 1869 and 1902 demonstrated the town's attempt to present itself regionally, nationally and internationally as a place where industries, education and arts were very much alive and on a sophisticated level.

Yet while the earlier event focused on rectifying the image of the Black Country and was regional in its scope, the aspirations of the later one to secure Wolverhampton's position as a commercial and cultural centre nationally and internationally were more prominent. Despite all the good effort to incorporate the basic features of great exhibitions, such as displays of industrial objects, art or the inclusion of entertainment, the success of Wolverhampton's events was limited.

The reasons are varied, but most significantly neither exhibition succeeded in attracting larger audiences and attention from outside the region – the very demographic that they needed to convince. This failure can be linked with their location in the geographical margins as well as with the timing of the exhibitions. Whereas in 1869, the more local event could still relate to much of the region's industrial heritage, the much more ostentatious exhibition of 1902 which included a larger investment and greater ambitions came at time when many of the industries either moved out or were in decline. Importantly, by this time national and international exhibitions elsewhere became even grander and were more of a commonplace, eventually leading to what may be called an 'exhibition fatigue'.[88]

Such development, from provincial and small to international and large, followed the more general trend in exhibitionary efforts around the world, which led to an increase in an interconnectedness of countries in a network of trade and production as well as the arts. Regional and provincial exhibitions can therefore in many ways be taken as a response to the economic situation both in the region (as exemplified by Wolverhampton's industrial decline) and to the increasing international competition that was so prominent at international and universal exhibitions in the capitals and large cities around the world.

Notes

1 'Great Exhibition: Note on a Building', *The Times*, 27 June 1850, 9.

2 *Wolverhampton Art and Industrial Exhibition 1902. Illustrated Guide to Wolverhampton and District. Including Places of Interest worth seeing and a description of the Art and Industrial Exhibition with Map of the Town and View of the Exhibition 1902* (Cheltenham, 1902), 5.

3 *Wolverhampton Art and Industrial Exhibition 1902*, 5.

4 Perilla Kinchin and Julie Kinchin, *Glasgow's Great Exhibitions 1888, 1901, 1911, 1938, 1988* (Wendlebury, 1988), 17–19; William Ernest Henley, *A Century of Artists: A Memorial of the Glasgow International Exhibition, 1888* (Glasgow, 1889.). On Manchester, see for example Bill Newton, *Manchester Royal Jubilee Exhibition Old Trafford, 1887* (Manchester, 1998).

5 Harold Parsons, *Portrait of the Black Country* (London, 1986), 16.

6 Cited in Walter Allen, *Black Country* (London, 1946), 7. A reference to 'Black country with its smoke, and blasts and furnaces' is first recorded by the *Oxford English Dictionary* in 1834, when it was probably not a new expression.

7 Cited in Charles J. L. Elwell, *Aspects of the Black Country* (Kingswinford, 1994), 143.

8 Charles Dickens, *The Old Curiosity Shop* (London, 1840).

9 Dickens, 241.

10 Allen, 25.

11 Nikolaus Pevsner, *Staffordshire: The Buildings of England* (Harmondsworth, 1974), 19.

12 George S. Barnsby, *Social Conditions in the Black Country, 1800–1900* (Wolverhampton, 1980), 238.

13 Parsons, 71.

14 Parsons, 73–74, 76.

15 Mark Shaw, 'The Ecology of Social Change: Wolverhampton 1851–1871', *Transactions of the Institute of British Geographers* 2, no. 3 (1977): 332–348, at 336.

16 R. Lawton, 'Peopling the Past', *Transactions of the Institute of British Geographers* 12, no. 3 (1987): 259–283, at 266.

17 Parsons, 75.

18 *South Staffordshire Industrial and Fine Arts Exhibition. Official Catalogue* (Wolverhampton, 1869), i, in Wolverhampton Archives and Local Studies, LS/L606/8.

19 *South Staffordshire*, 45.

20 Parsons, 44.

21 Paul Anderton, 'Staffordshire and the Setting up of the Great Exhibition 1851', *Staffordshire Studies* 9 (1997): 35–68; Carl Chinn, 'Making the Modern World: Smethwick', *Black Country Memories* 2 (2005): 103–110.

22 Alan Everitt, 'Country, County and Town: Patterns of Regional Evolution in England', *Transactions of the Royal Historical Society* 29 (1979): 79–108, at 80.

23 Edward Royle, 'Introduction: Regions and Identities', in *Issues of Regional Identity. In Honour of John Marshall*, ed. Edward Royle (Manchester and New York, 1998), 4. Cf. Benedict Anderson, *Imagined Communities: Reflections on the Origin and Spread of Nationalism* (London, 1991).

24 Royle, 9.

25 Victor Skipp, *The Making of Victorian Birmingham* (Studley, 1996), 7.

26 Roger Ward, *City-State and Nation. Birmingham's Political History c. 1830–1940* (Chichester, 2005), 5.

27 Richard H. Trainor, *Black Country Elites. The Exercise of Authority in an Industrialized Area 1830–1900* (Oxford, 1993), 24.

28 Trainor, 40.

29 Trainor, 56–57 and 72.

30 On George Wallis, see, for example, G. C. Boase, 'Wallis, George (1811–1891)', in *Oxford Dictionary of National Biography*, ed. R. C. Denis (Oxford, 2004), at http://www.oxforddnb.com/view/article/28570.

31 George Wallis, *Special Report on the Local Manufacturers of South Staffordshire and East Worcestershire as Represented in the South Staffordshire Exhibition Held at Wolverhampton, 1869* (Wolverhampton, 1869), 3.

32 *Showell's Dictionary of Birmingham*, ed. Thomas T. Harman (Wakefield, 1969), s.v. 'Exhibitions', 68.

33 *Showell's Dictionary*, 68.

34 *Showell's Dictionary*, 68.

35 *Showell's Dictionary*, 68.

36 Chris Upton, *A History of Birmingham* (Phillimore, 1993), 146. *Showell's Dictionary*, 68.

37 'South Staffordshire Industrial and Fine Art Exhibition, 1869', *The Birmingham Daily Post*, 17 August 1868, 6.

38 'The Wolverhampton Exhibition', *Art Journal* 8 (1869): 367; Wallis, 4.

39 'The Industrial & Fine Arts Exhibition at Wolverhampton', *The Wolverhampton Chronicle*, 26 May 1869, 3.

40 Catalogue, *The South Staffordshire Industrial and Fine Arts Exhibition 1869*, Wolverhampton Archive & Local Studies, DX-869/2/6.

41 Wallis, 21.

42 *South Staffordshire*, i.

43 *South Staffordshire*, ii.

44 Wallis, 3.

45 'Illustrated Interview. A Leader in Wulfruna', *The Wolverhampton Exhibition Pictorial* (Wolverhampton, 1902), 23.

46 Hilary Boucher, 'Wolverhampton Exhibition 1902', *Blackcountryman* 4, no. 1 (1971): 11–15, at 13.

47 Pamphlet, *A Proposal of Programme of Opening Ceremony*, Wolverhampton Archives & Local Studies, WOL-D-CE/12/6/2-6.

48 *Wolverhampton Art and Industrial Exhibition 1902*, 32.

49 *Wolverhampton Art and Industrial Exhibition 1902*, 9.

50 *Wolverhampton Art and Industrial Exhibition 1902*, 47.

51 *Wolverhampton Art and Industrial Exhibition 1902*, 10.

52 *Wolverhampton Art and Industrial Exhibition 1902*, 68.

53 *Wolverhampton Art and Industrial Exhibition 1902*, 22–23.

54 Boucher, 14.

55 Boucher, 14.

56 *South Staffordshire*, iii.

57 *South Staffordshire*, iv.

58 Joseph Jones, *Historical Sketch of the Art and Literary Institutions of Wolverhampton, from the year 1794 to 1897, 1897* (London, 1897), 147.

59 'N. Neal Solly, Vice-Chairman', in *South Staffordshire*, iii.

60 'Wolverhampton Exhibition', *Art Journal* (1869): 367.

61 Barnsby, 242–243.

62 Frederick Avery, ed., *The Book of the Century* (Manchester, 1948), 44.

63 Hugh Cunningham, *Leisure in the Industrial Revolution, c. 1780–c. 1880* (New York, 1980), 44.

64 *South Staffordshire*, v.

65 'Wolverhampton Exhibition', *Art Journal* (1869): 367.

66 'The South Staffordshire Exhibition', *The Illustrated London News*, 5 June 1869, 569.

67 *Wolverhampton Art and Industrial Exhibition, 1902*, 13.

68 *Wolverhampton Art and Industrial Exhibition, 1902*, 5.

69 *Wolverhampton Art and Industrial Exhibition, 1902*, 5.

70 *Wolverhampton Art and Industrial Exhibition, 1902*, 5.

71 For example 'The Wolverhampton Exhibition', *The Times*, 26 April 1902, 19; 'The Wolverhampton Exhibition', *The Times*, 2 May 1902, 6.

72 Paul Greenhalgh, *Ephemeral Vistas: History of the Expositions Universelles, Great Exhibitions and World's Fairs*, Studies in Imperialism (Manchester, 1988), 207.

73 Greenhalgh, 198.

74 Herbert Read, *Art and Industry: The Principles of Industrial Design* (London, 1966), 8.

75 Wallis, 4.

76 Wallis, 5–20.

77 Neal Solly, 'Guide to the Fine Arts Department of the South Staffordshire Exhibition', in *South Staffordshire Industrial and Fine Arts Exhibition* (Wolverhampton, 1869), 26.

78 Wallis, 21.

79 Wallis, 21.

80 Leaflets, *Wolverhampton and Staffordshire Fine Art & Industrial Exhibition, Historic Exhibitions (1986)*. Wolverhampton Archives & Local Studies, DX-869/2/6.

81 Leaflets, *Wolverhampton and Staffordshire.*

82 Report, *Wolverhampton and Staffordshire Exhibition (1884)*. Wolverhampton Archives and Local Studies, DX-634/198.

83 Greenhalgh, 208.

84 *Wolverhampton Exhibition Pictorial*, 6.

85 D. S. M., 'Editorial', *The Saturday Review*, 19 July 1902, 36. Wolverhampton Archive & Local Studies, Z806.36.

86 D. S. M.

87 'Second Day. Art Gallery Thrown Open', *Express and Star*, 2 May 1902, 5.

88 Alexander C. T. Geppert, *Fleeting Cities: Imperial Expositions in Fin-de-Siècle Europe* (Basingstoke, 2010), 4.

Bibliography

Printed Primary Sources

Catalogue, *The South Staffordshire Industrial and Fine Arts Exhibition 1869*, Wolverhampton Archive and Local Studies, DX-869/2/6.

D. S. M., 'Editorial'. *The Saturday Review*, 19 July 1902, 36. Wolverhampton Archive & Local Studies, Z806.36.

'Great Exhibition: Note on a Building', *The Times*, 27 June 1850, 9.

The Illustrated Guide to Wolverhampton and District Including Places of Interest and a Description of the Art and Industrial Exhibition 1902, 5. Wolverhampton Archives and Local Studies, DX-894/15.

'Illustrated Interview. A Leader in Wulfruna'. *The Wolverhampton Exhibition Pictorial*. Wolverhampton: Whitehead Bros, 1902.

'The Industrial & Fine Arts Exhibition at Wolverhampton'. *The Wolverhampton Chronicle*, 26 May 1869, 3.

Jones, Joseph. *Historical Sketch of the Art and Literary Institutions of Wolverhampton, from the year 1794 to 1897, 1897*. London: Alexander and Shepheard, 1897.

Leaflets, *Wolverhampton and Staffordshire Fine Art & Industrial Exhibition, Historic Exhibitions (1986)*. Wolverhampton Archives and Local Studies, DX-869/2/6.

Pamphlet, *A Proposal of Programme of Opening Ceremony*. Wolverhampton Archives and Local Studies, WOL-D-CE/12/6/2-6.

Report, *Wolverhampton and Staffordshire Exhibition (1884)*. Wolverhampton Archives and Local Studies, DX-634/198.

'Second Day. Art Gallery Thrown Open'. *Express and Star*, 2 May 1902, 5.

'The South Staffordshire Exhibition'. *The Illustrated London News*, 5 June 1869, 569.

'South Staffordshire Industrial and Fine Art Exhibition, 1869'. *The Birmingham Daily Post*, 17 August 1868, 6.

South Staffordshire Industrial and Fine Arts Exhibition. Official Catalogue. Wolverhampton Archives and Local Studies, LS/L606/8.

Wolverhampton Art and Industrial Exhibition 1902. Illustrated Guide to Wolverhampton and District. Including Places of Interest Worth Seeing and a Description of the Art and Industrial Exhibition with Map of the Town and View of the Exhibition. Cheltenham, 1902.

'The Wolverhampton Exhibition'. *Art Journal* 8 (1869), 367.

'The Wolverhampton Exhibition'. *The Times*, 2 May 1902, 6.

'The Wolverhampton Exhibition'. *The Times*, 26 April 1902, 19.

Secondary Sources

Allen, Walter. *Black Country*. London: Paul Elek, 1946.

Anderson, Benedict. *Imagined Communities: Reflections on the Origin and Spread of Nationalism*. London: Verso, 1991.

Anderton, Paul. 'Staffordshire and the Setting up of the Great Exhibition 1851'. *Staffordshire Studies* 9 (1997): 35–68.

Avery, Frederick, ed. *The Book of the Century*. Manchester: Grove Publishing Co., 1948.

Barnsby, George S. *Social Conditions in the Black Country, 1800–1900*. Wolverhampton: Integrated Publishing Services, 1980.

Boase, G. C. 'Wallis, George (1811–1891)'. In *Oxford Dictionary of National Biography*, edited by R. C. Denis. Oxford: Oxford University Press, 2004, at http://www.oxforddnb.com/view/article/28570, accessed 10 April 2013.

Boucher, Hilary. 'Wolverhampton Exhibition 1902'. *Blackcountryman* 4, no. 1 (1971): 11–15.

Chinn, Carl. 'Making the Modern World: Smethwick'. *Black Country Memories 2* (2005): 103–110.

Cunningham, Hugh. *Leisure in the Industrial Revolution, c. 1780–c. 1880*. New York: St Martin's Press, 1980.

Dickens, Charles. *The Old Curiosity Shop*. London: Chapman & Hall, 1840.

Elwell, Charles J. L. *Aspects of the Black Country*. Kingswinford Black Country Society, 1994.

Everitt, Alan. 'Country, County and Town: Patterns of Regional Evolution in England'. *Transactions of the Royal Historical Society* 29 (1979): 79–108.

Geppert, Alexander C. T. *Fleeting Cities: Imperial Expositions in Fin-de-Siècle Europe*. Basingstoke: Palgrave Macmillan, 2010.

Greenhalgh, Paul. *Ephemeral Vistas: History of the Expositions Universelles, Great Exhibitions and World's Fairs*, Studies in Imperialism. Manchester: Manchester University Press, 1988.

Kinchin, Perilla, and Julie Kinchin. *Glasgow's Great Exhibitions 1888, 1901, 1911, 1938, 1988*. Wendlebury: White Cockade, 1988.

Lawton, Richard. 'Peopling the Past'. *Transactions of the Institute of British Geographers* 12, no. 3 (1987): 259–283.

Parsons, Harold. *Portrait of the Black Country*. London: Robert Hale, 1986.

Pevsner, Nikolaus. *Staffordshire: The Buildings of England*. Harmondsworth: Penguin, 1974.

Read, Herbert. *Art and Industry: The Principles of Industrial Design*. London: Faber, 1966.

Royle, Edward. 'Introduction: Regions and Identities'. In *Issues of Regional Identity. In Honour of John Marshall*, edited by Edward Royle, 1–13. Manchester and New York: Manchester University Press, 1998.

Shaw, Mark. 'The Ecology of Social Change: Wolverhampton 1851–1871'. *Transactions of the Institute of British Geographers* 2, no. 3 (1977): 332–348.

Showell's Dictionary of Birmingham, ed. Thomas T. Harman. Wakefield: S. R. Publishers, 1969, s.v. 'Exhibitions', 68.

Skipp, Victor. *The Making of Victorian Birmingham*. Studley: Brewin Books, 1996.

Solly, Neal. 'Guide to the Fine Arts Department of the South Staffordshire Exhibition'. In *South Staffordshire Industrial and Fine Arts Exhibition*. Wolverhampton: Steen & Blacket, 1869.

Trainor, Richard H. *Black Country Elites. The Exercise of Authority in an Industrialized Area 1830–1900*. Oxford: Clarendon Press, 1993.

Upton, Chris. *A History of Birmingham*. Chichester: Phillimore, 1993.

Wallis, George. *Special Report on the Local Manufacturers of South Staffordshire and East Worcestershire as Represented in the South Staffordshire Exhibition Held at Wolverhampton, 1869*. Wolverhampton: Steen & Blacket, Steam Printing Works, 1869.

Ward, Roger. *City-State and Nation. Birmingham's Political History c. 1830–1940*. Chichester: Phillimore, 2005.

HISTORICITY AND MODERNITY

Nature and the Brazilian State at the Independence Centennial International Exhibition in Rio de Janeiro, 1922

Livia Rezende

Introduction

On his way to Brazil to inaugurate the University of São Paulo in 1935, Claude Lévi-Strauss was unimpressed by his first experience of urban life in the American continent.[1] The anthropologist observed that while European cities are the more highly regarded the older they become, 'in America, every year brings with it an element of disgrace. For [American cities] are not merely newly built; they are built for renewal, and the sooner the better'.[2] The perceived ephemerality of the American urban life and architecture, 'the flimsiness of the buildings, their preoccupation with effect, and their desire to catch the eye' were to him 'reminiscent of a Great Exhibition that had not been pulled down'.[3]

Lévi-Strauss' remarks, albeit uneasily Eurocentric, identify an anxiety for renewal that is central for the understanding of the experience of modernity in early twentieth-century Brazil. Since the 1900s, Rio de Janeiro, then national capital, suffered major and frequent urban reforms under a banner of modernization that aimed ultimately at hailing Brazil as a modern, civilized and export-oriented nation. This anxiety for renewal as a means to promote a sense of modernity within the country culminated in 1922 with the organization of an international exhibition in Rio.

The Independence Centennial International Exhibition [Exposição Internacional do Centenário da Independência] opened on 7 September 1922, one hundred years after Brazil gained independence from Portugal. Erected in downtown Rio de Janeiro, the Centennial Exhibition featured all the typical ingredients of an international exhibition on a small scale. The exhibition complex was divided into national and international sections made of several pavilions, palaces and monumental constructions. More than three million visitors enjoyed a large amusement park, cinemas, fancy restaurants and convenience bars. The Centennial Exhibition celebrated Brazil's sovereignty with new public monuments, musical and military parades, lavish parties, official speeches, and displays of the country's productive forces to date. The exhibition complex was built over two years, consumed much labour and capital and remained open for ten months until 24 July 1923.

Like previous World's Fairs, *expositions universelles* and international exhibitions, Brazil's Centennial Exhibition ascertained the constructive power of the nation that built it. Engineering, architectural and design wonders – from the Crystal Palace to the White City and the Eiffel Tower – displayed power. Even the ephemerality of exhibitions and the haste with which they vanished from site emphasized the human capacity to control and transform urban spaces. The international exhibition of Rio, however, went beyond this typical demonstration of material construction and destruction. The first and only exhibition ever organized on Brazilian soil also reconfigured the landscape of its host city and the history of its host nation. Thus, it is through a historical analysis of urban and national modernization and their articulation with discourses of modernity that this chapter discusses Rio's international exhibition.

Ephemeral Vistas in Rio

The Centennial Exhibition complex occupied a privileged area of 2,500 square metres right in between Rio's main boulevard and the Municipal Market [Mercado Municipal] next to the city harbour. This city-within-a-city started at the Main South Gate [Portão Principal Sul], a 33-metre-high monumental gate designed in eclectic style and located by the Rio Branco Avenue – a *belle époque* boulevard opened during the 1900s urban reforms. A reconstruction of the award-winning pavilion erected to represent Brazil in St Louis in 1904 during the Louisiana Purchase International Exposition crowned the exhibition entrance. Heading northwards along the sea-facing Avenue of Nations visitors passed by international pavilions, convenience shops, bars and an amusement park before reaching the Calabouço point by the Guanabara Bay. There, the Festivities Palace [Palácio das Festas] marked the beginning of the national section where nine pavilions housed displays of Brazil's commercial, political and social achievements (Fig. 7.1).

Two tall towers rose over the horizontality of the complex to offer complete views of the festival and promote the economy of vision and vistas that characterized that sort of spectacle.[4] The 45-metre-high Tower of Jewels [Torre das Jóias], whose name and electrification effect were borrowed from the Panama-Pacific International Exposition (1915), was on top of the States Pavilion [Pavilhão dos Estados]. Nearby, the 35-metre-high Calabouço Tower [Torre do Calabouço] at the Palace of Industries [Palácio das Indústrias]

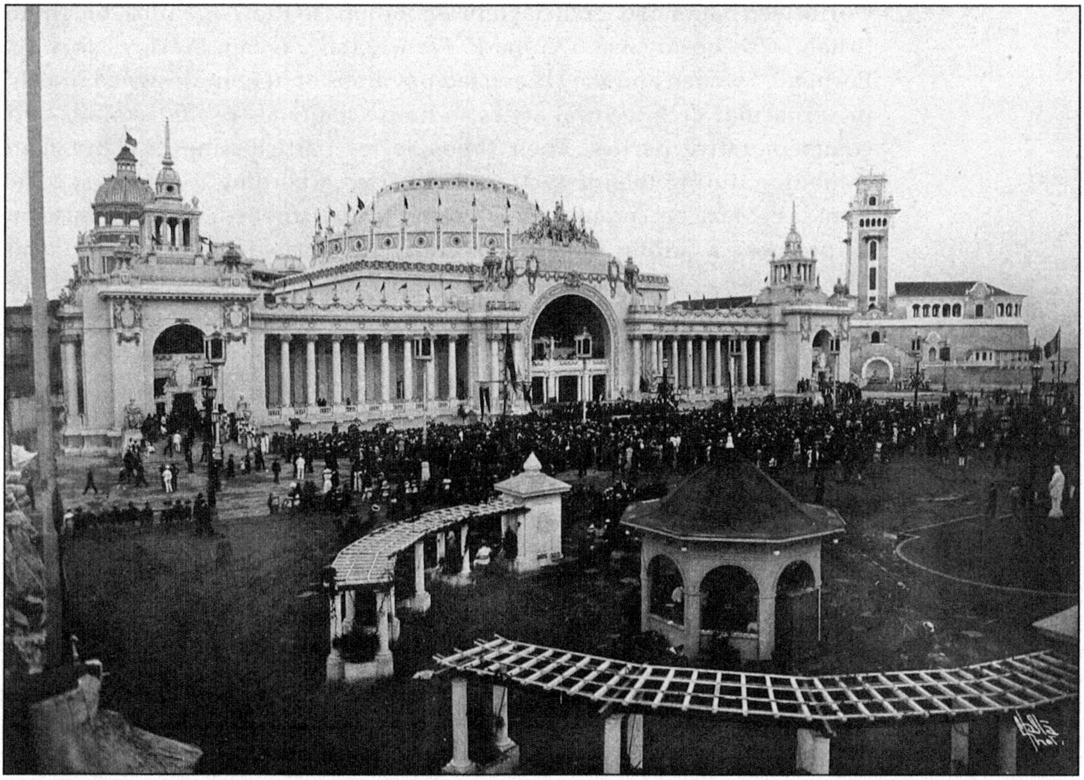

7.1 The Festivities Palace held an exhibition on hygiene (also known as the Hygiene Museum), photograph. National Historical Museum, Rio de Janeiro

worked as the best lookout onto the exhibition, the sea and the bay. The Centennial Exhibition ended by the North Gate [Portão Norte] with its two huge stanchions placed near the Municipal Market. The stanchions supported two large searchlights that transformed the exhibition's daylight architectural monumentality into a triumphant nocturnal spectacle.[5]

Three smaller shows complemented the Centennial Exhibition. Away from the seafront and the glitz of the main festival, live animals were displayed at a husbandry exhibition in the São Cristovão neighbourhood. The National Fine Arts School [Escola Nacional de Belas Artes] displayed Brazilian fine arts across two shows. The first on 'Retrospective Art' [Arte Retrospectiva] included historical, military, religious, decorative and applied arts from the past Brazilian centuries, including its colonial period. The second, on contemporary art, exhibited the latest Brazilian fine arts. Directors of the School organized a competition among artists for the creation of history paintings and sculptures that expressed the alleged greatness of the Brazilian nation. Besides promoting artistic patriotism and the imagining of the national past, the competition encouraged projections of the nation's future by awarding designs for 'remodelling and beautification alterations' of Rio.[6] The National Museum [Museu Nacional] hosted the third small exhibition with displays of botany, anthropology and zoology from the museum's ageing natural history collection.

Thirteen nations sent official representations to Rio. Argentina, Belgium, Britain, Czechoslovakia, Denmark, France, Italy, Japan, Mexico, Norway, Portugal, Sweden and the US erected pavilions of honour designed mostly in vernacular architectural styles to house national-specific exhibits and commemorative parties. Their rationale for participating in a Brazilian exhibition during the interwar period varied according to nation-specific agendas. Mexican commissioners came to Rio imbued with the mission to promote a strong alliance of Latin American nations.[7] Italian and Japanese immigrants had already been part of the Brazilian workforce for decades and their respective countries used the exhibition to strengthen bilateral ties.[8] Nations like Italy, Belgium, Portugal or the US erected large warehouses in Rio's docks area to show industrial exhibits. Multinational companies like Switzerland's Nestlé designed individual pavilions to popularize their products, while Chile, for example, was represented solely by its exhibitors of saltpetre.

The British Empire, initially hesitant about participating due to costs, finally accepted Brazil's invitation for 'fear' that it 'might fall behind others'.[9] 'There are few of the products of Brazil for which there is not a market in the United Kingdom', noted *The Times*, 'while its modern industrial development makes the country a magnificent field for British engineering firms, who are, we are glad to learn, strongly represented in the exhibition.'[10] Anxious aspirations for international economic recovery after the First World War accompanied conspicuous displays of power. In Rio, the British representation posed a show of commercial and military 'maritime greatness', its pavilion architecture and interior designed to convey possession of the Seven Seas. The funding required to send two warships to Rio, first vetoed by the House of Commons, was finally conceded when Parliament understood that the US, Japan and France, among others, were sending squadrons. 'The British Empire', said *The Times*, 'could not afford to be compared unfavourably with its rivals', even in hard economic times.[11]

In contrast to *The Times*' open rhetoric of rivalry, Brazilian exhibition organizers framed foreign participation in Rio's fair as acts of 'cordiality', 'friendship' and 'nobility'.[12] In return, they organized special nation's days, like Japan Day or Mexico Day, when dance, food, clothing and customs were promoted and propagandized on the streets of the complex and the pages of official illustrated magazines. In the gift-giving economy of exhibitions, more significantly and lasting was the legacy of buildings and infrastructure that visiting nations left in Rio. Almost all foreign pavilions were donated to the Brazilian local and federal governments and occupied by institutions and professional associations as the exhibition closed. France marked its prominent position in Brazilian cultural life by building a replica of Versailles' Petit Trianon, and was greatly commended by exhibition organizers for furnishing Rio with such 'a marvellous evocation of the past'.[13] This building – one of the few that survived posterior reoccupation of the exhibition site – is home today to the prestigious Brazilian Literary Academy [Academia Brasileira de Letras]. The US made a diplomatic and pragmatic move to establish a political presence in Brazil's federal capital. After buying land on the Avenue of Nations, the delegation led by Colonel David C. Collier erected a pavilion designed to become their embassy later.

Showing National Progress under the Republicans

The Centennial Exhibition was a matter of the state. It was conceived, organized and mostly financed by the local and federal governments. Proposals to commemorate the nation's centennial began to circulate between the Brazilian president, Epitácio Lindolfo da Silva Pessoa (1919–22) and the National Congress as early as 1920. Across the country, local authorities were to mark the occasion by organizing parades and solemnities. In the federal capital, the congress decided to celebrate the date first with a national exhibition that then became international when 'friendly' nations asked to participate, according to the official narrative.[14]

Despite the clear resolution from the president and the congress to organize a festival of this magnitude, at a glance conditions did not seem favourable for the erection of an international exhibition. This period in Brazilian political history is known as the 'coffee-with-milk politics' [política do café-com-leite], an oligarchic political system whereby presidential candidates from the coffee- and dairy-producing states of São Paulo and Minas Gerais alternated in power. Between 1925 and 1929, coffee alone provided 75 per cent of Brazil's foreign earnings.[15] The concentration of power and money in the hands of a few coffee plantation owners caused an uneven process of modernization in Brazil. São Paulo city and state experienced urban and economic growth driven by industrialization and coffee plantation revenue. In contrast, in the north and north-eastern regions pre-capitalistic relationships predominated. Rio and São Paulo were Brazil's largest cities where wealth and skilled workers concentrated. But even under urban expansion, during the 1920s Brazil was chiefly a rural country and 75 per cent of its population were illiterate.[16] Illiterates, as well as women, could not vote and as a result, popular participation in elections was minimal throughout Brazil's First Republic (1889–1930).

Commercially, the period also seems adverse for the building of an international exhibition. The First World War damaged Brazil's economy and marked the beginning of the end for Rio's *belle époque* and the 'coffee-with-milk' status quo. Import prices soared, importation and exportation numbers fell, and coffee revenue in particular suffered during the war period. Financial support and investments from Britain, a long-time financier of Brazil's infrastructure, faltered. Except for the US, the country's trading partners reduced their capacity to advance credit and import from Brazil. Politically, the presidential election of 1922 'revealed the growing regional tensions within the ruling class' of the oligarchic system as a 'bloc of intermediate states' challenged the status quo, albeit unsuccessfully.[17] Brazil's key economic pillar accompanied the slow demise of the oligarchic political system. The Brazilian 'black gold' drove the nation's export agenda and economy from the 1880s until the mid-1920s, when coffee receipts began to decline as well.[18] In a context of political uncertainty and economic constraint during the troubled interwar period, why did the Brazilian republican government deploy the costly medium of exhibitions to celebrate the nation's sovereignty?

The economic explanation resides in the immediate effect that exhibitionary promotion was likely to have on international investment in Brazil.[19] Between 1924 and 1929, while coffee receipts declined, a cyclical wave of foreign investment boomed, funding 'most of the banks, electric, telephone, and gas systems, port facilities, railways, steamship lines, and airlines'.[20] This boom suggests that the technologies of modernity flaunted at the capital during the exhibition paid back good dividends while Brazil showed to the world that its pre-war international position as raw materials exporter remained strong. The making of an international exhibition in Brazil in 1922 achieved results similar to those of earlier Brazilian participations abroad, when commissioners went to the *Exposition Universelle* of 1867 or the Louisiana Purchase International Exposition (1904), for instance, to attract foreign capital investment back into the country.[21]

In political terms, the organization of an exhibition in 1922 was part of a programme of public opinion formation, set by a political regime wishing to establish itself as the future of Brazil. Only 33 years earlier, Brazil had been the last monarchy in the Americas and the last American nation to adopt republicanism as a political solution to national problems. Thus, during its Centennial Exhibition, Brazilian government wished to project the image of a stable, civilized nation undergoing a process of modernization that would return international investments. João Luiz Alves, Brazilian minister and exhibition manager, dedicated the exhibition to those who 'came to admire the prodigious economic progress of Brazil over a hundred years'.[22] According to the official line, international approval of Brazil's development should come from foreign industrialists who would be able to 'say publicly "we have seen the Brazilian industrial display and verified that that Brazil is already an industrial country".'[23] Domestically, republican politicians mounted a predominantly visual display of the nation back to its inhabitants to convince a mostly illiterate audience how Brazil had progressed under their rule and how it would continue to grow, as we will soon see.

At the beginning of 1921, President Pessoa created the Independence Centennial Executive Commission [Commissão Executiva do Centenário da Independência] to plan and oversee the exhibition works. Posts for the Commission were initially appointed by Pessoa and then by Artur Bernardes (1922–26), elected president mid-way through the exhibition's duration. Chief posts were accumulated by politicians from the high echelons of the

Ministry of Justice and Internal Affairs [Ministério da Justiça e Negócios Interiores] and the Ministry of Agriculture, Industry and Commerce [Ministério da Agricultura, Indústria e Comércio], following the oligarchic political culture whereby public service appointments were made less on the basis of professional merit than to strengthen political agendas and alliances. This political culture coupled with the presidential change caused frequent overhaul in main exhibition posts. However, the financial and administrative operations of the exhibition were secured by the National Congress who passed bills authorizing cabinet ministers' acts and demands for funding.

The fair organizers sought funding also among the Brazilian public. During the planning and preparation stages of the Centennial Exhibition, they launched a financing scheme called Independence Bonus [Bônus da Independência] whereby future fairgoers bought coupons to help pay for the festival. Besides securing their entry after the opening ceremony, coupon-holders were allowed free admission to the construction site of the exhibition. There, visitors witnessed the investment of public and private money in a bid from the government to make the local public complicit in its exhibitionary efforts.

The Artifice of Nature

The actual preparations for the Centennial Exhibition began in 1920 with the razing of the Castelo Hill [Morro do Castelo], a mammoth feat of engineering that transformed not only Rio's landscape but also Brazil's past. The hill was significant to Brazil's colonial history. It held a fortress from where the region had been administrated and Rio's bay defended against foreign invasions since 1567.[24] Strategically located at the sea entrance to the city, the hill's surroundings included popular housing, forts, the oldest Jesuit college in Brazil and churches. Despite its historical and architectural significance, by the 1920s the hill was redefined by Rio's mayor, Carlos Cesar de Oliveira Sampaio, as 'an excrescence'.[25]

Sampaio was a firm believer in modernization as the inevitable consequence of technical progress. His worldview was formed during his graduate years as a civil engineer at the Polytechnic School [Escola Politécnica] in Rio and at the Engineering Club [Club de Engenharia], and informed by a positivist valorization of scientific reason as a means to life improvement.[26] In both institutions, Sampaio learnt and perpetuated the idea that national progress was a matter of modernization via technical solutions. During his electoral campaign for mayor, he proposed major urban reforms of the republican capital under the banners of beautification and sanitation.[27] As mayor, Sampaio promoted the razing of the hill for alleged sanitary reasons (Fig. 7.2). He claimed that the hill's location by the bay entrance blocked the circulation of air and encouraged the proliferation of miasmas and tropical diseases.[28] Sampaio was also one of the key players behind the Centennial Exhibition

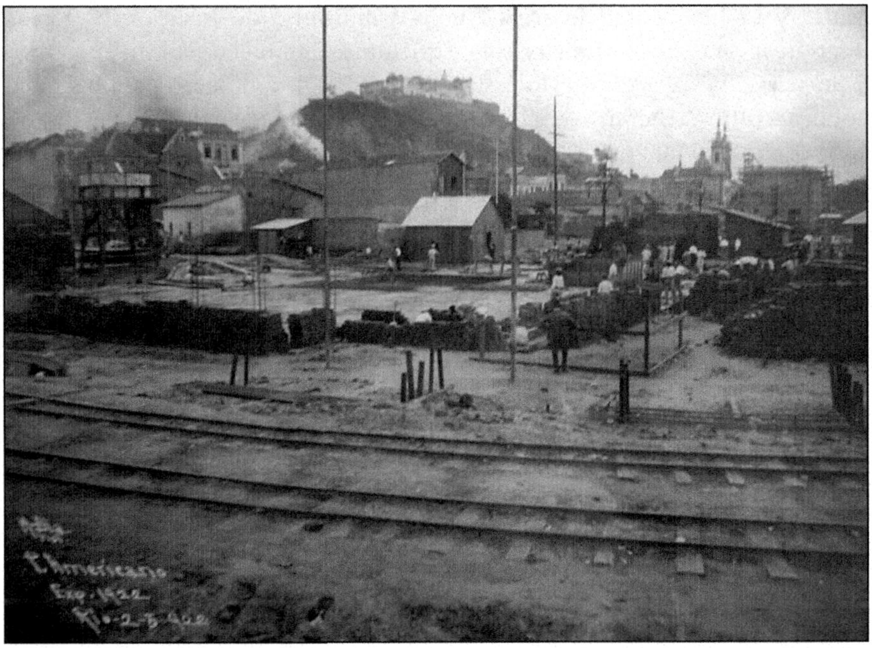

7.2 The land-filled area claimed from the sea (foreground) and the remains of the diminishing Castelo Hill (background), photograph. National Historical Museum, Rio de Janeiro

acting as Exhibition General Commissioner. In this capacity, he used the earth remains of the hill to landfill the shore by the bay entrance and thus create a brand new space on which to erect the exhibition. In this way, the old literally paved the way for the new.

The elimination of the hill, however, aimed not only at eradicating tropical diseases. The razing removed 'the remnants of the Portuguese colonial period as well as rundown and insalubrious housing' where poor dwellers lived in the space earmarked for modernization.[29] Sampaio framed the problem of insalubrious housing in downtown Rio, mostly a socioeconomic consequence of the lack of popular housing, as a matter of sanitation to be solved by the removal of a supposed topographic obstacle to ventilation. The removal of this vast portion of land resulted in home destitution for several families, and promoted further occupation of other hills by improvised housing, known today as Rio's 'favelas' (slums). Sampaio developed an ambitious reform programme for Rio in the 1920s, one that began with the elimination of the past and the poor and culminated with the realization of an exhibition, understood as a harbinger of modernity. By radically transforming Rio's landscape through modern technology, Sampaio was helping forge an identity for the city and for Brazil aligned with the idea that modernization alone formed an experience of modernity.

Sampaio wanted to 'show foreigners who still do not know us what we are capable of'.[30] His ambition and project found strong allies on the pages of widely circulated illustrated magazines like the *Revista da Semana* or *Illustração Brasileira* (Fig. 7.3). To celebrate Rio's topographic modifications these magazines frequently and convincingly displayed photos of the engineering wonders capable of transforming 'a mountain [*sic*] into a plain' and elongating 'the city over the waters of the Guanabara Bay'.[31] For a nation suffering from what was perceived as a lack of tradition, the razing of the hill served paradoxically to create ruins out of Brazil's colonial past. As the Castelo hill was gradually lowered to landfill the bay for the future exhibition site, a magazine spread compared the destruction of Rio's colonial remnants to spectacular ruins.[32] The disappearance of parts of 'old' Rio was naturalized to the wide public in unsentimental photographs and promoted as the desirable and inevitable consequence of becoming modern.[33]

Parallel to the eradication of the old (and the poor) was a conspicuous display of manpower in conquering nature and ushering in the new. In another instance of persuasive visual communication, a 'vision of the city of the future' was carefully drawn 'hovering over the spoils of the historical hill'.[34] The argument made visually by these images, in accordance to the official view of modernity, spoke of a future that was not just the succession of the past. The future was the necessary demolition of the past – an appropriate illustration to Lévi-Strauss' observations about the frantic renewal of American cities.

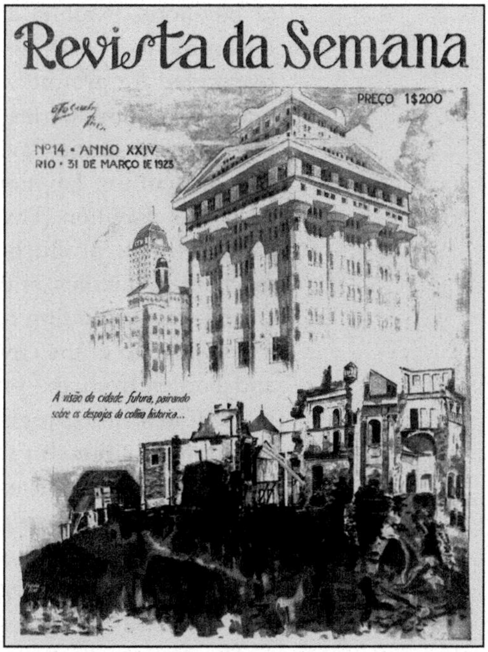

7.3 Cover of the *Revista da Semana* (Week Review) depicting a 'vision of the future city hovering above the ruins of the historical hill', magazine. National Library of Brazil

The Naturalization of the State

Besides promoting an official view that conflated the concept of modernity with that of urban modernization, the exhibitionary complex built in Rio was designed to instil abstract civic sentiments in local visitors and render visible an organization of society proposed by the republican state. Tony Bennett's Gramscian analysis of the exhibitionary complex sustains that exhibitions like the Brazil's Centennial, entirely planned by governmental agencies, operated not as a coercive political force but though the 'organisation of consent'.[35] The creation of a 'voluntarily self-regulating citizenry' is key here, and it is by analysing the typology of pavilions that formed the exhibition's national section that this project is best observed.[36]

The national section of the Centennial Exhibition consisted of nine purpose-built pavilions and reclaimed edifices near the Municipal Market. Some prominent national industries like Antarctica and Brahma breweries or the Matarazzo United Industries [Industrias Reunidas Matarazzo] built their own pavilions on site. Rio's exhibition, however, was mostly made up of pavilions, exhibits and displays designed and occupied by the federal government. The Ministry of Agriculture, Industry, and Commerce alone responded for around 75 per cent of the national exhibition. Four pavilions managed by ministries – the Palace of Industries [Palácio das Indústrias], the Fish and Game Pavilion [Pavilhão de Caça e Pesca], the Transportation and Agriculture Pavilion [Pavilhão de Viação e Agricultura] and the Small Industries Pavilion [Pavilhão das Pequenas Indústrias] – displayed Brazil's most saleable products in sections dedicated to agriculture, mineralogy, mechanized textiles, crafts and foodstuff, among others.[37]

The transformation of Rio's landscape and the subsequent occupation of a new portion of the city by the exhibition accompanied a 'transfer of objects and bodies from the enclosed and private domains', be they domestic or institutional, 'into progressively more open and public arenas where, through the representations to which they were subjected, they formed vehicles for inscribing and broadcasting the messages of power ... throughout society.'[38] Still in the national section, a specific group of pavilions and exhibits, here termed 'nation-building pavilions', were designed to educate fairgoers on how Brazil had progressed under the republic's rule and how it would continue to grow.

The type of nation-building pavilion that best promoted the republican rule was the Pavilion of Statistics [Pavilhão de Estatísticas]. Following Benedict Anderson's proposal that 'communities are to be distinguished by the style in which they are imagined', the organization of a Pavilion of Statistics for Rio's exhibition offers an invaluable insight into how Brazil was being imagined and displayed during the First Republic.[39] Dedicated to show statistical works, this pavilion is an unusual feature in the history of exhibitions, although statistical congresses were common in nineteenth-century world's fairs. Exhibition organizers intended to display 'numerical studies on territory, demographics, the economy' and the 'intellectual' and 'moral' advances of Brazil during the last hundred years.[40] These achievements were demonstrated, above all, through techniques of visualization like photographs, posters, murals, maps, graphs and diagrams that formed, according to the official guide, 'a complete exhibition on the statistics of all works from Brazil', especially designed for those who could not read (Fig. 7.4).[41]

While commercial pavilions undertook a census of Brazilian artefacts and productive forces to date, the Pavilion of Statistics offered a visualization of the political and social states of the nation. Above all, as Bennett noted, the Pavilion of Statistics was 'ordering objects for public inspection and ordering the public that inspected'.[42] Bennett, however, did not list statistical works or designed displays among the disciplinary apparatuses that form the exhibitionary complex.[43] The study of cultures of International Exhibitions expands this kind of methodological approach to include local forms of societal ordering and organization of consent. These forms had to adjust to Brazil's large illiterate population, anti-democratic electoral system and multicultural dimension. If in 1851 the Great Exhibition operated as a series

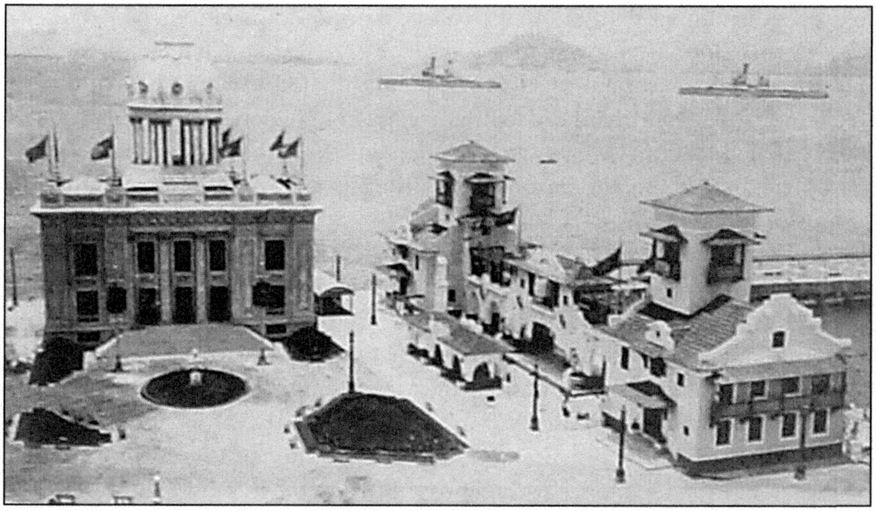

7.4 The Pavilion of Game & Fish (right), Pavilion of Statistics (left) and international warships (background), photograph. National Historical Museum, Rio de Janeiro

of 'object lessons', at the Centennial Exhibition of 1922 the object to be learned by Brazilians was the nation itself.

Furthering Anderson's proposal to investigate the ways in which communities imagine themselves, the study of the Rio exhibition also demonstrates how the statistical exhibit materialized the 'abstract quantification/serialization' of race, mortality rates, diseases, gender, ageing or foreigners in Brazil.[44] Abstract numbers were made visible to help establish the national political space and to persuade Brazilians of their cohesion and progress as a nation under the aegis of the republican state.[45] Beyond the crucial visual aspect of this persuasion, the forging of a national community through a nation-centred exhibition in 1922 substantiates Anthony D. Smith's 'modernist' form of explaining nation-building and nationalism.[46] Exhibition organizers were self-acclaimed 'nation-builders' ['construtores da nacionalidade'][47] or elites that 'mobilized and united populations in novel ways to cope with modern conditions and modern political imperatives', such as the works of sanitation and beautification that validated national progress.[48] The clear discourse of renewal distance this exhibition from a 'reviving of the past' event or the 'ethno-symbolist contention that most nations are formed on the basis of pre-existing ethnic ties and sentiments', and approximate it to explanations of a 'creative' version of nationalism.[49] This creativity, for Smith, explains a sort of 'developmental nationalism' where 'the "nation" is defined, not as a given ascriptive solidarity, but as realized in and through the "project" of transcending its own backward situation in the effort of development'.[50] 'It is the creativity of the plan and the energy of the effort of self-transcendence that marks off a nationality', and it was in the Pavilion of Statics where visitors could locate, measure and see the alleged results of this effort.[51]

Another nation-building pavilion, the Festivities Palace helped naturalize the presence of the republican state in peoples' lives. Beyond its primary purpose of hosting parties and official ceremonies, the Festivities Palace housed educational exhibits that exposed fairgoers to the ways of modern life. Its first floor was given to the eminent Dr Carlos Chagas, director of the Brazilian Department of Health, just nominated for the Nobel Prize in medicine. The Department of Health organized a hygiene exhibit, also called the Hygiene Museum [Museu da Higiene], which, besides showing advances in tropical disease control, endorsed scientifically the programme of sanitation proposed by Sampaio.[52]

Appeals to citizenry and organization of consent and control were attempted also via 'softer' ways. Exhibition organizers distributed to fairgoers hundreds of reproductions of Brazilian history painting free of charge in the name of 'civic and patriotic education'.[53] Complementing the promotion of an alleged greatness of the nation, they distributed maps, posters, stamps and postal cards of Brazil, many intended for foreign visitors and translated into six languages. The Brazilian public were summoned to contribute to this plethora of visual mementos by creating, for example, hymns, coins or theatre plays that boasted civic sentiment and praised 'Brazilianness'. Screening of films and displays of photographs specially commissioned for the exhibition helped the visualization of Brazil's agricultural forces, natural riches and rural workforce, usually distant from the gaze of Rio's urban dwellers and foreign visitors. The Campinas Cinema, for example, erected by the government of the prosperous São Paulo state, provided a leisurely appreciation of that region's industrial power through carefully chosen propaganda films.[54]

The celebration of national sovereignty would not be complete without the establishment and show of a new pantheon of heroes that linked the men behind the First Republic and the Centennial Exhibition to those responsible for Brazil's independence in 1822. The pantheon proposed by exhibition organizers and circulated on newspapers, illustrated magazines and fair souvenirs aggrandised historical actors like Pedro I (Brazil's first emperor who claimed independence from Portugal), the Andrada brothers or Tiradentes and placed them in a political genealogy that ended with Pessoa, Sampaio and other politicians-turned-exhibition commissioners.[55] This proposed pantheon aimed to legitimize the Centennial Exhibition and the First Republic as parts of the greater project of nation-building Brazil. Notably, in this political historical genealogy construed during the exhibition, a long silence fell over Brazil's Second Reign (1840–89), the period governed by Pedro II immediately prior to the advent of the Republic and against which the First Republic set its process of national identification.

Two Exhibitions

Apart from exposing Brazilians to scientific novelties and technologies that brought with them a sense of a modern world, the Centennial Exhibition was also the locus of another experience of modernity. This understanding of modernity is less conceptualized as the gain of technical progress and more attuned to the major changes in the perception of time and space experienced by people across the globe. This perceptual change was prompted, for example, when hydroplanes took fairgoers to experience the exhibition and the razing of the Castelo Hill from above. Also, news broadcasting technologies made it possible for citizens in distant cities to hear the opening speech of President Pessoa as it happened in Rio. The experiences of simultaneity of events and collapse of spatial distances are two of the most fundamental aspects of the experience of modernity when the concept is understood less as the spectacle of modernization and the ephemeral result of constant renewal, and more as a worldwide phenomenon prompted by technological and cultural changes.[56]

Besides promoting perceptual changes, the Centennial Exhibition offered novel opportunities for mass consumption and leisure. The focal point for these new practices of public life was the Amusement Park, erected as an integral part of Rio's exhibitionary complex by its organizers. Merry-go-rounds, seesaws, swings, electric cars and mock airplane rides sided with food and drinks concessions, from the popular 'caldo-de-cana' (a sugar cane syrup) and numerous beer huts to expensive silver-service restaurants like Bol's or Falconi. Rio's exhibition is historically located in between a nineteenth-century industrial and instructional exhibition and a twentieth-century popular entertainment fair.[57] The Centennial Exhibition participated in the economy of vision typical of that historical period whereby a visit to the fair's industrial exhibits was as significant as to see and to be seen in an environment of entertainment and pleasure.[58]

Benjamin Costallat, a popular Brazilian author writing for a contemporary local newspaper, visited the fair and voiced these experiences of consumption, leisure and technologies of vision, adding yet another dimension to the experience of a modern world brought about by the Rio exhibition:

> I dined last night at the Falconi Restaurant in the exhibition. The exhibition was on fire, all illuminated! The pavilions looked surreal at night! Searchlights intersected one another in space, whitening the sky and clouds with their silver dust. And as if in a bacchanal of lights and colours, the palaces seemed suspended, floating, ethereal like a dream that will be dissipated. In front of it all, the sea. Over the dark waters, gigantic forms could be guessed. These were the 'dread-naughts' [*sic*], steel giants, modern dragons, the fabulous of the sea, blood dragons, fire dragons, who slept there peacefully, their steam-boilers cold, their cannons silent, their iron plethora in rest. The contrast was shuddering. In land – light, joy, pleasure; at sea – darkness, silence, threats! This is when I realised the existence of two exhibitions![59]

Anchored in Guanabara Bay, dozens of international warships reminded Costallat and fairgoers of the enmities and cruelties recently practised during the First World War. The sending of foreign warships for naval reviews had long been a tradition of international exhibitions. However, during the interwar exhibition in Rio this assertion of military power became even more pressing. Reaching beyond the rhetoric of progressive technical development sustained by fair organizers and participating nations, Costallat grasps the essential contradiction of the project of modernity itself. Technical progress will both improve and destroy lives:

> And then I realised all the uselessness of the exhibition on land, falsely colourful, falsely happy, falsely illuminated, made of plaster, made of fantasy, transitory, passing, weak, theatrical in front of those armours made of real iron, those howitzers made of real steel, all that organisation for death and extermination, eternal like humanity, and which humanity eternalises for its own disgrace ….[60]

The Rio exhibition exposed the bleak and destructive potential of material and technical progress during a critical historical period when national differences recrudesced in an increasingly internationalized world. The military parade and the spectre of another war haunted the old rhetoric of exhibitions whereby nations peacefully competed by comparing displays of machinery, objects and art. Costallat's use of the term 'disgrace' eerily echoes that of Lévi-Strauss, for whom the passage of time to American cities was an 'element of disgrace'.[61] Both considered accelerated and impetuous expansion – from the feverish expansion of American urban life to European belligerent imperialism – the disgraceful dimension of modernity.

Conclusion

The Independence Centennial International Exhibition held in Rio de Janeiro in 1922 was at the same time the culmination of an anxiety for renewal to grasp modernity via modernization and the creative project of a state wanting to establish consent among its incipient citizenry.

Domestically, the Centennial Exhibition exposed contradictions in the national project. The dimension of creativity – or 'construction' – of a nation 'transcending its own backward situation in the effort of development', as noted by Smith, reached an apex when both history (the colonial architecture) and nature (the hill) were framed as obstacles to be transcended via modernization.[62] However, while Rio underwent constant renewal to be transformed into the national locus of modernity, Brazil overall remained a rural, conservative, illiterate country.

Despite its small size and the timid international participation, the Centennial Exhibition also revealed in full the cleavages in the experience of modernity across the globe during the interwar period. While the rhetoric of peaceful competition that sustained nineteenth-century exhibitions still lingered, a novel, 'disgraceful' rationale based on international rivalry for participating in exhibitions developed, as seen with the British discourse.

The erection of a dedicated Pavilion of Statistics reiterate both Bennett's and Anderson's approach to the power of visualization in the processes of national persuasion, popular consent organization and imagination of communities. An exhibition permitted the visualization of the nation through performance, spectacle and materialization of abstract values, codes and structures that would be otherwise difficult to be communicated to a largely illiterate population. Following both premises, the analysis just undertaken demonstrates that if the 1851 Great Exhibition helped form a new public and inscribe it in new relations of sight and vision, the Centennial Exhibition helped form a new object for public appreciation, the nation itself.

Acknowledgements

This research has been supported by the CAPES Foundation and the National Council for Scientific and Technological Development (CNPq), Ministry of Science and Technology, Brazil.

Notes

1 Claude Lévi-Strauss, *Tristes Tropiques* (New York, 1961), 36.

2 Lévi-Strauss, 100.

3 Lévi-Strauss, 36.

4 Tony Bennett, 'The Exhibitionary Complex', *new formations* 4 (1988): 73–102. See also Paul Greenhalgh, *Ephemeral Vistas: A History of the Expositions Universelles, Great Exhibitions and World's Fairs, 1851–1939* (Manchester, 1988).

5 Ruth Levy, *A Exposição do Centenário e o Meio Arquitetônico Carioca no Início dos Anos 1920* (Rio de Janeiro, 2010). Levy's work on the architectural and urban aspects of Rio's exhibition offers an exhaustive survey of the complex and its eclectic style as an alternative narrative for the establishment of a modern, but not modernist style in Rio.

6 'c) – Um plano geral relativo às possíveis modificações de remodelação e embelezamento da cidade do Rio de Janeiro'. *Official Guide to the Exposition*, ed. Bureau Official de Informações (Rio de Janeiro, 1922), 229. All translations from Portuguese to English are the author's.

7 Mauricio Tenorio-Trillo, 'A Tropical Cuauhtémoc: Celebrating the Cosmic Race at the Guanabara Bay', *Anales del Instituto de Investigaciones Estéticas* 65 (1994): 93–137.

8 Tenorio-Trillo, 84–85.

9 'Brazil's Centenary Exhibition', *The Times*, 6 September 1922, 11.

10 'Brazil's Centenary Exhibition', 11.

11 'The Brazil Centenary', *The Times*, 29 July 1922, 9.

12 Terms used by minister João Luiz Alves in his closing speech: 'Foi esta cordialidade que mais uma vez tivemos ocasião de apreciar, com o nobre concurso de paizes amigos, nesta Exposição'. *Relatório dos Trabalhos da Exposição Internacional do Centenário*, ed. Ministério da Justiça e Negócios Interiores da República do Brasil (Rio de Janeiro, 1926), 42.

13 'Como poderemos esquecer essa maravilhosa evocação do passado ... que é o Petit Trianon, aqui reproduzido pela França na Avenida das Nações?'. *Relatório dos Trabalhos*, 42.

14 *Relatório dos Trabalhos*, 29 and 42. The exhibition gained its official international status on 22 July 1922.

15 Warren Dean, 'Economy', in *Brazil: Empire and Republic, 1822–1930*, ed. Leslie Bethell (Cambridge, 1993), 227.

16 Boris Fausto, 'Society and Politics', in *Brazil: Empire*, ed. Bethell, 278.

17 Fausto, 272 and 295–296.

18 Dean, 251.

19 The term 'exhibitionary' is used after Bennett's seminal work on the 'exhibitionary complex', or the institutions, discursive formations and technologies of vision that articulated power and knowledge seeking to transform the problem of order from disciplinary into one of culture. Bennett, 73–76.

20 Dean, 239.

21 From its first official representation in the 1862 International Exhibition of London to the organization of the 1922 Centennial in Rio and beyond, Brazilian participation in international exhibitions, world's fairs and *expositions universelles* was frequent and relevant. For the Brazilian exhibition of 1867 see Livia Rezende, 'The Raw and the Manufactured: A Discussion of Brazilian National Identity Projected in International Exhibitions (1862–1867)', in *Another Name for Design: Words for Creation*, ed. Haruhiko Fujita (Osaka, 2008), 354–357. For a discussion of Brazilian craft display, see Livia Rezende, 'Crafting the Nation: Brazilian "Civilised Exoticism" at World Exhibitions', in *Design and Craft: A History of Convergences and Divergences*, ed. Javier Gimeno-Martinez and Fredie Floré (Brussels, 2010), 136–139. For a thorough design historical analysis of the period comprising the first 60 years of Brazilian participation, see Livia Rezende, *The Raw and the Manufactured: Brazilian Modernity and National Identity as Projected in International Exhibitions (1862–1922)* (PhD thesis, Royal College of Art, 2010).

22 'Foi, sim, para os que aqui vieram admirar o prodigioso progresso econômico do Brasil em cem anos de existência como Nação'. *Relatório dos Trabalhos*, 41.

23 'Tão notável foi a lição colida, que ella ... levou industriais estrangeiros a dizerem de publico: "Vimos os mostruários da indústria brasileira e verificamos que o Brasil já é um país industrial ..."'. *Relatório dos Trabalhos*, 41.

24 Jorge Czajkowski, *Guia da Arquitetura Colonial, Neoclássica e Romântica no Rio de Janeiro* (Rio de Janeiro, 2000).

25 '... à extirpação dos três pólipos que tanto concorrem para dificultar a respiração da nossa capital, pólipos fáceis de serem extraídos, aliás sem o perigo de reprodução, e que são formados por estas excrescências do solo, conhecidas sob a denominação de morro do Senado, morro de Santo Antonio e morro do Castelo.' See Carlos Kessel, *A Vitrine e o Espelho: o Rio de Janeiro de Carlos Sampaio* (Rio de Janeiro, 2001), 49.

26 Kessel, 26–27.

27 Kessel, 50–51.

28 Kessel, 49–50.

29 Bruno Bontempi Jr and Noah W. Sobe, 'Rio de Janeiro, 1922–1923', in *Encyclopaedia of World's Fairs and Expositions*, ed. John E. Findling and Kimberly D. Pelle (London, 2008), 228.

30 '… eu não vejo que melhor forma possamos dar a essa comemoração [do Centenário] do que realizando obras de saneamento, de instrução, de assistência, de embelezamento, que completadas com uma Exposição Internacional no próprio local em que estejamos realizando as obras do arrasamento, possam mostrar ao estrangeiro, que ainda não nos conhece, de quanto somos capazes.' See Kessel, 52.

31 'A Montanha e a planície. Cidade prolongada sobre as águas da Guanabara', *Revista da Semana*, 19 August 1922, 18–19.

32 'A Engenharia e a Montanha. Como se tem operado, em dois anos, a demolição do Morro do Castelo', *Revista da Semana*, 19 August 1922, 14–15.

33 'O Rio que desaparece. Os Trabalhos no Morro do Castelo', *Illustração Brasileira*, 17 January 1922, 79–80.

34 'A visão da cidade futura, pairando sobre os despojos da colina hitórica …', *Revista da Semana*, 31 March 1923, cover page.

35 Bennett, 99.

36 Bennett, 76.

37 *Official Guide to the Exposition*, 153–159.

38 Bennett, 74.

39 Benedict Anderson, *Imagined Communities: Reflections on the Origin and Spread of Nationalism* (London, 1991), 6.

40 'A Sub-Comissão de Estatística … optou por expor estudos numéricos territoriais, demográficos, econômicos, intelectuais (no que diz respeito à educação e cultura) e morais (justiça e religião); murais com estatísticas comparativas; mapas, cartogramas e diagramas; quadros e gráficos'; Thais Rezende da Silva de Sant'Anna, 'A Exposição Internacional do Centenário da Independência: Modernidade e Politica no Rio de Janeiro do Inicio dos Anos 1920' (Master's dissertation, Universidade Estadual de Campinas, 2008), 50–51.

41 'No Pavilhão da Estatística, a cargo do ilustre Diretor de Estatística Dr. Bulhões Carvalho, se fará por meio de quadros uma exposição completa de estatística de todos os trabalhos do Brasil'. *Official Guide to the Exposition*, 159.

42 Bennett, 74.

43 Bennett, 73–74.

44 Anderson, xiv.

45 Anderson, xiv.

46 Anthony D. Smith, *Nationalism and Modernism. A Critical Survey of Recent Theories of Nations and Nationalism* (London, 1998).

47 *Relatório dos Trabalhos*, 41–42. Term used by João Luiz Alves during the closing ceremony of the Rio exhibition on 24 July 1923.

48 Smith, *Nationalism and Modernism*, 223–224.

49 Anthony D. Smith, *Theories of Nationalism* (London, 1971), 199.

50 Smith, *Theories of Nationalism*, 199.

51 Smith, *Theories of Nationalism*, 199.

52 *Official Guide to the Exposition*, 7 and 157.

53 'O Governo fará distribuir em profusão por todo o Brasil, por preço reduzido, ou gratuitamente, ou ainda como prêmio, ... reproduções de quadros históricos, gravuras e retratos ... destinados à educação cívica e patriótica ...'. *Relatório dos Trabalhos*, 14.

54 Film was a central medium for the dissemination of the exhibition itself. The film *1922: A Exposição da Independência* by Silvino dos Santos offers great views of the exhibitionary complex, technological advances, the hydroplane concession and the amusement park. Santos produced his film under the Batoque Cinematografia company. Digital copies are held at the Image and Sound Museum [Museum da Imagem e do Som – MIS] in Rio de Janeiro.

55 Tiradentes, or Joaquim José da Silva Xavier, led a revolutionary movement for Brazilian independence from the Portuguese colonial power in 1789. The First Republic boasted Tiradentes as a hero and the pioneer of republicanism in Brazil. The Andrada brothers – José Bonifácio de Andrada e Silva, Martim Francisco Ribeiro de Andrada and Antônio Carlos Ribeiro de Andrada Machado – had played crucial political roles during and right after independence in 1822.

56 Stephen Kern, *The Culture of Time and Space, 1880–1918* (Cambridge, MA, 1983).

57 Peter van Wesemael, *Architecture of Instruction and Delight: A Socio-Historical Analysis of World Exhibitions as a Didactic Phenomenon, 1798–1851–1970* (Rotterdam, 2001).

58 Jonathan Crary, *Techniques of the Observer: On Vision and Modernity in the Nineteenth Century* (Cambridge, 1990). See also Bennett.

59 'Jantei ontem, comigo mesmo, no Restaurante Falconi da Exposição ... A Exposição ardia, toda em luz! Os pavilhões pareciam irreais dentro da noite! ... Holofotes se entrecortavam no espaço, embranquecendo o céu e as nuvens com sua poeira de prata! E numa bacanal de luzes e de coloridos, os palácios pareciam suspensos, flutuantes, aéreos como um sonho que se vai dissipar! Em frente a tudo isso, o mar ... No escuro das águas havia formas gigantescas que se adivinhavam. Eram os "dread-naughts", os gigantes de aço, os dragões modernos e fabulosos do mar, dragões de sangue, dragões de fogo, que dormiam ali, pacatamente, as suas caldeiras frias, os seus canhões silenciosos, todas a sua pletora de ferro, em repouso... O contraste era tremendo. Em terra – luz, alegria, prazer; no mar – trevas, silêncio, ameaças! Então eu percebi a existência das duas exposições.' See Benjamim Costallat, 'As Duas Exposições', *Jornal do Brazil*, 17 September 1922, 5.

60 'E então compreendi tosa a inutilidade da exposição de terra, falsamente colorida, falsamente alegre, falsamente iluminada, feita de gesso, feita de fantasia, transitória, passageira, fraca, teatral, diante das couraças bem de ferro, dos obuses bem de aço, de toda aquela organização de morte e de extermínio, eterna como a humanidade e que a humanidade eternizou para a sua própria desgraça ....' See Costallat, 5.

61 Lévi-Strauss, 100.

62 Smith, *Theories of Nationalism*, 199.

Bibliography

Printed Primary Sources

'A Engenharia e a Montanha'. *Revista da Semana*, 19 August 1922.

'A Montanha e a Planície. Cidade Prolongada sobre as Águas da Guanabara'. *Revista da Semana*, 19 August 1922.

'The Brazil Centenary'. *The Times*, 29 July 1922.

'Brazil's Centenary Exhibition'. *The Times*, 6 September 1922.

Costallat, Benjamin. 'As Duas Exposições'. *Jornal do Brazil*, 17 September 1922, 5.

Official Guide to the Exposition. Edited by the Bureau Official de Informações. Rio de Janeiro: Typographia do Anuario do Brasil, 1922.

'O Rio que Desaparece: os Trabalhos no Morro do Castelo'. *Illustração Brasileira*, 17 January 1922.

Relatório dos Trabalhos da Exposição Internacional do Centenário. Rio de Janeiro 1922–1923. Edited by the Ministério da Justiça e Negócios Interiores da República do Brasil. Rio de Janeiro: Imprensa Nacional, 1926.

'Visão da Cidade Futura'. *Revista da Semana*, 31 March 1923.

Secondary Sources

Almeida, Annamaria Rego de. 'Uma Exposição Internacional no Rio de Janeiro'. *Anais do Museu Histórico Nacional* 34 (2002): 371–384.

Anderson, Benedict. *Imagined Communities: Reflections on the Origin and Spread of Nationalism*. London: Verso, 1991.

Bennett, Tony, 'The Exhibitionary Complex'. *new formations* 4 (1988): 73–102.

Bontempi, Jr, Bruno, and Noah W. Sobe. 'Rio de Janeiro, 1922–1923'. In *Encyclopaedia of World's Fairs and Expositions*, edited by John E. Findling and Kimberly D. Pelle, 228–229. Jefferson, NC and London: McFarland, 2008.

Crary, Jonathan. *Techniques of the Observer: On Vision and Modernity in the Nineteenth Century*. Cambridge, MA: The MIT Press, 1990.

Czajkowski, Jorge. *Guia da Arquitetura Colonial, Neoclássica e Romântica no Rio de Janeiro*. Rio de Janeiro: Casa da Palavra, 2000.

Dean, Warren. 'Economy'. In *Brazil: Empire and Republic, 1822–1930*, edited by Leslie Bethell. Cambridge: Cambridge University Press, 1993.

Elkin, Noah C. 'Promoting a New Brazil: National Expositions and Images of Modernity, 1861–1922'. PhD thesis, Rutgers University, 1999.

Fausto, Boris. 'Society and Politics'. In *Brazil: Empire and Republic, 1822–1930*, edited by Leslie Bethell. Cambridge: Cambridge University Press, 1993.

Greenhalgh, Paul. *Ephemeral Vistas: A History of the Expositions Universelles, Great Exhibitions and World's Fairs, 1851–1939*. Manchester: Manchester University Press, 1988.

Kern, Stephen. *The Culture of Time and Space, 1880–1918*. Cambridge, MA: Harvard University Press, 1983.

Kessel, Carlos. *A Vitrine e o Espelho: o Rio de Janeiro de Carlos Sampaio*. Rio de Janeiro: Arquivo Geral da Cidade do Rio de Janeiro, 2001.

Lévi-Strauss, Claude. *Tristes Tropiques*. New York: Criterion, 1961.

Levy, Ruth. *A Exposição do Centenário e o meio arquitetônico carioca no início dos anos 1920*. Rio de Janeiro: Editora EBA/UFRJ, 2010.

Rezende, Livia. 'Crafting the Nation: Brazilian "Civilised Exoticism" at World Exhibitions'. In *Design and Craft: A History of Convergences and Divergences*, edited by Javier Gimeno-Martinez and Fredie Floré, 136–139. Brussels: Universa, 2010.

Rezende, Livia. 'The Raw and the Manufactured: Brazilian Modernity and National Identity as Projected in International Exhibitions (1862–1922)'. PhD thesis, Royal College of Art, 2010.

Rezende, Livia. 'The Raw and the Manufactured: A Discussion of Brazilian National Identity Projected in International Exhibitions (1862–1867)'. In *Another Name for Design: Words for Creation*, edited by Haruhiko Fujita, 354–357. Osaka: Osaka University, 2008.

Sant'Anna, Thaís Rezende da Silva de. 'A Exposição Internacional do Centenário da Independência: Modernidade e Política no Rio de Janeiro do Início dos Anos 1920'. Master's dissertation, Universidade Estadual de Campinas, 2008.Smith, Anthony D. *Nationalism and Modernism. A Critical Survey of Recent Theories of Nations and Nationalism*. Routledge: London, 1998.

Smith, Anthony D. *Theories of Nationalism*. London: Duckworth, 1971.

Tenorio-Trillo, Mauricio. 'A Tropical Cuauhtémoc: Celebrating the Cosmic Race at the Guanabara Bay'. *Anales del Instituto de Investigaciones Estéticas* 65 (1994): 93–137.

Wadsworth, James E., and Tamera L. Marko. 'Children of the Pátria: Representations of Childhood and Welfare State Ideologies at the 1922 Rio de Janeiro International Centennial Exposition'. *The Americas* 58 (2001): 65–90.

Wesemael, Peter van. *Architecture of Instruction and Delight: A Socio-Historical Analysis of World Exhibitions as a Didactic Phenomenon, 1798–1851–1970*. Rotterdam: 010 Publishers, 2001.

8

The Ghent Universal and International Exhibition of 1913: Reconciling Historicism, Modernity and Exoticism

Davy Depelchin

... this Exhibition, because of its international character, will be a new and enlightened indication of direction on the path of humanity, moving towards progress, coming to a halt, contemplating, examining, judging the state of the global civilisation, to go better oriented to new and decisive triumphs.[1]

Universal exhibitions have always been a celebration of the achievements of mankind. Displaying state-of-the-art technology, countless products from all over the world, and including some of the most exquisite works of art, such exhibitions formed three-dimensional surveys of the world economy and culture. These temporary international exhibitions reaped the fruits of the past and translated them into ambitions for the future. They were a demonstration of optimism and a catalyst for progress.

Although united in their claims of internationalism – a fair for the benefit of mankind! – many of the organizers of universal exhibitions, as well as the participating nations, had objectives that were less altruistic and far more concrete. In an era of competing nation-states, this type of exhibition provided a platform for symbolic demonstrations of influence. Instead of fighting it out on the battlefield, countries competed upon the stage of the universal exhibition, which became a new way of measuring power. The creation of an elaborate pavilion, filled with agricultural, industrial and commercial displays, was an ideal way for a country to enhance its reputation as a stable, wealthy nation. Success also bred success: the organization of such a large-scale event could be taken as a proof of competence. Moreover, universal exhibitions contributed to the prestige of the host cities and countries. For this reason, most of the exhibitions had an urban and architectural programme highlighting the elements that were important to the creation, or re-creation, of communal, regional or national identities. Each exhibition, however, had a very different focus.[2]

A cursory look at documents related to the Ghent exhibition of 1913 might give the impression that the architectural design of the fair was dominated by historicism and traditionalism, at the expense of exoticism. It would certainly seem, at first glance, as though the organizers of the event in the provincial capital of East Flanders were not particularly interested in the inclusion of the exotic fancies that usually featured so prominently in universal exhibitions. This impression, however, would be erroneous. Exotic features did appear on the fairgrounds in Ghent, albeit less obviously. The emphasis placed by the organizers upon local history, and the future, appears to have prevented any such exoticism from taking centre stage or being used for exhibition marketing. Was this, however, the result of an inability to reconcile the local agenda with the universality inherent in the notion of a universal exhibition? Or can it be considered, on the other hand, to be an expression of an evolution in taste?

Belgian Exhibitions in the Nineteenth Century

The World Exhibition of 1958, held in Brussels between 17 April and 19 October, is the only universal exhibition that lives on in Belgian popular culture; the pre-war and the inter-war exhibitions organized on Belgian soil have largely been forgotten. The enduring popularity of Expo '58, as it is also known, is mainly due to the post-war optimism of the period and with the simple fact that the event is still within living memory. However, the role of historiography should not be underestimated. Ever since the fiftieth anniversary celebrations in 2008, accessible publications and documentaries on the event have become widespread.[3] Although Belgian universities have taken the lead in terms of researching and celebrating the centennials of the earlier universal exhibitions, the interest in these fairs is largely confined to an academic or, at best, local audience.[4]

The phenomenon of the universal exhibition was extremely important in pre-war Belgium. Nevertheless, Belgium's first international exhibition only took place in 1885. It was organized in Antwerp, in the context of the completion of major urban improvement works on the riverbanks, and became the first showcase for the King's new acquisition in Central Africa: Congo Free State.[5] Although the Antwerp exhibition was the first to be associated with the label 'universal' it was, in fact, a tributary of the National Exhibition that had been organized in Brussels five years earlier. This latter exhibition, which was actually a festival celebrating the fiftieth anniversary of Belgian independence, can be viewed as a national rehearsal for Antwerp 1885 and the exhibitions that followed it.[6] It marked the start of an impressive series of international showcases: Brussels in 1888, Antwerp in 1894, Brussels in 1897, Liège in 1905, Brussels in 1910 and, finally, Ghent in 1913. Although it is possible, in retrospect, to contest the use of 'universal' or 'international' to describe all of these exhibitions,[7] they did share the same objective: to exceed national interests. Furthermore, the scale of the fairs required enormous financial investments that could only be justified by offering financiers an international platform. Even so, the list of fairs cited above reveals that seven international exhibitions were held in Belgium over a period of 28 years. In other words, an international exhibition was organized in Belgium every four years, on average, between 1885 and the outbreak of the First World War. This level of frequency is extremely high, especially for a small country like Belgium.[8] Why, then, did Belgium feel the need to organize so many exhibitions?

In his 1993 article, Paul Greenhalgh identified four main politico-economic motives that help to explain the high frequency of universal exhibitions in certain countries: industry, trade, national cohesion and empire.[9] These parameters can easily be applied to the Belgian case. At the time of the international exhibitions, Belgium was a progressive nation with a highly developed and powerful industrial base: mining (coal) in the Borinage, heavy industry (steel) in the Sambre and Meuse valleys and textiles in the regional capitals of the North. Belgian national output exceeded national demand and, as a result, foreign trading partners were considered to be a condition *sine qua non* for the future growth of the gross domestic product (GDP), and for the wealth of the nation itself.[10] Where better to display these products and to establish commercial relationships than in an international exhibition? On the other hand, Belgium's interest in the world fair phenomenon coincided with a series of internal political tensions. The political unity that characterized the early days of the country had given way to two centres of power, with the Catholics on the one side and the Liberals on the other. Furthermore, both parties had to deal with the rise of socialism and with the emancipation of the Flemish-speaking part of the country. Greenhalgh argues that it is precisely these factors that encouraged those looking for a new, shared identity to undertake the preliminaries for an international exhibition.[11]

The Ghent Universal and International Exhibition of 1913

The *Exposition Universelle et Internationale de Gand* opened on 26 April 1913 and closed on 3 November, a total of 192 days. The exhibition grounds, with a total surface area of 1,250,000 square metres, consisted of three interconnected terrains in the south of Ghent, near the newly finished Saint Peter's train station (Fig. 8.1). National and international trains stopped at the main gates of the exhibition, thereby allowing a fluid and efficient stream of visitors in and out of the fair. On the other hand, from the northern gate of the site, it was only a 15-minute walk to the historic city centre. The proximity of the exhibition site to the heart of city was vitally important to the organizers, as will be discussed further on.

Twenty countries, excluding Belgium, participated in the event. Not all of the contributions, however, were made through official channels. The German pavilion, which stood alongside those of the other major exhibitors (Argentina, Canada, Chile, France, Great Britain, Italy, the Netherlands, Persia, Spain and Tunisia), was, in fact, a private initiative of the *Deutsche*

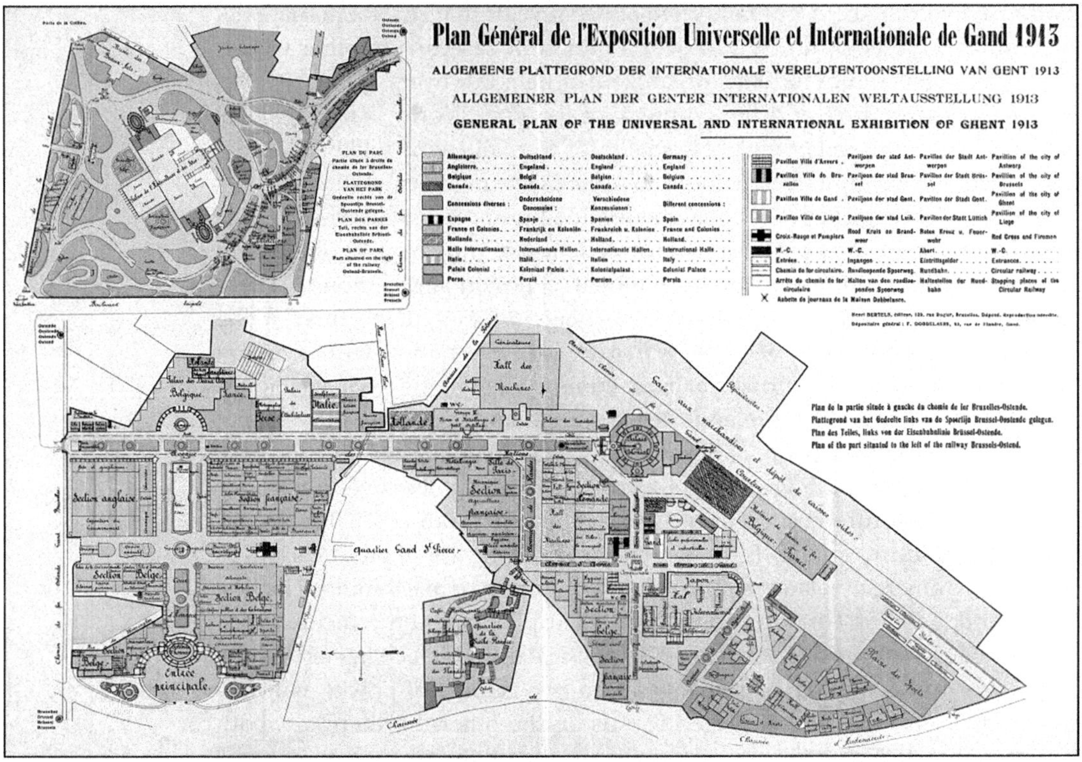

8.1 General plan of the Universal and International Exhibition of Ghent 1913, published by Henri Bertels (Brussels), circa 1913. Ghent University Library, BHSL. HS.III.0003. 001182. © Reproduction right Ghent University Library

Werkbund, a pressure group of German industrialists founded in 1907.[12] Some of these participants were even represented in more than one section of the fair, or in different locations. In addition to its vast palace on the Avenue of Nations, France took part in the railway exhibition and dominated the Fine Arts salon. France also built a pavilion in the Beaux Arts style to represent the city of Paris and constructed seven more pavilions for its colonies, protectorates and trade posts. The other participating countries (Austria, Bolivia, Brazil, Denmark, California and the United States of America, Guatemala, Japan, Romania and Russia) had more modest displays in the international hall in the southern part of the fair. The number of foreign exhibitors varies according to the source consulted, but it is possible to estimate that between 14,000 and 20,000 people were involved.[13]

During the six months that the Ghent exhibition was open to the public, it attracted 9.5 million visitors. Almost 2 million single tickets, 86,000 season tickets and 500,000 passes were issued to personnel. If we relate these figures to those compiled during the 1910 Brussels exhibition, three years earlier, it is possible to ascertain that the number of participating countries had decreased (20 compared to 26), as had the total number of visitors (a difference of 3.5 million). The surface area of the exhibition ground in Ghent was, however, 25 per cent larger. Does this mean that the exhibition was a failure? Not in the least. Compared to the earlier exhibitions in Antwerp,

Brussels and Liège, Ghent did very well.[14] Moreover, the cost of producing the 1913 exhibition (16,500,000 Belgian francs) was lower than those for the Brussels event (18,000,000 Belgian francs).[15]

An Agenda and a Promotion Strategy

Although the Ghent exhibition is internationally recognized as having been genuinely universal, it tends to be marginalized within the history of the world fairs because of its location and objectives. What, therefore, motivated the provincial city of Ghent to host an international event?

From 1889 onwards, there was a growing body of public opinion that Ghent should organize an international exhibition. The city had proved its efficiency in arranging large-scale public events with the very successful Provincial Exhibition of 1898. Nevertheless, Ghent was not a capital city like Brussels, nor was it one of the world's leading harbours like Antwerp. Only when the people of Liège demonstrated that they were capable of turning their 1905 exhibition into a huge success did a group of local representatives – from the worlds of industry, commerce, finance and the arts – join forces to establish a society dedicated to examining the possibilities. The ambitions of the initiators were also shared by a number of people working in the highly influential textile industry that dominated the local economy and led to Ghent being nicknamed 'the Belgian Manchester'. The society immediately assessed the commercial opportunities that would be created by a universal exhibition. As mentioned earlier, an investment in a universal exhibition was viewed as an investment in the future. Moreover, the city could also incorporate some of its most popular events into a universal exhibition, such as the five-yearly horticulture show, also known as the *Floralies*, and the three-yearly *Salon des Beaux-Arts*. Ghent's central location and close proximity to both the capital city and the ports of Antwerp and Ostend were important factors in terms of attracting British and other overseas visitors. It was also easily reached by rail from the neighbouring countries of France, Germany and the Netherlands.

Looking back today, it is evident that the 1913 exhibition was not so much a catalyst for progress, but the end of a line. Fortunately, at the time, the people behind the Ghent exhibition did not know that all their efforts and long-term ambitions would be curtailed, only six months later, by the outbreak of the First World War. Although dark clouds were already gathering over South-East Europe (the Balkan Wars), the 1913 Ghent exhibition was very much a festival of optimism. It focused on the future and on the notion of modernity but this did not, however, preclude historical references. On the contrary, the exhibition was firmly rooted in the city's glorious past. In the Middle Ages, Ghent was a large, wealthy city at the junction of continental trade routes and enjoyed a high level of autonomy – which, if challenged, it protected fiercely. Fostered by nationalism, and later also by an urge to legitimize a country that had only been invented in 1830, nineteenth-century historians were unafraid to mine the past and to emphasize the importance of cities like Ghent. History played a significant role in the establishment of the Belgian nation, and also in the re-establishment of the regions and cities.[16] During the first half of the nineteenth century, Ghent became an important commercial and industrial centre once more; a position it retained up till 1913.

It is necessary to contextualize the organization of Ghent exhibition by looking at the local politics of the day and, more precisely, at the city's urban vision. From the 1880s onwards, Ghent had invested in the redevelopment of the historical centre by liberating the medieval monuments from their so-called 'parasitic' structures. New buildings were inserted into the urban fabric that supported the 'picturesque' modern image that the local government was striving towards. In the heart of the historical centre, just in front of the thirteenth-century Saint Nicholas Church, for instance, the architect Louis Cloquet built a vast new post office (1898–1909) using an eclectic version of the neo-Gothic style. Nor was the city administration afraid to historicize pre-existing iconic elements within the urban landscape. The medieval appearance of the Belfry, which had been rebuilt over the centuries, was 'restored' (1911–13) following a Eugène Viollet-le-Duc inspired logic. Town planners did not hesitate to historicize and to beautify the city, driven by a wish to create order in the chaos that they ascribed to the medieval city grid. A particularly fine illustration of this can still be seen today in the vista that was artificially created by knocking down a whole city block and constructing a bridge across the River Leie (Saint Michael's Bridge). This resulted in a spectacular view that showcases Ghent's three famous medieval towers (belonging to the Saint Nicholas Church, the Belfry and the Cathedral of Saint Bavo). The image of the city has been synonymous with this postcard-like vista ever since.

Leaflets and posters for the 1913 exhibition document that one of the organizers' primary objectives was the promotion of cultural and city tourism (Fig. 8.2). The city's historical monuments were thus an extra selling point for the temporary event. It is no coincidence that the Cathedral of Saint Bavo and the newly built post office both featured on the exhibition poster, albeit discreetly. The main element of the poster is a purposeful young woman on a white horse covered in a ceremonial, medieval-looking blanket. In her right hand, she carries a bicolour flag featuring the words *Gloire au Travail* ['Glory in Work'], while in her left hand she holds a laurel wreath incorporating ribbons in the Belgian colours of black, yellow and red. These are references to youth, history, the nation and work. The past and the future united; an encapsulation of the self-image that Ghent projected upon itself.

Despite this emphasis placed on local history and culture, the 1913 exhibition was also presented as a truly international event. As mentioned above, the full title of the Ghent exhibition included the words 'universal' and 'international'.[17] In the absence of any international convention – this would only be formulated in 1928 in Paris – this explicit labelling was a way of stressing that the event was not to be mistaken for a local happening, and

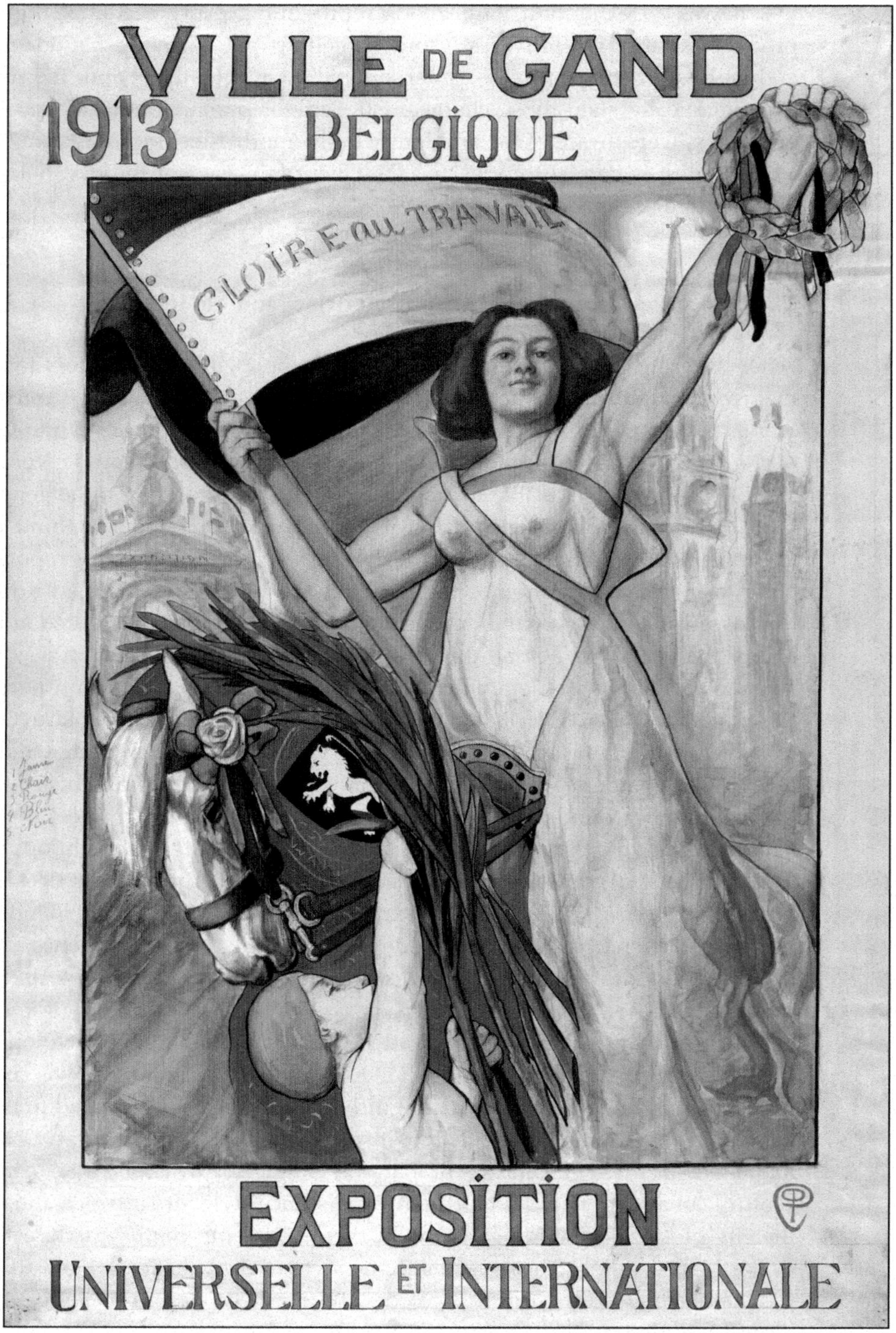

8.2 Promotion poster for the Ghent Universal and International Exhibition, artist unknown, circa 1913. Ghent University Library, BIB. AFF.C.000170. © Reproduction right Ghent University Library

that it was an exhibition with goods representing every sector of human production (that is, 'universal') and a melting pot of many nationalities and national displays (that is, 'international'). On the other hand, the need to emphatically state those characteristics might perhaps indicate that the organizers were aware of the ambiguity in the motives behind the exhibition – although it is crucial to note that the Ghent fair was not the first Belgian exhibition to use this nomenclature.[18]

The Exhibition Grounds: Between Historicism and Modernity

When entering the Ghent exhibition through the main gates, visitors found themselves right at the heart of the fair grounds. The majestic Court of Honour, onto which the portico gave out, was modelled on the White City of the 1893 Chicago Columbian Exhibition. Architect Oscar Van de Voorde (1871–1938) was responsible for the design, and also worked on the general layout of the exhibition and many of the individual pavilions. Behind the Court of Honour's white, Beaux Arts-style façades (praised by contemporary critics as being a 'symphony of snow')[19] lay the Belgian, French and British displays. The Fine Arts pavilion, executed in similar style, was located at the end of the visual axis. Of all the sections in the Ghent exhibition, it was the Avenue of Nations that epitomized modernity. It contrasted with a part of the exhibition called 'Old Flanders', a picturesque, idealized 'reconstruction' of a historical Flemish city, designed by Valentin Vaerwyck and Armand Heins. 'Old Flanders' was certainly the most obvious reference to history on the exhibition grounds. It was a private initiative designed to meet visitor expectations – this type of scenery had become something of a tradition. The 'Old Flanders' display was the successor of the 'Old Antwerp' of 1894, the *Bruxelles-Kermesse* of 1897 and 1910 and the *Vieux Liège* of 1905. 'Old Flanders' was an attraction in its own right, and was conceived as total experience, with stand holders dressed up as historical figures, performances, contests and processions.[20]

A degree of historicism also pervaded the area containing the pavilions of the Belgian cities. It is remarkable that each of the participating towns chose to erect a building in a clearly distinguishable historical style related to their heyday: a thirteenth-century Gothic-inspired pavilion for Ghent, an edifice modelled on fifteenth-century Brabantine Gothic for Brussels, a sixteenth-century construction in the Northern Renaissance style for Antwerp and, for the city of Liège, a building typical of the seventeenth-century architecture of the Meuse valley. These very specific style choices, which looked to the past rather than to the future, accord with the thesis described above: that municipal identity was vitally important and could be shaped by key historical buildings.

It is not true, however, that there was an inability to adapt historical characteristics to the present. The 'Modern Village', a prototype settlement with architecture that accommodated the 'modern way of life', and the (terminological) counterpart of 'Old Flanders', illustrates this. Although the 'modern touch' is instantly perceptible in this part of the exhibition, it cannot be denied that these buildings stood for regionalism and traditionalism. As Tim Dickel has already indicated in his work on these sections of the exhibition, both 'Old Flanders' and the 'Modern Village' implicitly promoted architectural regionalism.[21] The organizers and designers of the Ghent exhibition saw the origins of modernity in their own local history – and they were not afraid to exploit this notion.

Ghent 1913 and the Exotic

> The general disposition of the Ghent Exhibition testifies abundantly that the leading idea was all about symmetry and method. In earlier and similar manifestations, apart from some exceptions, the hand of fate had scattered dazzling halls and beautiful parcels in a more or less picturesque chaos without trying to create 'ensembles', which would have emphasized the buildings of which they were composed. Without a doubt was there an Avenue of Nations, a principal alley, stereotypical expressions in the dictionary of exhibitions; but these, far from presenting a homogeneity of style and architecture, gave, on the contrary, a right of asylum to the most unbridled fantasy and to the widest cosmopolitanism. In Ghent it was different; the innovation was fortunate.[22]

For the author of the above quote, taken from *Le livre d'or* [Golden Book] that accompanied the exhibition, it seems to have been a relief that Ghent did not focus on one of the most persistent and typical aspects of universal exhibitions: the exotic. While numerous exotic pavilions lined the paths of the 1905 Liège exhibition, the 1910 Brussels exhibition, and their international predecessors, the Ghent exhibition seemingly lacked this exotic accent. It could be argued that the organizers identified more with the glorious past of the city, and its promising industrial future, than with the clichés of Western popular culture. This interpretation is not, however, entirely accurate. Non-Western pavilions in exotic styles were to be found within the exhibition grounds, but instead of being given prominent places on the Avenue of Nations, the contributions of Morocco, Tunisia and the unofficial Algerian presentation were relegated to a wooded part of the site. There was also a Philippine village, but it was well hidden on the outskirts of 'Old Flanders', and visitors were required to walk through half of this section in order to reach its only entrance. The very opposite was true, however, for the Belgian colonial display dedicated to the Congo. This was not only accessible but given great prominence at the end of the Avenue of Nations. Although there has been a tendency to associate the architecture of this pavilion with Orientalism,[23] the style is, in fact, very difficult to categorize. At best, it could possibly be labelled 'eclectic exoticism' or, in the words of Johan Lagae, an 'orientalized classicism'.[24] The indeterminate nature of the architecture can be explained by the architects' search for a telling, authentic Congolese style. Unable to find a distinctive and

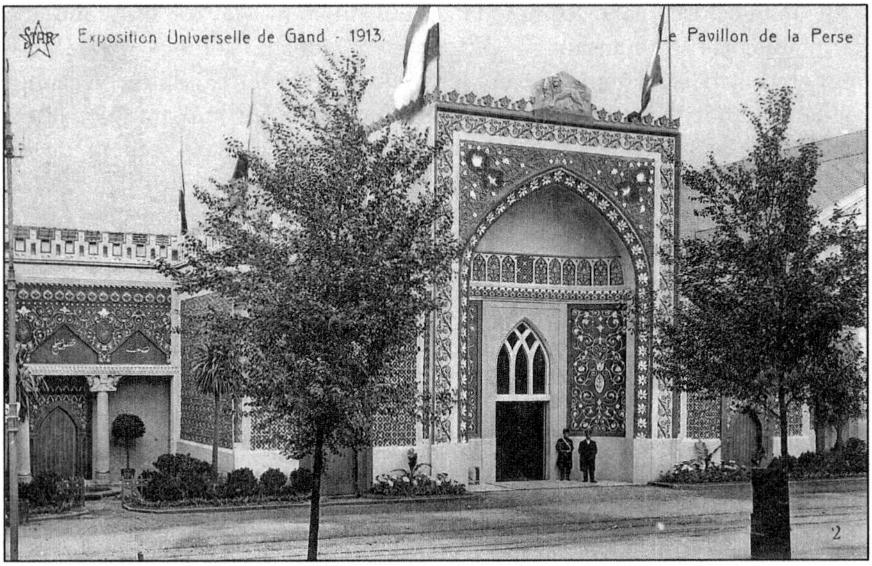

8.3 Postcard of the Persian Pavilion, published by Héliotypie De Graeve (Ghent), 1913. Author's collection

easily identifiable theme in the vernacular architecture of Central Africa, they adopted a hybrid style.

Persia was the only country to possess an eye-catching 'exotic' pavilion in a prime location (Fig. 8.3). As a result, it received a fair amount of attention: 'In this total symphony of snow, not one discordance, not one conflict, not one friction; over there, only the Persian pavilion played its shrill note; one would think this a rosette of the Legion of Honour, aiming to glorify architecture, upon the ensemble of his oeuvre.'[25] The general architect of exhibition, Oscar van de Voorde, was responsible for the elegant design: the mosques of Isfahan and Tabriz inspired the façades and the interior was reminiscent of pre-Islamic Persepolis. The brightly coloured building was considered to be a good example of a well-considered architectural attraction. The Persian section of the exhibition came about through the devoted efforts of a committee consisting of three officials, whose ambition it was to strengthen the ties between the two countries: Georges Van Oost, consul in Ghent, Louis Coetermans, consul in Antwerp, and Mahmoud Khan, the plenipotentiary Persian minister in Belgium. Persia's increasing exposure to the West had led to a growing demand for Western products. Belgium, hitherto a negligible trade partner, had the potential to become an important market for Persian goods. Belgians, for their part, were interested in buying Persian wood, cotton, rice, wheat, raisins, dates, almonds, pistachio nuts, attar of roses and tobacco. On the other hand, they also saw an opportunity to export the products of their own industry, that is to say: cotton yarn, cloth, iron, steel, glass, paper and flour. Although Belgian products were already available in Persia, they were imported through British and German intermediaries. In an attempt to improve the trading conditions for both partners, Coetermans seized the initiative and decided to eliminate the go-betweens. In 1906, he convinced the Hamburg-America Line to establish a direct connection between Antwerp and Persia and trade became more lucrative as a result. In 1913, for the first time in Belgian exhibition history, Persia invested in a freestanding pavilion instead of opting for a display in one of the larger exhibition halls. In so doing, the country emphasized the importance of its partnership with the exhibition's host country.[26] It is interesting to note, however, that Persia was the only nation to claim an orientalist construction on the Avenue of Nations. In the light of this, it is worth examining the ambitions, and the 'exotic' contributions, of the other participants.

A little further to the north, on the border between the Avenue of Nations and the commercial Avenue du Belvédère, where tasting pavilions and theme restaurants were located, the Indian delegation erected a pavilion in the Moghul style. An initiative of the Bhumgara trading company, its location was not prominent enough for it to be taken seriously as an official pavilion. The Indian palace was enclosed by a restaurant called Zillerthal and a sanitary block. It is significant, therefore, that out of all the plans drawn up of the exhibition site, of which there were many, only a handful indicate the building. And it was not only the cartographers who paid so little attention to the Indian pavilion. It is also very difficult to find descriptions of India's participation in the publications of the day.[27]

Of all the participating nations, France had the most potential to be 'exotic'. No less than seven pavilions were constructed to house the country's colonial display(s): a pavilion *des colonies françaises*, one for the colonial commerce, an ambulant construction with offices and a room for the colonial press, another one for the French Sangha-Oubanghi Company, a pavilion for Morocco, one for Tunisia and a showcase for Algerian products (Fig. 8.4). In contrast to

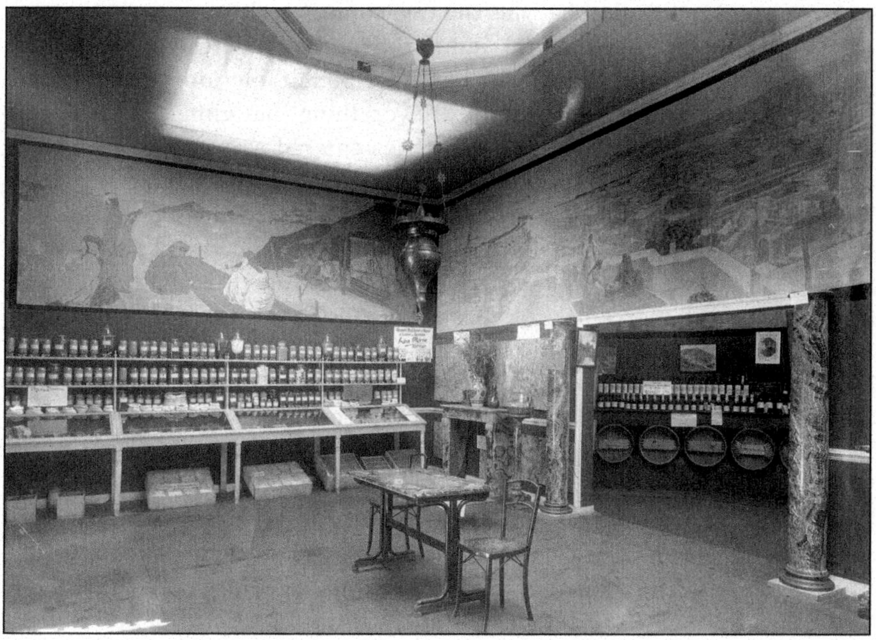

8.4 The Algerian Pavilion, interior view, author unknown, 1913, photograph. City of Ghent, Municipal Archives, De Zwarte Doos, Collection Expo 1913, MN1913_FO_0248. © Reproduction right Municipal Archives Ghent

Morocco and Tunisia, Algeria was not officially represented and the latter was an initiative of a group of Algerian entrepreneurs. It was also Morocco's first participation in a universal exhibition in its capacity as a French protectorate, a status it had obtained on 30 March 1912. The pavilion was located next to that of Tunisia. Both were designed in an orientalist style, and included horseshoe arches, a minaret and a crescent-topped dome. The total surface area of the pavilions was approximately 10,000 square metres and they were located in the park where the Ghent citadel had once stood.

The French pavilions dedicated to the Maghreb were mostly conceived as modern showcases designed to develop commercial ties with Belgium and the world. The showcases focused on geographical maps, samples of the natural resources and products of local workmanship. Inside, the exoticism was diluted in favour of a less fanciful, more objective and mercantile approach. The Tunisian pavilion may have contained a souk, and the Moroccan one an exhibition of the *Société des Peintres orientalistes français*,[28] but they no longer seemed designed to appeal because of their picturesque 'otherness'. The core business of the pavilions, and the displays within them, was to stimulate international trade. In the *Livre d'or*, Gustave Drèze noticed that the objective of earlier exhibitions was 'to impress the visitor's mind and to evoke the sight of these distant territories and the customs of their inhabitants by establishing palaces in exotic styles'.[29] He goes on to remark that 'in Ghent, the [French] colonial Exhibition was a natural history; it was documentary, commercial rather than picturesque ....'[30] This contemporary testimony accords with the analysis made by Zeynep Çelik in *Displaying the Orient*.[31] She writes that the Moroccan and Tunisian pavilions in the Paris exhibition of 1867 were 'a deliberate escape from the economic and political upheavals the two countries [Tunisia and Morocco] were experiencing. Agricultural, industrial, and commercial spheres were omitted: instead, there were palaces with exuberant interiors and furniture, royal costumes and armour – simply "everything that glitters, shines, and adorns"'.[32] However, when these countries entered a system of European tutelage based on neo-mercantilist politics, there was a tendency among the organizing committees to adopt new, rationalized display techniques in order to present the North African regions as developed and trustworthy trading partners.[33] While the free trade principle that had governed international commerce in the 1860s and 1870s had, by that time, been set aside in favour of protectionism, the international markets had expanded. During the last quarter of the nineteenth century and the pre-war years of the twentieth century, the development of European industry compelled the nation-states to look for new export markets, while an improvement in the standard of living fuelled a European demand for goods. European interventionist politics in North Africa, and elsewhere, were dictated by this need. The resulting policies aimed to improve the efficiency of production process in these countries and, at the same time, to control any subsequent trade. In addition to this politico-economic factor, the imperial powers considered their colonies, protectorates, spheres of influence and foreign concessions to be instruments through which to gain prestige.[34]

Besides the change in scope and objectives, there was also an evolution in the characteristics and visibility of the orientalist pavilions. The colonial displays of the Ghent exhibition seem to mirror the tendencies that Zeynep Çelik traces back to the Paris exhibition of 1900 – although many authors consider the 1900 *Exposition Universelle* to be the very paroxysm of exoticism:[35]

> Along with the return to classicism, a stricter control was exercised in planning the fairgrounds in 1900, and picturesque effects were minimized. Although the buildings on the Rue des Nations were designed in many different styles, they were neatly aligned and their regularized siting complemented the symmetry and axiality across the river. Moreover, in the turn-of-the-century exhibition, the haphazardness of the indigenous villages was tactfully hidden from immediate view.[36]

This is very true of Ghent, where the most explicit architectural manifestations of exoticism were in less prominent locations: the Moroccan, Tunisian and the unofficial Algerian pavilion were situated in the wooded area of the site. Although this setting might, at first glance, appear to have been rather picturesque, the contrast between the Eastern-style buildings and the European broadleaf and coniferous trees produced the opposite effect. Meanwhile, and despite the tendency to objectify the showcases, the pavilions of the French colonies were located in the zone where the attractions were to be found. An exhibition perennial, entitled 'The Cairo Street', containing a display of stereotypical scenes from Middle Eastern daily life, was situated in close proximity to the French colonial section.

Conclusion: Reconciling Traditionalism, Exoticism and Modernity

One immediate question that arises, given the elements described above, is why there was such a big difference in the exotic display between the Ghent exhibition of 1913 and its immediate Belgian predecessor, the 1910 exhibition in Brussels. The question becomes even more intriguing if one takes into account the fact that the preparations for the Ghent fair began as early as 1905. This means that, to a certain extent, the two exhibitions were planned simultaneously.

Obviously, an analysis that uses the notion of 'changing times' to explain the objectification of the colonial displays, as well as the fact that exotic pavilions became less visible, does not tell the whole story. However, if one correlates this reading with the exhibition organizers' focus on traditionalism, an interesting explanation can be found. Brussels, the capital of Belgium, had less need to anchor itself in local history since it was already relatively well positioned as a modern, cosmopolitan metropolis. In Ghent – without denying the chronological, market-driven changes in colonial displays – the desire to emphasize local history explains the suppression of exoticism. Instead of creating a typical, eclectic Avenue of Nations (a staple of world exhibitions), the organizers of the Ghent exhibition tried to obscure the exotic pavilions by banishing them to a less visible part of the fair. The lavishness of the genuine orientalist architecture was considered to be discordant. Since modernity was linked with the historical past of the city, the common juxtaposition of primitivism and/or exoticism with modernity that was so typical of other exhibitions was redundant in Ghent.

The Ghent exhibition of 1913 illustrates how the concept of the universal exhibition could be rethought to meet local demands and ambitions. As the latter were so prominent in Ghent, it is tempting to view the 1913 event as a project that exceeded its initial, locally anchored goals. Indeed, it was through the consistent execution of the central tenets and ideas of their programme that Ghent exhibition organizers managed to create such a quirky exhibition.

Notes

1 All quotations translated from the French by the author. '… cette Exposition, de par son caractère international, sera un nouveau et lumineux jalon planté sur la route où l'humanité, en marche vers le progrès, s'arrête, se recueille, s'examine, juge de l'état de la civilisation mondiale pour aller, mieux orientée, à de nouveaux et décisifs triomphes.' Albert Dutry, *Coup d'oeil sur l'Exposition de Gand* (Brussels, 1913), 16.

2 Paul Greenhalgh, 'The Tradition of Expositions Universelles', in *De panoramische droom. Antwerpen en de wereldtentoonstellingen 1885–1894–1930 /The Panoramic Dream. Antwerp and the World Exhibitions 1885–1894–1930*, ed. Mandy Nauwelaerts et al. (Antwerp, 1993), 33–34; Pieter Van Wesemael, *Architecture of Instruction and Delight. A Socio-Historical Analysis of World Exhibitions as a Didactic Phenomenon (1798–1851–1970)* (Rotterdam, 2001), 19–26.

3 The Department of Architecture and Urban Planning of the University of Ghent is noted for its research on this topic and for the dissemination of the study results. Two publications of importance are Rika Devos and Mil De Kooning, eds, *Moderne architectuur op Expo 58: voor een humaner wereld* (Brussels, 2006); Gonzague Pluvinage, ed., *Expo 58: tussen droom en werkelijkheid* (Tielt, 2008).

4 To celebrate the Brussels exhibition of 1910, the University of Brussels organized an exhibition accompanied by a catalogue (Serge Jaumain and Wanda Balcers, eds, *Bruxelles 1910. De l'Exposition universelle à l'Université* (Brussels, 2010). In Ghent, several books were published on the occasion of the centennial of the 1913 exhibition: Wouter Van Acker and Christophe Verbruggen, eds, *Gent 1913. Op het breukvlak van de moderniteit* (Ghent-Kortrijk, 2013); Jo Van Herreweghe, ed., *Impact. Wereldtentoonstelling Gent 1913* (Ghent, 2013); Guido Everaert, *Opgeknapt. 10 monumenten in de Gentse Kuip en de Expo van 1913* (Ghent, 2013); Joost Vandommele, *Wereldtentoonstelling Gent 1913* (Ghent-Kortrijk, 2013); and Peter Jacobs and Erwin De Decker, *Wandelen door Gent, 1913–1918. Van Wereldtentoonstelling tot Wereldoorlog* (Tielt, 2013). Other academic articles on the Ghent exhibition include Leen Wyffels, 'De wereldtentoonstelling te Gent in 1913. Een architektonische evaluatie van het werk van Oscar Van de Voorde', MA thesis, Ghent University, Department of Art History, 1979; André Capiteyn, *Gent in weelde herboren: wereldtentoonstelling 1913* (Ghent, 1988); Norbert Poulain,

'Op zoek naar (h)erkenning. Een openbaar onderzoek naar de inhoud van het paviljoen van de Gentse Academie op de wereldtentoonstelling in 1913', *Interbellum* 18, no. 2 (1998): 7–30; Tim Dickel, 'Wereldtentoonstelling Gent 1913. Oud Vlaendren en het Modern Dorp: schijnbare tegenpolen?', MA thesis, Ghent University, Department of Art History, 2004; Christiane Heiser, 'Der Deutsche Werkbund auf der Weltbühne oder wie Deutschland 1913 doch noch an der Weltausstellung teilnahm', *Archiv für Kulturgeschichte* 92 (2010): 363–398.

5 Ruddy Doom, 'De wereldtentoonstellingen en de koloniale propaganda/ The World Exhibitions and Colonial Propaganda', in *De panoramische droom*, ed. Mandy Nauwelaerts et al., 195–207; Thomas Coomans, 'Wereldtentoonstellingen', in *Repertorium van de architectuur in België: van 1830 tot heden*, ed. Anne Van Loo (Antwerp, 2003), 571.

6 Wanda Balcers and Serge Jaumain, 'La Belgique et les expositions universelles' in *Bruxelles 1910*, ed. Jaumain and Balcers, 11–37.

7 One should especially question the 1888 Brussels Exhibition or the *Grand Concours International des Sciences et de l'Industrie*. In national historiography it is regarded as a genuinely international exhibition, while international publications such as the *Encyclopaedia of World's Fairs and Expositions* do not to include this event.

8 A census carried out in 1910 counted 7,423,784 people living in Belgium.

9 Greenhalgh, 'Tradition of Expositions Universelles', 34; also Greenhalgh, *Ephemeral Vistas. The Expositions Universelles, Great Exhibitions and World's Fairs, 1851–1939* (Manchester, 1988).

10 Rik Coolsaet, *België en zijn buitenlandse politiek, 1830–2000* (Leuven, 2011), 137–181.

11 Greenhalgh, 'Tradition of Expositions Universelles', 33–34.

12 Heiser, 363–398.

13 Two different figures even appear in *Le livre d'or*: 19,139 and 14,156. Gustave Drèze, *Le livre d'or de l'Exposition universelle et internationale de Gand en 1913* (Ghent, 1913), 45 and 112.

14 Antwerp 1885 had 3.5 million visitors, Antwerp 1894 5 million, Brussels 1897 6.2 million, and Liège 1905 had 7 million visitors. Drèze, *Le livre d'or*, 45.

15 Drèze, *Le livre d'or*, 36 and 44–45.

16 Jo Tollebeek, 'Enthousiasme en evidentie in de negentiende-eeuwse Belgisch-nationale geschiedschrijving', *Leidschrift* 8, no. 2 (1992): 61–81; Jo Tollebeek, 'The Hyphen of National Culture. The Paradox of National Distinctiveness in Belgium and the Netherlands, 1860–1918', *European Review* 18, no. 2 (2010): 207–225; Jo Tollebeek, 'At the Crossroads of Nationalism: Huizinga, Pirenne and the Low Countries in Europe', *European Review of History – Revue européenne d'histoire* 17, no. 2 (2010): 187–215.

17 The title was not *Universal and Industrial Exhibition*, as Findling notes in his article in the *Encyclopaedia of World's Fairs and Expositions*. John E. Findling, 'Ghent 1913', in *Encyclopedia of World's Fairs and Expositions*, ed. John E. Findling and Kimberly D. Pelle (Jefferson, NC and London, 2008), 213–215.

18 The Brussels exhibition of 1910 – or, in full, *Exposition Universelle et Internationale de Bruxelles 1910* – had the same title.

19 Drèze, *Le livre d'or*, 103.

20 Dickel, 66–75.

21 Dickel, 109–114.

22 See also Davy Depelchin, 'Tussen exotisme en neomercantilisme. Oriëntalistische paviljoenarchitectuur met een politieke boodschap', in *Gent 1913. Op het breukvlak van de moderniteit*, ed. Wouter Van Acker and Christophe Verbruggen (Ghent-Kortrijk, 2013), 140–153.

23 Iwein Meyskens, 'Architectuur van de koloniale paviljoenen op wereldtentoonstellingen', MA thesis, Ghent University, Department of Architecture and Urban Planning, 1997, 58–59.

24 Johan Lagae, '"Le Congo tel que les Belges l'ont trouvé et le Congo tel qu'ils l'ont outillé" ? Het palais colonial en de nationalistische verbeelding van de Belgische kolonie', in *Gent 1913. Op het breukvlak van de moderniteit*, ed. Wouter Van Acker and Christophe Verbruggen (Ghent-Kortrijk, 2013), 62–79.

25 'Dans toute cette symphonie de neige, aucun désaccord, aucun heurt, aucun froissement; là-bas, le pavillon de la Perse claironnait seul sa note plus aiguë; mais on eut dit une rosette de la Légion d'Honneur, tranchant à la gloire de l'architecture, sur l'ensemble de son oeuvre.' Drèze, *Le livre d'or*, 103.

26 Levaho, 'La manifestation en l'honneur de M. Louis Coetermans, consul général de Perse à Anvers', *Gand-Exposition* 2, no. 23 (1912): 277–278; 'Nog eene reeks belangrijke tijdingen', *Gand-Exposition* 2, no. 25 (1912): 306–307; 'Autour de l'exposition', *Gand-Exposition* 2, no. 26 (1912): 314–315; 'La visite de Sa Majesté le Roi Albert, le vendredi 6 juin 1913', *Gand-Exposition* 3, no. 17 (1913): 197; 'L'exposition au jour le jour', *Gand-Exposition* 3, no. 20 (1913): 235–236; Gustave Drèze, 'La Perse à l'exposition de Gand', *Gand-Exposition* 3, no. 29 (1913): 357–364; Drèze, *Le livre d'or*, 123–126; Dutry, 16.

27 The most detailed description can be found in Drèze, *Le livre d'or*, 183–184.

28 Léonce Bénédite (1856–1925), an art historian and curator at the Musée de Luxembourg in Paris, founded the society in 1893 to promote the oriental and 'Orient inspired' – that is, orientalist – arts in the West. See Roger Benjamin, *Orientalist Aesthetics: Art, Colonialism and French North Africa, 1880–1930* (Berkeley, 2003), 57–77; Pierre Sanchez and Stéphane Richemond, *La Société des Peintres orientalistes français. Répertoire des exposants et liste de leurs oeuvres, 1889–1943* (Dijon, 2008), 13–38.

29 '… impressionner l'esprit du visiteur et lui faire apparaître l'aspect de ces contrées lointaines et les moeurs de leurs habitants, en édifiant des palais de style exotique, destinés à la présentation de collections extraordinaires ou à l'exhibition de naturels de colonies.' Drèze, *Le livre d'or*, 174.

30 'A Gand, l'Exposition coloniale [française] fut une leçon de choses; elle fut documentaire, commerciale plutôt que pittoresque ….' Drèze, *Le livre d'or*, 174.

31 Zeynep Çelik, *Displaying the Orient. Architecture of Islam at Nineteenth-Century World's Fairs* (Berkeley, Los Angeles and Oxford, 1992).

32 Çelik, 122.

33 Çelik, 125–138.

34 R. R. Palmer, Joel Colton and Lloyd S. Kramer, *A History of the Modern World* (New York, 2007), 629–638; Coolsaet, 171–181.

35 Sylviane Leprun, *Le théâtre des colonies. Scénographie, acteurs et discours de l'imaginaire dans les expositions, 1855–1937* (Paris, 1986), 106–108; Greenhalgh, *Ephemeral Vistas*, 67–68; Van Wesemael, 362–394.

36 Çelik, 93.

Bibliography

Printed Primary Sources

'Autour de l'exposition'. *Gand-Exposition* 2, no. 26 (1912): 314–315.

Drèze, Gustave, 'La Perse à l'exposition de Gand'. *Gand-Exposition* 3, no. 29 (1913): 357–364.

Drèze, Gustave, *Le livre d'or de l'Exposition universelle et internationale de Gand en 1913*. Ghent: Vanderpoorten, 1913.

'La visite de Sa Majesté le Roi Albert, le vendredi 6 juin 1913'. *Gand-Exposition* 3, no. 17 (1913): 197.

Levaho, 'La manifestation en l'honneur de M. Louis Coetermans, consul général de Perse à Anvers'. *Gand-Exposition* 2, no. 23 (1912): 277–278.

'L'exposition au jour le jour'. *Gand-Exposition* 3, no. 20 (1913): 235–236.

Secondary Sources

Balcers, Wanda, and Serge Jaumain 'La Belgique et les expositions universelles'. In *Bruxelles 1910. De l'Exposition universelle à l'Université*, edited by Serge Jaumain and Wanda Balcers, 11–37. Brussels: ULB/Racine/Dexia Banque, 2010.

Benjamin, Roger. *Orientalist Aesthetics: Art, Colonialism, and French North Africa, 1880–1930*. Berkeley: University of California Press, 2003.

Capiteyn, André. *Gent in weelde herboren: wereldtentoonstelling 1913*. Ghent: Stadsarchief, 1988.

Çelik, Zeynep. *Displaying the Orient. Architecture of Islam at Nineteenth-Century World's Fairs*. Berkeley, Los Angeles and Oxford: University of California Press, 1992.

Coolsaet, Rik. *België en zijn buitenlandse politiek, 1830–2000*. Leuven: Van Halewyck, 2011.

Coomans, Thomas, 'Wereldtentoonstellingen'. In *Repertorium van de architectuur in België: van 1830 tot heden*, edited by Anne Van Loo, 570–577. Antwerp: Mercatorfonds, 2003.

Depelchin, Davy, 'Tussen exotisme en neomercantilisme. Oriëntalistische paviljoenarchitectuur met een politieke boodschap'. In *Gent 1913. Op het breukvlak van de moderniteit*, edited by Wouter Van Acker and Christophe Verbruggen, 140–153. Ghent-Kortrijk: Snoeck, 2013.

Devos, Rika, and Mil De Kooning, eds. *Moderne architectuur op Expo 58: voor een humaner wereld*. Brussels: Mercatorfonds, 2006.

Dickel, Tim. *Wereldtentoonstelling Gent 1913. Oud Vlaendren en het Modern Dorp: schijnbare tegenpolen?* Master's dissertation. Ghent University, Department of Art History, 2004.

Doom, Ruddy. 'De wereldtentoonstellingen en de koloniale propaganda/The World Exhibitions and colonial propaganda'. In *De panoramische droom. Antwerpen en de wereldtentoonstellingen 1885 – 1894 – 1930/The Panoramic Dream. Antwerp and the World Exhibitions 1885–1894–1930*, edited by Mandy Nauwelaerts et al., 195–207. Antwerp: Antwerpen 93, 1993.

Dutry, Albert. *Coup d'oeil sur l'Exposition de Gand*. Brussels: J. Goemaere, 1913.

Everaert, Guido. *Opgeknapt. 10 monumenten in de Gentse Kuip en de Expo van 1913*. Ghent: City of Ghent, 2013.

Findling, John E. 'Ghent 1913'. In *Encyclopedia of World's Fairs and Expositions*, edited by John E. Findling and Kimberly D. Pelle. Jefferson, NC and London: McFarland, 2008.

Greenhalgh, Paul. *Ephemeral Vistas. The Expositions Universelles, Great Exhibitions and World's Fairs, 1851–1939*. Manchester: Manchester University Press, 1988.

Greenhalgh, Paul, 'The Tradition of *expositions universelles*'. In *De panoramische droom. Antwerpen en de wereldtentoonstellingen 1885–1894–1930/The Panoramic Dream. Antwerp and the World Exhibitions 1885–1894–1930*, edited by Mandy Nauwelaerts et al., 23–34. Antwerp: Antwerpen 93, 1993.

Heiser, Christiane. 'Der Deutsche Werkbund auf der Weltbühne oder wie Deutschland 1913 doch noch an der Weltausstellung teilnahm'. *Archiv für Kulturgeschichte* 92 (2010): 363–398.

Jaumain, Serge, and Wanda Balcers, eds. *Bruxelles 1910. De l'Exposition universelle à l'Université*. Brussels: ULB/Racine/Dexia Banque, 2010.

Lagae, Johan. '"Le Congo tel que les Belges l'ont trouvé et le Congo tel qu'ils l'ont outillé" ? Het palais colonial en de nationalistische verbeelding van de Belgische kolonie'. In *Gent 1913. Op het breukvlak van de moderniteit*, edited by Wouter Van Acker and Christophe Verbruggen, 62–79. Ghent-Kortrijk: Snoeck, 2013.

Leprun, Sylviane. *Le théâtre des colonies. Scénographie, acteurs et discours de l'imaginaire dans les expositions, 1855–1937*. Paris: l'Harmattan, 1986.

Meyskens, Iwein. *Architectuur van de koloniale paviljoenen op wereldtentoonstellingen*. MA thesis, Ghent University, Department of Architecture and Urban Planning, 1997.

'Nog eene reeks belangrijke tijdingen'. *Gand-Exposition* 2, no. 25 (1912): 306–307.

Palmer, R. R., Joel Colton and Lloyd S. Kramer, *A History of the Modern World*. New York: McGraw-Hill International, 2007.

Pluvinage, Gonzague, ed. *Expo 58: tussen droom en werkelijkheid*. Tielt: Lannoo, 2008.

Poulain, Norbert. 'Op zoek naar (h)erkenning. Een openbaar onderzoek naar de inhoud van het paviljoen van de Gentse Academie op de wereldtentoonstelling in 1913'. *Interbellum* 18, no. 2 (1998): 7–30.

Sanchez, Pierre, and Stéphane Richemond. *La Société des peintres orientalistes français. Répertoire des exposants et liste de leurs oeuvres, 1889–1943*. Dijon: l'Echelle de Jacob, 2008.

Tollebeek, Jo. 'At the Crossroads of Nationalism: Huizinga, Pirenne and the Low Countries in Europe'. *European Review of History – Revue européenne d'histoire* 17, no. 2 (2010): 187–215.

Tollebeek, Jo. 'Enthousiasme en evidentie in de negentiende-eeuwse Belgisch-nationale geschiedschrijving'. *Leidschrift* 8, no. 2 (1992): 61–81.

Tollebeek, Jo. 'The Hyphen of National Culture. The Paradox of National Distinctiveness in Belgium and the Netherlands, 1860–1918'. *European Review* 18, no. 2 (2010): 207–225.

Van Acker, Wouter, and Christophe Verbruggen (eds). *Gent 1913. Op het breukvlak van de moderniteit*. Ghent-Kortrijk: Snoeck, 2013.

Vandommele, Joost. *Wereldtentoonstelling Gent 1913*. Ghent-Kortrijk: Snoeck, 2013.

Van Herreweghe, Jo (ed.). *Impact. Wereldtentoonstelling Gent 1913*. Ghent: City of Ghent, 2013.

Van Wesemael, Pieter, *Architecture of Instruction and Delight. A Socio-Historical Analysis of World Exhibitions as a Didactic Phenomenon (1798–1851–1970)*. Rotterdam: 010, 2001.

Wyffels, Leen. *De wereldtentoonstelling te Gent in 1913. Een architektonische evaluatie van het werk van Oscar Van de Voorde*. Master's dissertation, Ghent University, Department of Art History, 1979.

9

Old London, Old Edinburgh: Constructing Historic Cities

Wilson Smith

From the middle years of the 1880s great European exhibitions began to embrace a new type of attraction which can be termed the 'historic city'.[1] Ancient buildings which had once graced the exhibition's host city were faithfully and accurately reconstructed and assembled into a spectacular three-dimensional representation of a historical townscape, re-creating the distinctive features of local architecture and its urban setting and evoking a vision of the city's past. Surrounded by the exhibition's displays of modernity, of advanced technology, industrial innovation and new consumer products, the historic city looked backwards to a popular but authentic imagining of the host city's history. At the same time the definition and presentation of this past was determined by the authors of the exhibit and the story they wished to tell about their city: the historic city displayed a vision of the past mediated by contemporary discourses of power, belief and politics. This chapter will examine two of these historic city reconstructions: Old London, the first of the genre, exhibited annually between 1884 and 1886, and its immediate successor and imitator, Old Edinburgh of 1886. Like all historic city exhibits, their appeal rested on the specifics of past and present time and place; their appearance and content reflected the circumstances of their origin, their authorship, and the character of their host cities and exhibitions.

Old London: Realistic and Truthful

Picking their way through the labyrinthine geography of the South Kensington buildings, visitors to the International Health Exhibition in the summer of 1884 passed through a reproduction of Bishopsgate, one of the ancient entries to the City, to find themselves in 'Old London' (Fig. 9.1).[2] Re-creations of some 25 historic but long-since-demolished buildings had been grouped together into an open-air representation of a street as it might have

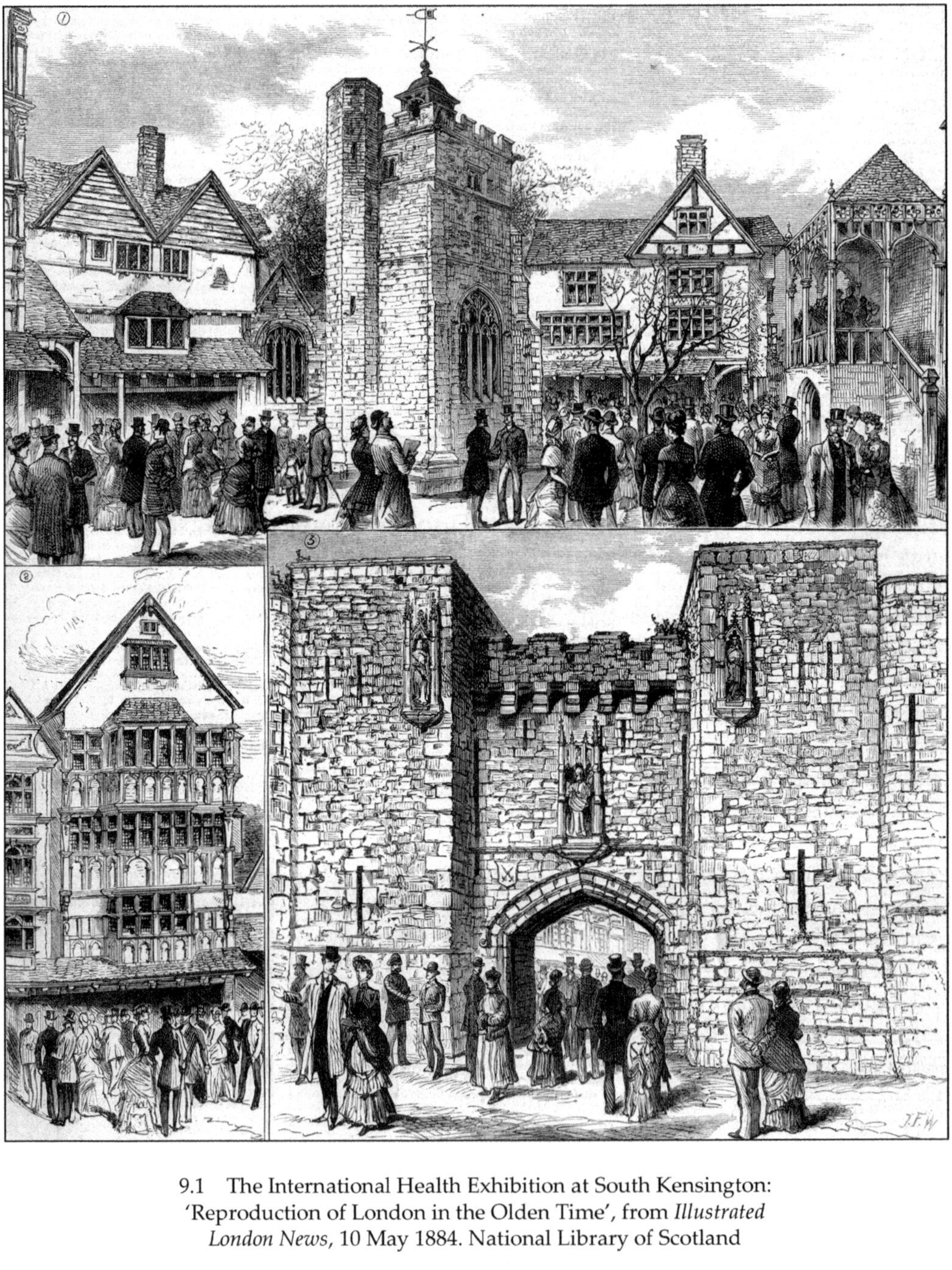

9.1 The International Health Exhibition at South Kensington: 'Reproduction of London in the Olden Time', from *Illustrated London News*, 10 May 1884. National Library of Scotland

existed before the Great Fire, 'that swift furnace of flame of 1666', destroyed much of the ancient city.[3] The reproductions included hostelries – the Rose Inn and the Cock Tavern – the tower of a medieval church, a Guild Hall with its stained glass, and such examples of domestic architecture as Isaac Walton's House, 'a magnificent specimen of an ordinary citizen's house in Elizabeth's reign'.[4] The centrepiece depicted Whittington's Palace, a lavish re-creation of the richly decorated residence associated with Richard Whittington, the semi-mythical medieval Lord Mayor (Fig. 9.2). The reproduction buildings had been assembled to create the effect of an ancient cityscape: from a broad High Street the passage narrowed to Elbow Lane 'in order to obtain that sinuosity so characteristic of London streets'.[5]

Old London's designer, George H. Birch, was at pains to stress the realism and truthfulness of his creation: it was 'realistic, in that these houses are no pasteboard and painted canvas delusions, but honest structures. Truthful, in that they represent no fanciful restorations from written records, but are faithful delineations from actual drawings derived from authentic sources.'[6]

Birch, an antiquarian and an architect specializing in church restoration, had originally contemplated the faithful reproduction of a complete London street, a section of historical Cheapside.[7] Lacking sufficiently accurate source material to satisfy his desire for authenticity, he hit instead on the device of juxtaposing 'true' representations of individual lost buildings in a seamless assemblage: an approach which, in any case, allowed for greater variety and artistic control in which necessary reductions in scale could be adjusted to form a satisfactory whole. The materials of wood, plaster and paint were worked into realistic but illusionary finishes. Surfaces were distressed to give the appearance of age: 'In order to give reality to the

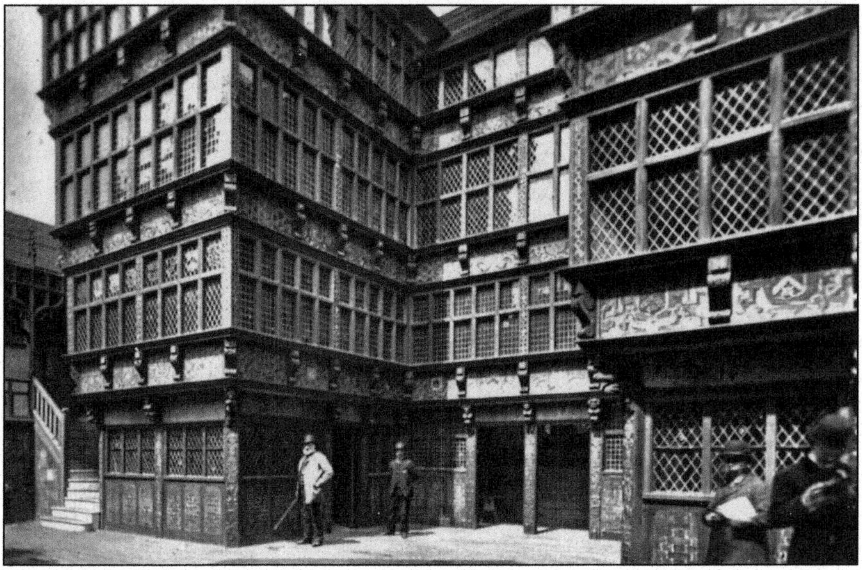

9.2 Whittington's House, from *A Souvenir Containing Six Permanent Photographs of the International Health Exhibition*. London Stereoscopic Company. David A. Hanson Collection. Sterling and Francine Clark Art Institute Library, Williamstown, Massachusetts

general effect the plaster itself is covered with different shades of brown and green while, to increase the illusion, lichen and mosses are inserted into the old walls.'[8] The resulting 'reality' was not that of the pristine original, but of undisturbed ageing: the illusion, that the buildings had survived undisturbed into the present.

Old London was characteristic of a new approach to exhibitions at South Kensington. After the triumph of 1851 the unsuccessful outcome of the second London International Exhibition in 1862 had constrained the ambitions of Sir Edward Cole and the London exhibition organizers. Doubtful of the prospects of future universal exhibitions Cole now proposed a turn to smaller-scale, more comprehensible themed shows. However, the dry treatment of the resulting South Kensington series of 1871–74 failed to excite public enthusiasm: the projected run of ten annual exhibitions closed after four, having attracted sparse audiences and run up a deficit of £150,000.[9] For the next South Kensington development Sir Philip Cunliffe-Owen, Cole's protégé and his successor as Director of the South Kensington Museum, adopted a more visitor-friendly approach influenced by his experience as a Commissioner for the 1876 Philadelphia Centennial and the Paris Exposition of 1878.[10] Cunliffe-Owen took the unassuming project of an International Fisheries Exhibition scheduled for 1883 and transformed it into a fashionable success, broadening its scope – '"The Fisheries", to all intents and purposes, became an exhibition of things in general "from a fishy point of view"' – and adding outdoor attractions: promenades under electric light, illuminated fountains and visual effects and musical entertainments.[11]

The popular success of the Fisheries Exhibition was taken up by Edward Prince of Wales, in an echo of his father Prince Albert's achievement of 1851. A programme of three subsequent annual events was projected under the Prince's patronage and directed by Cunliffe-Owen: Health in 1884; Inventions in 1885, and the Colonial and Indian Exhibition of 1886; all with the same formula of broadly interpreted themes and added spectacle.

In this setting Birch's reconstructed city was greeted with acclaim. Old London was predicted to be one of the hits of the Health Exhibition, and the exhibition-going public duly responded with enthusiasm to this authentic vision of the historic city. Visitors and commentators alike revelled in the immersive nature of the experience; of the illusion of the past, of its dream-like quality:

> Nothing can be more delightful, more perfectly illusive, more full of charm and interest, than this admirable work … the stroller in the street … must be dull indeed if he does not feel his imagination quicken by the suggestion and store of poetry about him.[12]

> The walls are green with moss and age, wind-beaten, storm-discoloured, sometimes rickety … There is indeed no part of this fine reproduction of the past which does not inspire the old world sentiment, and help fill the mind with the idea of old world customs and bygone scenes.[13]

The audience was encouraged to suspend disbelief, to imagine themselves carried back in time to pre-Fire London in pleasurable contrast to the displays of modern production and consumption around them. 'The surprise of the visitor when he finds himself transported from the region of furnished apartments and model bakeries to the heart of Old London cannot fail to be very pleasing.'[14] Anything which disturbed this illusionary absorption in the picturesque, the quaint and the romantic – such as 'the telegraph-wires high overhead against the blue sky' – could only be condemned as the jarring, anachronistic intrusion of the 'real', modern world of the surrounding exhibition.[15]

The Past in the Modern Present

Old London became the prototype of the historic city exhibit, this re-creation of the historical townscape of the host city which became a standard feature of succeeding exhibitions. With its characteristic reproduction of a past urban space the historic city occupied its own place in the typology of reconstructed architecture at great exhibitions. It could be distinguished from the signature building, such as the *Castello di Fenis* reproduced at the 1884 Turin Exhibition; from the assemblage of national styles in the *Rue des nations* at Paris in 1878 and 1900; or from ethnographic village reconstructions such as those at Vienna in 1873; or the evocation of the exotic exemplified by the *Rue du Caire* first shown at Paris in 1889 – or indeed from the depiction of *foreign* cities such as Old Vienna at the 1893 Chicago Columbian Exposition or *Venedig in Wien* of 1895. After Old London, historic city exhibits developed in scale and ambition but were constructed on similar principles and provoked the same audience delight. Old Edinburgh in 1886, discussed below, was followed amongst others by Old Manchester and Salford in 1887, *Oud Antwerpen* in 1895, *Alt-Berlin* in 1896, *Gamla Stockholm* in 1897, and most spectacularly by *Vieux Paris* at the Paris Exposition of 1900.[16]

The disjunction between these immersive chronotopic reconstructions of the urban past and their situation in the great exhibition, the epitome of modernity with its commitment to the values of progress through technological advance and the accumulation of commodities, is striking.[17] This disjunction can be taken as an instance of modernity's constant recourse to history, of the interplay between past and present contained within the modern, where for Lynda Nead, drawing on the work of writers such as Benjamin and de Certeau, '[t]he present remains permanently engaged in phantasmatic dialogue with the past'.[18] For Allan Pred, in his Benjaminian montage exploration of the 1897 Stockholm Exhibition, *Gamla Stockholm* was one of the exhibition's 'spectacular articulations of modernity'. Within an escapist and illusory space, '[a] city of the past that was a pastiche of the past, present and future', a commodified and sanitized past legitimated and idealized the present.[19]

As well as providing a pleasurable contrast to the modern world of the exhibition, the historic city existed within it, the dream-world of the past within the dream-world of the present: and like Pred's Stockholm, in its representation of the past revealing the circumstances of its creation within the host exhibition and the host city. For ultimately the exhibit's fascination rested on its specificity as an evocation of a particular historical place: either as an image of the city attractive to the outside visitors drawn to the exhibition, or more tellingly as an invitation to its own citizens to identify with a shared and familiar past – prompting a *frisson* of recognition of historic civic identity.

The reproduction historic city can therefore be seen as an opportunity for the exercise of the contemporary popular historical imagination. In the case of Old London, the reconstructed cityscape showed many of the characteristics identified by Peter Mandler in his study of the Victorian tourist experience of real historic buildings.[20] Chronologically Old London was situated in Mandler's Olden Time of popular history, the comfortable period of Tudorism and Stuartry ended by the spectacular cataclysm of the Great Fire which provided a satisfying boundary between Old London and the city's contemporary realities. Again, Birch associated at least some of his reconstructed buildings with historical figures – the Gunpowder Plotters and Oliver Cromwell as well as Isaac Walton and Richard Whittington – although the attributions were often fugitive and anecdotal.[21] However, it was Birch's concern for authenticity and truthfulness which was central to the audience experience of Old London. In an elision of the distinction between simulated and surviving ancient buildings, his artfully distressed and antiqued finishes evoked the 'age-value' so appreciated by the Victorian tourist.[22]

Old London, and the historic city exhibit generally, took a place in 'a dense web of representation', 'the riot of new representational technologies' which for Mandler characterize the Victorian period.[23] Graphic images reproduced by new technical processes, a flood of print in the form of historical novels and guidebooks, theatrical productions, panoramas and dioramas all fed 'the insatiable appetite manifested for historical material of all types'.[24] So the historic city exhibit was not without precedent or parallels. Mandler recalls a spectacular predecessor a generation before, another facsimile London with the same historical punctuation mark: 'in 1844, a painted model of the whole of "London in the Olden Time", covering five acres, was on display in the Royal Surrey Zoological Gardens, the evening exhibition of which culminated in a pyrotechnic display simulating the Great Fire of London.'[25]

Much more recently an analogous vision of ancient England had been promoted in the Albert Hall, itself part of the South Kensington exhibition complex. The Olde Englyshe Fayre held in June 1881 was a fashionable charity event in aid of the Chelsea Hospital for Women which featured much mock-medieval entertainment, jollification and dressing-up by its upper-class lady stall-holders.[26] Its re-creation of an Olde English Marketynge Plaice was just the sort of 'pasteboard and canvas delusion' against which Birch had set his face: 'The great error to be carefully avoided in my mind is to give that air of insubstantiality to the whole suggestive of those fearful theatrical scenic representations of "Old England" familiar to us now by these modern bazaars and fancy fairs.'[27] The scholarly construction of Old London was intended to be much less frivolous, but as spectacularly enjoyable.

Politics and Power in Old London

The vision of the historic city presented to visitors to the Health Exhibition and with which contemporary Londoners were being invited to identify was, however, the product of modern relations of power and influence. The construction of Old London had been extravagantly funded as a separate project by the City of London Corporation and a number of the City of London Livery Companies, at a distance from the official South Kensington administration.[28] The Corporation sat at the apex of the largely unreformed structure of London local government that had been under attack by Liberals and municipal reformers for most of the century; the Livery Companies had outlived their origins as medieval craft guilds and survived as wealthy, self-perpetuating charitable foundations.[29] The moving spirit behind Old London was George Shaw, Master of the Plumbers' Company and the projector of a new role for his guild in overseeing a scheme for the registration of plumbers.[30] Within the framework of the Health Exhibition, Old London had ostensibly been intended to confirm the modern progress of Shaw's plumbers by depicting the insanitary nature of the real historical London; but this rather unappealing aim had, unsurprisingly, not been carried through with any conviction. The exhibit's purpose was to provoke pleasure rather than disgust.[31] Instead, Old London conjured up a more positive view of the Olden Time. Its apartments were used to display artefacts from private collections or from the rich resources of the Livery Companies; in its workshops nominated firms demonstrated the craft skills with which the guilds were once associated.

In the South Kensington exhibitions' growing polarization between traditional 'practical instruction' and worldly 'public entertainment', Old London was clearly on the side of entertainment.[32] '[T]he erection is, in fact, a most charming and costly toy': a toy which, however, might allow its corporate sponsors to present themselves as the representatives of Old London at a time when their wealth and privileges were under attack.[33] A Royal Commission had been investigating the Livery Companies since 1882. Its report, issued in May 1884 as the Health Exhibition was under way, contained further detail on the fortunes amassed by the Companies and their lack of public accountability.[34] George Augustus Sala, a veteran commentator on exhibitions, wrote caustically in the *Illustrated London News*:

> 'Old London' should be regarded not only as a generous boon to the public, but also as a masterstroke of policy on behalf of those who wish well to the Corporation and the Guilds of the City of London ... virtually a retort courteous, by anticipation, to the

portentous blue-book of the City of London Livery Companies Commission; to say nothing of Sir William Vernon Harcourt's Government of London Bill. 'What right-minded Briton would wish to tamper with such a "lovely London" ... ?'[35]

Shaw himself was in no doubt as to the political purpose of his project:

The proposed representation at S. Kensington should be used as a means of imparting real information regarding the City of London and impressing on the public mind ... the value of the Guilds in past and present times. The object being to popularize the facts of history in opposition to the fancies of modern so-called reformers.[36]

And so Old London used spectacle and display in the setting of the modern great exhibition to expound a favourable view of the traditional arrangements for the governance of London. This exposition involved the same mobilization of history and re-animation or outright invention of tradition evident in, for example, the more prestigious observances of the Lord Mayor's Show.[37] The reconstruction of Whittington's Palace which featured so centrally in Old London could only serve to jog historical memories of the poor boy made good (at least in the folk tale of Dick Whittington and His Cat) and his legendary philanthropy, the chief justification of the contemporary Livery Companies.

Such was the popular appeal of this 'lovely London' that the attraction was retained complete for the next event in the South Kensington series, the International Inventions Exhibition of 1885. Now under the direct control of the exhibition executive committee a few remaining exhibitors from the previous year were joined by a haphazard selection of industrial products, modern crafts and curios 'in order to demonstrate, as far as possible, the contrast between ancient and modern industries'.[38] In an exhibition devoted to 'Apparatus, Appliances, Processes and Products, invented or brought into use since 1862', Old London seemed even more incongruous than at the Health Exhibition.[39] By 1886 and the Colonial and Indian Exhibition, the last in the South Kensington programme, the organizers had given up making any connection at all: it 'was found to be such a popular feature during the last two years that it was determined to retain it although it has, properly speaking, nothing to do with the subject matter of the existing Exhibition' and Old London, retained once more by popular demand, was used as office accommodation.[40] Juxtaposed with the Orientalist splendours of the Gwalior Gateway, a European-conceived 'make-work monument' to Indian stone-carving skills, Old London persisted as a hollow if still spectacular sideshow.[41]

Exhibitionary Edinburgh

1886 was a year of exhibitions. The first major English provincial international exhibition – of Shipping and Navigation – took place in Liverpool at the same time as the South Kensington Colonial and Indian Exhibition and the final appearance of Old London; and that summer also saw the realization of

Scotland's first universal international exhibition, the International Exhibition of Industry, Science and Art in Edinburgh. Although smaller international themed shows had recently been held in the city – Fisheries, in 1882, the year before the corresponding South Kensington event, and Forestry in 1884 – the 1886 Edinburgh Exhibition was to be the first of a series of large-scale exhibitions organized in Scotland before 1914. The first Glasgow International Exhibition followed in 1888, another Edinburgh event was held soon after in 1890, and the second Glasgow International Exhibition took place in 1901. The series continued with two significant Scottish National Exhibitions: the first in Edinburgh in 1908 was followed by a Glasgow response in 1911. This alternating sequence might be taken as an expression of the continuing rivalry between Scotland's contrasting urban centres: Edinburgh, 'Scotland's metropolis' and its historic capital; and Glasgow, 'the Second City of the Empire' at the heart of Scottish industrial production. If so, it was a competition Glasgow won convincingly. Mobilizing the city's industrial strength and resources, the Glasgow exhibitions were bigger, more ambitious and more successful in terms of attendance and financial receipts. In the process, Glasgow became known as Britain's premier exhibition city outside London.[42] The Edinburgh exhibitions, in contrast, have remained in comparative obscurity.

The 1886 Exhibition, however, not only demonstrated the feasibility of a Scottish international exhibition; it was a considerable success in its own right. The project was initiated by local business, commercial and industrial interests who succeeded in gaining the support of Edinburgh Town Council, the city's municipal authority, with its commitment to developing the city as a location for prestigious events.[43] The Exhibition was a popular triumph: it dominated Edinburgh's social life that summer, attracted a great number of visitors and excursionists – attendances reached a creditable 2,750,000 – and made a moderate surplus of around £5,500.[44] The show's substance belied its 'international' aspirations; it had struggled to attract foreign exhibitors and overt colonial and imperial content had been largely scooped by the contemporary London exhibition. On the other hand, the organizers took every opportunity to present the event as an outward-looking showcase for Scottish industry, commerce and culture: 'It would be difficult to conceive any more perfect means of instructing a foreigner in the habits and pursuits of Scotchmen in their native land than a visit to this Exhibition, so full is it with everything tending to reflect the national life.'[45]

This assertion of a distinctive Scottishness typified the consciousness of national identity that characterized late Victorian Scotland. In an era of largely autonomous municipal government by self-confident urban bourgeois elites, the local patriotism which characterized English provincial life was in Scotland given a national rather than regional character by the distinctive institutions whose existence had been guaranteed by the Union settlement of 1707: the Presbyterian religion and the Scottish education system, Scots law, Scottish banking, and the structure of burgh government itself. Edinburgh was central to this sense of identity as the stateless nation's capital and the seat of these defining institutions: the assemblies of Scotland's three competing Presbyterian Churches, the higher law courts, a range of educational provision at whose summit lay the University of Edinburgh with its world-famous medical school, and the city's burgeoning financial services industry.[46] The authority of these institutions was performed or expressed through the creation of symbols and observances inviting identification with the imagined nation: an instance of the much-observed 'invention of tradition'.[47] In Edinburgh this invention centred round the figure of Sir Walter Scott, the Edinburgh citizen whose poetry and novels had opened up a vision of Scotland to the European gaze and to the industry of tourism; and who, to a large extent, was held to have crystallized the identification of Scottishness with Highlandism in his choreography of the ceremonial visit of George IV to the city in 1822.[48] Though his Tory patriarchalism was out of sympathy with the modern Liberal political ethos of the city, half a century after his death Scott still provided a reference point for national and, for Edinburgh, municipal observances.

This assertion of national and cultural difference prompted no significant demands for political separation: Victorian Scots were in general profusely loyal to the Union and the Empire, and, in particular, their personification in the monarchy. Victoria's espousal of an imagined Highland life at Balmoral and her affection for Scottish imagery and ceremonial reinforced her popularity in Scotland. The Queen's visit to the Exhibition in August 1886 would be greeted with the customary obsequious ceremonial and expressions of loyalty.[49] The avowal of Scottish nationhood was rather a competitive claim to equality – and equal treatment – with England within the United Kingdom and the Empire in which Scotland's modern progress and prosperity were seen to be founded. However, as urban government became more complex and the very remoteness of the Imperial Westminster Parliament became a focus for suspicions about the centralizing tendencies of London administration such assertions of the distinctiveness of Scottish institutions, and Scottish civil society, could only grow more strident.[50]

Honest and True: Old Edinburgh

So the Edinburgh Exhibition took on the characteristics of a national enterprise, and the show was replete with Scottish symbolism. The spectacle of the products of modern industry was accompanied by statuary, banners and other historical artefacts, paintings and jewellery alluding to a distinctive Scottish history and distinctive, if often imagined, Scottish traditions. At a time when the universal exhibition, in South Kensington practice at least, had been supplanted by specialized, themed events, the theme of the Edinburgh Exhibition could be taken as Scottishness; and its focal point was another reconstructed historic city. Old Edinburgh was designed by the young architect Sydney Mitchell (Fig. 9.3). Like Birch in Old London

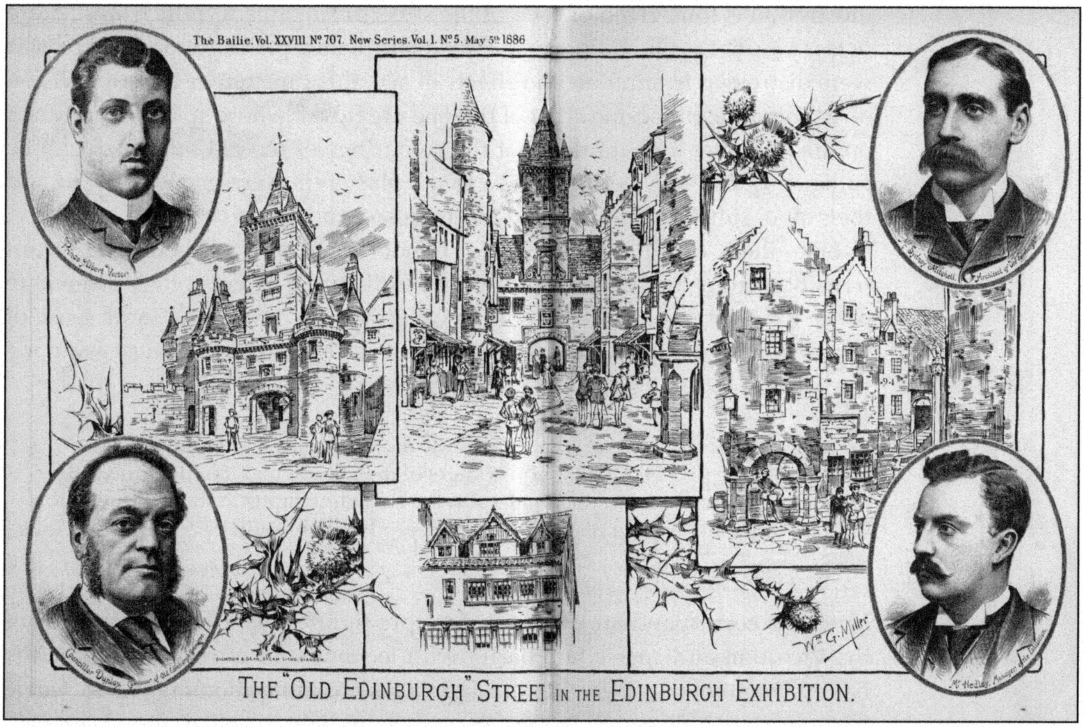

9.3 The 'Old Edinburgh' Street in the Edinburgh Exhibition, from *The Bailie*, 5 May 1886. Portraits of Sydney Mitchell, top right, and J. C. Dunlop, bottom left. By permission of the University of Glasgow Library, Special Collections

Mitchell was an architect and an antiquarian, but one of a different political hue. His reconstruction of Edinburgh's Mercat Cross, a gift to the city from W. E. Gladstone, had been unveiled with great ceremonial in November 1885, a Liberal offering in an antiquarian wrapping.[51]

Old Edinburgh was plainly based on Old London, 'which I have seen and minutely studied', Mitchell declared in his competition-winning prospectus.[52] The Edinburgh organizers had nevertheless learned from the popularity of their predecessor. Rather than being tucked away in the maze-like South Kensington complex, Old Edinburgh occupied a central position in the Edinburgh Exhibition building. Through the Grand Hall with its colossal statue of William Wallace,[53] along the Central Court and past the movement, noise and smell of modern Machinery in Motion, the exhibition visitor was drawn into the historic city. Beyond another reproduced city gate – this time the Nether Bow, which had marked the boundary of Edinburgh's High Street and the burgh of Canongate – lay a re-created street of some twenty 'lost' buildings authentically reproducing the distinctive features of the Edinburgh, and Scottish, vernacular. Domestic tenements were joined by the Cunzie-House or Royal Mint, the Tolbooth or city jail – Scott's 'Heart of Midlothian' – and a replica of an earlier version of Mitchell's Mercat Cross. As in Old London, scale had been adjusted to produce a convincing ensemble reproduction of the historical cityscape; again, an attempt had been made to represent ancient street patterns, this time of Edinburgh's distinctive wynds and closes: 'An illustration is thus given of each of the street arrangements which were more or less peculiar to Old Edinburgh.'[54] And once again, pristine wood and plaster were distressed to simulate the effects of age, this time under the supervision of 'Mr Smythe, the scenic artist of the Theatre Royal, whose splendid canvases on the stage are a guarantee that the colourings of the old buildings will be sufficiently realistic, yet refined in tone'; plasterers were encouraged to add their own, suitably historical, graffiti to the reproduction surface.[55]

Like their counterparts in Old London, visitors relished the abrupt transition from the displays of modern industry and technology elsewhere in the Exhibition into this re-creation of Edinburgh's past. The same sense of immersion, of delight and wonderment was produced, aided, once again, by the modern technology of electric light:

> the moon (electric of course) ... produces the most fantastic and vivid effects of light and shade upon the nooks and corners of the ancient and quaint edifices. The Rembrandtesque scenes in the booths and the twinkling lights in the windows aloft lend aid to the making up of a scene more deceptive than could be attempted upon a stage.[56]

Here the correspondent of the *Builder* pre-figures Mandler's discussion of the Victorian audience's familiarity with technologies of representation: the happy conjunction of glaring arc lighting simulating moonlight and feeble incandescent bulbs giving the effect of lamplight recalled the *chiaroscuro* effects of graphic reproduction and of Victorian spectacular theatre.[57]

But once again, the historic city display encapsulated its creators' interpretation of its host city's history. The *Book of Old Edinburgh* was produced as a handbook to the exhibit. Co-written by J. C. Dunlop, chair of the organizing committee and a prominent town councillor, and his sister Alison Hay Dunlop this attractive publication was produced in large numbers for sale in the Exhibition while leather-bound presentation copies were bestowed on visiting dignitaries.[58] Informed by the Dunlops' Radical Liberal and United Presbyterian Church background as much as by their considerable antiquarian knowledge, the *Book* portrays Old Edinburgh from the ruling municipal Liberal and Presbyterian point of view. Edinburgh's distinctiveness as a place and as a capital city is linked to a fervid assertion of its central role in a moral history of the Scottish nation:

> Besides her dower of beauty, the Capital of the North has ever possessed an individuality more marked than any other city in the Empire. ... [W]e offer some representation of the scenes where ... Scottish history was lived and enacted; for in so far as that history was pure and honest, fearlessly God-fearing and true, it has given our country its place among the nations.[59]

To this end the Dunlops situate the re-created buildings of Old Edinburgh within a Romantic/Tragic emplotment of this tumultuous history. The reconstructed historical spaces are associated with historical characters such as the Stuart kings, Cardinal Beaton, Mary of Guise, and of course Mary

Queen of Scots, but the stories are told from a contemporary Presbyterian perspective: while Walter Scott's role in creating Edinburgh's past is gratefully acknowledged, his political views, particularly his tendencies to Jacobitism, are rejected. Vivid and intense passages describe the fate of the Covenanters, the seventeenth-century Presbyterian fundamentalists treated with disdain by the Enlightenment (and unsympathetically by Scott) but whose rehabilitation provided a martyrology for the Radical and working-class movements of Victorian Scotland.

Improvement and Nostalgia

Edinburgh was felt to be particularly suitable for this Romantic historical treatment. 'During the nineteenth century, the city reconstructed itself as an elaborate symbol for Scotland.'[60] Processes of renovation had monumentalized what J. C. Dunlop elsewhere described as Edinburgh's 'Three Historic Boulders' of Palace, Kirk and Castle.[61] By 1886, the Royal Palace of Holyroodhouse had become a visitor attraction, the High Kirk of St Giles had been transformed into the semblance of a Presbyterian cathedral, and the Castle, under development as a monument since the 1840s, was undergoing a similar process of 'authentic' reconstruction.[62] Old Edinburgh, with its own painstaking re-creation of a historical cityscape and its appeal to the popular imagination of a distinctive national history, fitted into this Romantic re-imagining of modern Edinburgh.

Conversely, Old Edinburgh evoked feelings of loss. The monuments which comprised Dunlop's 'Three Boulders', and many of Edinburgh's other institutions of power – the law courts, the City Chambers, the ecclesiastical assemblies – were to be found in the heart of the historic city, the backdrop for the spectacular urban pageantry for which Mitchell's restored Mercat Cross itself provided a significant site. However, they also shared a space with many of Edinburgh's poorest inhabitants within this now-decaying central core. With the flight of upper- and middle-class residents to the New Town after 1766, Edinburgh's characteristic ancient tenements had deteriorated into picturesque squalor as once-grand mansions were endlessly subdivided or 'made down' into crowded lodgings. What was now the Old Town became synonymous with deprivation, disease and high mortality; a crumbling urban fabric subject to the processes of Improvement, slum clearance and demolition.[63] As Lord Provost Thomas Clark proclaimed in 1885: 'I confess to what some may look upon as a sentimental difficulty in the removal of historic tenements which have been famous as the residence of men, and women too, notable in bygone days. Still, the health of the community must ever be paramount to sentiment.'[64] Nevertheless many citizens felt that these often ineffective incursions of modernity had caused irreversible damage to the city's distinctive architectural and urban legacy: 'Old Edinburgh is being fast improved off the face of the earth.'[65] The reconstructed Old

Edinburgh provided, in the historical imagination at least, a respite from this improvement: albeit one which, as in Old London, ignored or romanticized the actualities of pre-modern plumbing and sanitation.[66]

In contrast to Old London, set in the comfortable past of the Olden Time, many of Old Edinburgh's re-creations had stood until very recently. The distinctive tenement at the Bow Head in the heart of the Old Town had been 'unnecessarily and unfortunately demolished by civic authority in 1878, an irreparable loss to the antiquities of the city'.[67] Re-incarnated as one of the exhibit's set-pieces it was the subject of one of the Dunlops' most effective passages of historical imagination (Fig. 9.4).[68] Such a reminder of recent loss – even in the name of progress – could provoke nostalgia and reinforce Edinburgh's antiquarian predilections. It could also be a spur to more positive approaches to architectural preservation and conservation. Patrick Geddes, whose move to the Old Town in 1887 and whose subsequent developments of 'conservative surgery' represent the beginnings of architectural conservation in Edinburgh, was fascinated by the Old Edinburgh exhibit which he saw as 'at once stimulating and popularizing the historic spirit, and helping to the recovery of the fallen, yet highest art of civilized production of permanent wealth – that of rational, fitting and beautiful architecture, civic and domestic'.[69]

Geddes went on to employ Sydney Mitchell on his Ramsay Garden development, partly, as Volker Welter surmises, on the strength of Mitchell's design for the Old Edinburgh Nether Bow gateway which Geddes proposed to reconstruct along the ridge of the Royal Mile as part of a scheme intended to restore 'the missing elements of the ideal medieval Edinburgh'.[70] More than 25 years later, Geddes displayed photographs of Old Edinburgh in the Outlook Tower exhibition, representing the Romantic side of his dichotomy between Edinburgh's Romance and (still very present) Squalor. For him, Old Edinburgh represented 'probably the most admirable reconstruction of an ancient city yet effected, and a suggestion of what may yet be done in some of our old quarters in permanent form'.[71]

The Historic City Performed

To heighten and dramatize the illusion of the past it was an obvious step to provide the historic city with inhabitants. To realize the scenographic possibilities of the three-dimensional experience and its appeal to the popular historical imagination it was necessary that Old London and Old Edinburgh should be 'populated by real human beings doing ordinary and familiar things'.[72] The treatment of this element of performance in the two historic cities reflected, once again, the different aims of their creators. In Old London's original showing at the Health Exhibition in 1884, the Livery Company-sponsored demonstrations of craft work in the exhibit's booths were intended to be performed by craftsmen dressed appropriately in historical costume. However, such dressing-up was uncoordinated and disorganized: 'to flippant persons, like ourselves, the incongruous dresses adorning some of the artizans [*sic*] in the Old London Street seem excessively funny'.[73] By its very nature, this was an exclusively male affair. In the following year's incarnation at the Inventions Exhibition, Old London's fitfully costumed craftsmen were joined by the female textile workers of Mrs Ernest Hart's Donegal Industrial Fund 'At Ye Signe of Ye Rose and Shamrocke' demonstrating the techniques of Irish domestic industry in traditional peasant dress: a rural idyll superimposed on the supposedly urban re-creation.[74]

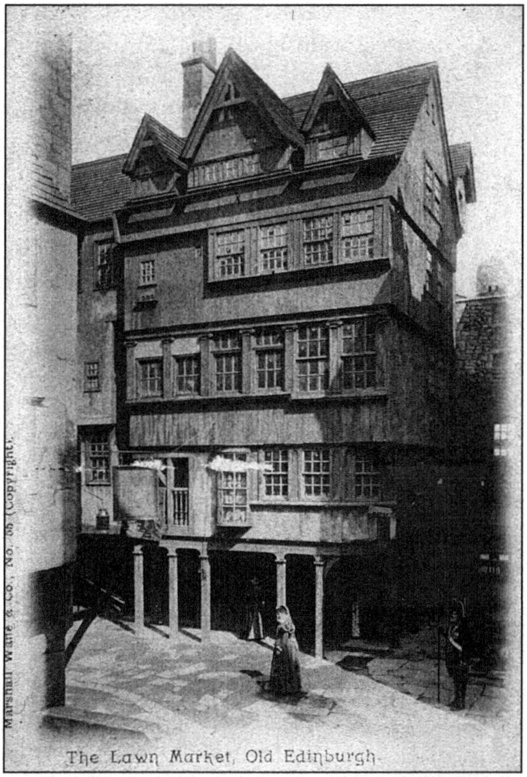

9.4 The Lawn Market, Old Edinburgh. Souvenir postcard by Marshall Wane, from the Papers of Sir Patrick Geddes, University of Strathclyde. © University Archives, University of Strathclyde. Licensor www.scran.ac.uk

Old Edinburgh took its costume and performance more seriously from the outset. This was in large part due to matters of finance. Old London, funded by Corporation and guild largesse, operated under the South Kensington ethos with its prohibition of over-the-counter sales of any kind. Old Edinburgh on the other hand reflected the more commercial instincts of its organizers. The exhibit was conceived as a self-funding venture: Mitchell's facades concealed retail premises rented out at premium rates.[75] The organizers' original intention was to replicate Old London's worthy display of craft activity, and this was indeed carried through in a few cases. The display of Edinburgh printers Ballantyne, Hanson and Co., the descendants of Walter Scott's first publisher, featured 'An old Hand Press at work showing Printing in the olden time' producing (for sale) a souvenir biography of Scott, 'with various specimens of old Printing, and Curiosities and Implements connected with the Art'. However, commercial imperatives prompted other stall-holders to offer a more diverse, and sometimes incongruous, range of wares: 'Souvenirs of Exhibition, Scotch Dolls and Figures, Mechanical Toys'; 'Fancy and Japanese Goods', or 'Oriental Curios and Novelties'.[76]

In this retail environment, costume played a prominent, and gender-specific, role. A detailed specification of the dress to be worn by the largely female staff of shop assistants and sales attendants was issued by the Exhibition organizers: the 'Sumptuary Committee' as one account put it, remarking on 'the summary and almost disrespectful way in which the "journeymen" are hustled out of the way … compared with the curious and tender care with which each article of female attire is discussed'.[77] The female inhabitants of Old Edinburgh were thus displayed as part of the historical illusion, while costume and performance were part of their conditions of employment: 'Must be good Saleswomen, and wear the Fancy

Dress provided'.[78] Their very visible presence added to the commercial hubbub, the animation and excitement of the re-created street scene. In the bazaar-like atmosphere of over-enthusiastic selling which permeated the Exhibition, the shopkeepers of an Old Edinburgh 'degraded by squeaking dolls and india-rubber toys' tried to recoup their considerable outlay with the sale of souvenirs and kitsch.[79] The exhibit came to resemble a down-market and commercialized version of the aristocratic fancy fair which had so horrified Birch. Geddes, for all his admiration of Old Edinburgh's authentically reconstructed buildings, may well have had the exhibit in mind in his condemnation of this aspect of the modern exhibition, 'the sale of mere gewgaws to holiday-makers'.[80]

Conclusion: Historic Cities, Forward to the Past?

The examination of Old London and Old Edinburgh, as early examples of the historic city exhibit, reveals three apparent paradoxes which became increasingly evident as the genre proliferated into a widely popular exhibition attraction.

Firstly, the historic city was backward-looking, in a context – the great exhibition – whose very existence had hitherto been predicated on a vision of progress, on technological innovation and the growth of industry and trade as means of social improvement.[81] Originating in a desire for antiquarian reconstruction, the historic city instead provided a pleasurable immersion in the Olden Time, appealing to its audience's vision of popular history: the past in the present, in the midst of modernity.[82]

Secondly, the historic city became a popular generic attraction at great exhibitions, yet was by its nature site-specific. In Alexander Geppert's analysis the multiplication of historic cities is an example of 'exhibitionary networks' in action: as information and personnel moved between sites, great exhibitions came to share similar attractions and become increasingly alike.[83] So beneath their plaster skins, the historic cities might make use of similar mechanisms to achieve their effects. Thus the device of performance and costume, so prominent in Edinburgh, proved irresistible to the organizers of future exhibitions. Succeeding historic city exhibits – Manchester, Antwerp, Berlin, Stockholm – developed the theme, which reached its peak in *Vieux Paris* in 1900 where 500 actors were engaged to provide an air of verisimilitude to the reconstructed city.[84] These shared devices notwithstanding, it is the specificity of the historic city exhibit which, however, remains its defining feature. Characteristic (and character-ful) details of the architectural and urban vernacular of the host city were lovingly re-created for recognition by the audience. But more profoundly, the different stories which the exhibits' creators felt moved to tell about their cities' pasts framed the discourse behind the attractions. To Old London's justification of Livery Company privilege, or Old Edinburgh's evocation of Scotland's place amongst the nations, can be added the elaboration of a history for the paradigmatic 'shock city' of industrialism (Manchester); the celebration of a bourgeois mercantile elite (Antwerp), or a corrective to the 'inferiority complex' of a new Imperial capital (Berlin).[85]

Finally, there was a tension between the desire for scholarly architectural authenticity in recreating the distinctive cityscapes of Birch's Old London or Mitchell's Old Edinburgh and their successors and the commercialism which came to accompany their popularity. The construction of Old London fitted into Cunliffe-Owen's vision of a revived South Kensington; however, at the same time the South Kensington ethos restricted its function to the educational. With popularization came the demands of commerce, the intrusion of elements of the bazaar and the 'fancy fair' into the historic city. This is perhaps a specific example of the general tension between education and amusement in the great exhibitions as they developed as sites of mass entertainment, increasingly incorporating elements of the fairground into their spectacular attractions.[86] The 'authentic' setting of the historic city was used as an opportunity to sell novelties and souvenirs at Edinburgh, and increasingly afterwards as a picturesque site for commerce and catering – 'history could be eaten, drunk or bought'.[87] 'Consuming the past' in the historic city also came to mean consumption in the present.

Notes

1 No standard terminology seems to have emerged for this genre of exhibit. For Paul Greenhalgh, *Ephemeral Vistas: History of the Expositions Universelles, Great Exhibitions and World's Fairs* (Manchester, 1988), they are 'old cities' (p. 108); for Alexander Geppert, *Fleeting Cities: Imperial Expositions in Fin-de-Siècle Europe* (Basingstoke, 2010), 'domestic' or 'retrospective sections'.

2 Annmarie Adams, 'The Healthy Victorian City: The Old London Street at the International Health Exhibition of 1884', in *Streets: Critical Perspectives on Public Space*, ed. Zeynep Çelik, Diane Favro and Richard Ingersoll (Berkeley, CA, 1994), 203–212. Adams's concern with the health aspects of Old London differs from the emphasis here. The exhibit was variously described as 'Old London' or the 'Old London Street': as with the Edinburgh equivalent, I have used the shorter form throughout.

3 International Health Exhibition, 'Official Catalogue', *The Health Exhibition Literature* XVIII (London, 1884), 172.

4 International Health Exhibition, 'Official Catalogue'.

5 International Health Exhibition, 'Official Catalogue'.

6 International Health Exhibition, 'Official Catalogue'.

7 For Birch's career see Margaret Richardson, 'George Henry Birch (1842–1904)', in *Oxford Dictionary of National Biography*, online edn, accessed 26 March 2013.

8 *Burgess's Illustrated and Descriptive Guide to the International Health Exhibition* (London, 1884), 24.

9 Kenneth Luckhurst, *The Story of Exhibitions* (London, 1951), 130–135; Greenhalgh, *Ephemeral Vistas*, 32; Paul Greenhalgh, 'The Art and Industry of Mammon: International Exhibitions, 1851–1901', in *The Victorian Vision*, ed. John M. MacKenzie (London, 2001), 265–279.

10 H. Trueman Wood and R. C. Denis, 'Sir Francis Philip Cunliffe-Owen (1828–1894)', in *Oxford Dictionary of National Biography*, online edn, accessed 26 March 2013.

11 *Glasgow Herald*, 31 October 1884, 6.

12 *Saturday Review*, 24 May 1884, 679.

13 *Illustrated Sporting and Dramatic News*, 17 May 1884, 228.

14 *Burgess's Guide*, 24.

15 *Saturday Review*, 24 May 1884.

16 Alan J. Kidd, 'The Industrial City and Its Pre-Industrial Past: The Manchester Royal Jubilee Exhibition of 1887', *Transactions of the Lancashire and Cheshire Antiquarian Society* 89 (1995): 54–73 [my thanks to Malcolm Cooper for this reference]; Mandy Nauwelaerts, ed., *De panoramische droom: Antwerpen en de wereldtentoonstellingen, 1885, 1894, 1930* (Antwerp, 1993), 256–269; Katja Zelljadt, 'Presenting and Consuming the Past – Old Berlin at the Industrial Exhibition of 1896', *Journal of Urban History* 31, no. 3 (March 2005): 306–333; Allan Pred, 'Spectacular Articulations of Modernity: The Stockholm Exhibition of 1897', *Geografiska Annaler. Series B, Human Geography* 73, no. 1 (January 1991): 45–84; Robert W. Brown, 'Albert Robida's Vieux Paris Exhibit', *Yearbook of Interdisciplinary Studies in the Fine Arts*, 2 (1991): 421–445; Geppert, 52–56.

17 Geppert, 245–246.

18 Lynda Nead, *Victorian Babylon* (New Haven, 2000), 8; see also Simon Gunn, *History and Cultural Theory* (Harlow, 2006), ch. 5.

19 Pred, 68.

20 Peter Mandler, '"The Wand of Fancy": The Historical Imagination of the Victorian Tourist', in *Material Memories*, ed. Marius Kwint, Christopher Breward and Jeremy Aynsley (Oxford, 1999), 125–141.

21 Thomas St Edmund Hake, *Ye Historical Sketch of Ye Olde London Streete* (London, 1884).

22 Mandler, 137; see also M. Christine Boyer, *The City of Collective Memory* (Cambridge, MA, 1994), 143–144.

23 Mandler, 135.

24 Mandler, 131.

25 Mandler, 135.

26 *A Booke of Ye Olde English Fayre* (Manchester, 1881); see also Tom Hughes, 'June 8, 1881 – Ye Olde Fancy Fayre', *Victorian Calendar*, at http://victoriancalendar.blogspot.co.uk/2011/06/june-8-1881-ye-olde-fancy-fayre.html, accessed 5 December 2012.

27 Birch to George Shaw, 18 February 1884, London Metropolitan Archives CLC/261/1/83.

28 Total costs of £14,853 were reported at 7 October 1884, LMA CLC/261/2/305.

29 David Owen, *The Government of Victorian London 1855–1889* (Cambridge, MA, 1982), ch. 11.

30 George Shaw, *Revived Guild Action* (London, 1889). This second edition asserts Shaw's responsibility for Old London.

31 The sanitization of the past – cf. Pred's analysis of *Gamla Stockholm* – was a universal feature of historic city exhibits: see, for example, Zelljadt, 314, for Old Berlin.

32 'British Exhibitions', *Illustrated London News*, 2 August 1884, 94.

33 *Builder*, 17 May 1884, 687.

34 City of London Livery Companies' Commission, *Report and Appendix*, vol. 1 (London: Eyre and Spottiswoode, 1884) [C.4073].

35 G. A. Sala, 'The Health Exhibition, a Look Round', *Illustrated London News*, 2 August 1884, 94.

36 George Shaw, ms. note LMA CLC/261/1/32.

37 Andreas Fahrmeir, 'The London Lord Mayor's Show in the Nineteenth Century: Celebrating Achievement or Glorifying Tradition?', in *Political Rituals in Great Britain 1700–2000*, ed. Jörg Neuheiser and Michael Schaich (Augsburg, 2006), 55–75.

38 International Inventions Exhibition, *Official Guide* (London, 1885), lix.

39 International Inventions Exhibition, *Prospectus* (London, 1884).

40 Colonial and Indian Exhibition, *Official Guide* (London, 1886), 7.

41 Jeffrey L. Spear, 'A South Kensington Gateway from Gwalior to Nowhere', *SEL Studies in English Literature 1500–1900* 48, no. 4 (2008): 911–921, at 913.

42 Perilla Kinchin and Julie Kinchin, *Glasgow's Great Exhibitions 1888, 1901, 1911, 1938, 1988* (Wendlebury, 1988); Greenhalgh, 'Mammon', 275–276.

43 Richard Rodger, 'The "Common Good" and Civic Promotion: Edinburgh 1860–1914', in *Cities of Ideas*, ed. Robert Colls and Richard Rodger (Aldershot, 2004), 144–177.

44 Comparative figures for the South Kensington exhibitions: 1884: 4,153,390 visits, £15,580 profit; 1885: 3,760,581 visits, £18,089 loss; 1886: 5,550,745 visits, £34,642 profit.

45 *Black's Guide to Edinburgh, the International Exhibition and the Environs* (Edinburgh, 1886), x.

46 Graeme Morton and R. J. Morris, 'Civil Society, Governance and Nation, 1832–1914', in *The New Penguin History of Scotland*, ed. R. A. Houston and William Knox (London, 2001), 355–416; Lindsay Paterson, *The Autonomy of Modern Scotland* (Edinburgh, 1994), ch. 4.

47 E. J. Hobsbawm and T. O. Ranger, eds, *The Invention of Tradition* (Cambridge, 1992).

48 For George IV's 1822 visit see Caroline McCracken-Flesher, *Possible Scotlands: Walter Scott and the Story of Tomorrow* (Oxford, 2005), 73–110; Ian Duncan, *Scott's Shadow: The Novel in Romantic Edinburgh* (Princeton, NJ, 2007), 3–19; for a review of literature on the invention of Highland tradition see Matthew P. Dziennik, 'Whig Tartan: Material Culture and Its Use in the Scottish

Highlands, 1746–1815', *Past and Present* 217, no. 1 (January 2012): 117–147, as an update (and a corrective) to Hugh Trevor-Roper, 'The Invention of Tradition: the Highland Tradition of Scotland', in *Invention*, ed. Hobsbawm and Ranger, 15–41.

49 Richard J. Finlay, 'Queen Victoria and the Cult of Scottish Monarchy', in *Scottish History: The Power of the Past*, ed. Edward J. Cowan and Richard J. Finlay (Edinburgh, 2002), 209–224; on royal visits to Edinburgh generally see Rodger, 'Common Good', 156–158.

50 Graeme Morton, *Unionist-Nationalism: Governing Urban Scotland, 1830–1860* (East Linton, 1999), ch. 8; Richard J. Finlay, 'National Identity, Union, and Empire, c.1850–c.1970', in *Scotland and the British Empire*, ed. John M. MacKenzie and T. M. Devine (Oxford, 2011), 280–316.

51 *The Scotsman*, 24 November 1885, 5.

52 'Tolbooth' [Sydney Mitchell], [Tender for Old Edinburgh Street], 23 October 1885. Edinburgh City Archives Acc.423/9/6.

53 For Wallace's contemporary significance see Graeme Morton, 'The Most Efficacious Patriot: The Heritage of William Wallace in Nineteenth-Century Scotland', *Scottish Historical Review* 77 no. 2 (October 1998): 224–251.

54 Mitchell, 2.

55 *Daily Review*, 9 March 1886, 2; *Cameron's Guide Through the International Exhibition of Industry, Science, and Art* (Edinburgh, 1886), 55–56.

56 *Builder*, 31 July 1886, 172.

57 Mandler, 135–136.

58 In fact Alison Hay Dunlop was largely responsible for the book's authorship: J. C. and Alison Hay Dunlop, *The Book of Old Edinburgh, and Hand-Book to the Old Edinburgh Street* (Edinburgh, 1886); Hake, *Ye Historical Sketch*, its London equivalent, was much less ambitious both in content and production.

59 Dunlop and Dunlop, 5.

60 Morton and Morris, 399.

61 J. C. Dunlop, *Official Penny Guide to the Old Edinburgh Street* (Edinburgh, 1886).

62 Katherine Haldane Grenier, *Tourism and Identity in Scotland, 1770–1914: Creating Caledonia* (Aldershot, 2005), ch. 4; Rosalind K. Marshall, 'Scotland's Westminster Abbey: The Memorials in St Giles' Cathedral, Edinburgh', *Review of Scottish Culture* 23 (2011): 94–117; Robert J. Morris, 'The Capitalist, the Professor and the Soldier: The Re-Making of Edinburgh Castle, 1850–1900', *Planning Perspectives* 22, no. 1 (2007): 55–78.

63 Richard Rodger, *The Transformation of Edinburgh* (Cambridge, 2001), ch. 12; P. J. Smith, 'Slum Clearance as an Instrument of Sanitary Reform: The Flawed Vision of Edinburgh's First Slum Clearance Scheme', *Planning Perspectives* 9, no. 1 (1994): 1–27.

64 *The Scotsman*, 7 November 1885, 6. Clark was a leading promoter of the Exhibition, and became an active Vice-President of its Executive Council.

65 J. B. Gillies, *Edinburgh Past and Present* (Edinburgh, 1886), 162.

66 But see 'Gardy-loo', illus. by W. Hole in Dunlop and Dunlop, 70, as an example of Edinburgh's continuing fascination with this aspect of its sanitary past.

67 Bruce J. Home, 'Provisional List of Old Houses Remaining in the High Street and Canongate of Edinburgh', *Book of the Old Edinburgh Club* 1 (1908): 1–30, at 5.

68 Dunlop and Dunlop, 23–32.

69 Patrick Geddes, *Industrial Exhibitions and Modern Progress* (Edinburgh, 1887), 9, quoted in Volker Welter, 'History, Biology and City Design – Patrick Geddes in Edinburgh', *Architectural Heritage* 6, no. 6 (1995): 60–82, at 64; Lou Rosenburg and Jim Johnson, '"Conservative Surgery" in Old Edinburgh, 1880–1940', in *Edinburgh: The Making of a Capital City*, ed. Brian Edwards and Paul Jenkins (Edinburgh, 2005), 131–49.

70 Welter, 75.

71 Patrick Geddes, *The Civic Survey of Edinburgh* (Edinburgh, 1911), 565.

72 Mandler, 136.

73 *Fun*, 3 September 1884, 108.

74 International Inventions Exhibition, *Official Catalogue*, lx–lxii; Janice Helland, 'Working Bodies, Celtic Textiles, and the Donegal Industrial Fund 1883–1890', *Textile* 2, no. 2 (2004): 134–155.

75 Although in the event rentals only covered £1,843 of Old Edinburgh's £4,897 building costs: *The Scotsman*, 7 December 1887, 10.

76 International Exhibition of Industry, Science, and Art, *Official Catalogue* (Edinburgh, 1886), 293–294.

77 *Edinburgh Evening Dispatch*, 8 April 1886, 2.

78 Advertisement, *The Scotsman*, 3 May 1886, 1.

79 *Edinburgh Evening Dispatch*, 1 June 1886, 2. The accompanying article 'The Rupture at the Exhibition' provides an analysis of the economics of exhibiting at Edinburgh.

80 Geddes, *Industrial Exhibitions*, 12.

81 Greenhalgh, *Ephemeral Vistas*, 23–24; Geppert, 204–206.

82 Cf. David Lowenthal, *The Past Is a Foreign Country* (Cambridge, 1985), 96–105.

83 Geppert, 5; as a (failed) example of networking see 'The "Old Edinburgh" Buildings', *The Scotsman*, 10 December 1886, 4, reporting an unsuccessful attempt to sell Old Edinburgh to the organizers of the 1889 Paris Exposition.

84 Geppert, 55.

85 Walter Tomlinson, *The Pictorial Record of the Royal Jubilee Exhibition, Manchester, 1887* (Manchester, 1888), 127–132; Alfons Thijs, 'Old Antwerp 1894, a Bourgeois Dream World', in *De panoramische droom*, ed. Nauwelaerts, 267–269; Zelljadt, 307.

86 Paul Greenhalgh, 'Education, Entertainment and Politics: Lessons from the Great International Exhibitions', in *The New Museology*, ed. Peter Vergo (London, 1989), 74–98.

87 Zelljadt, 328; see also Pred, 69–70.

Bibliography

Printed Primary Sources

Black's Guide to Edinburgh, the International Exhibition and the Environs. Edinburgh: A. & C. Black, 1886.

A Booke of Ye Olde English Fayre. Manchester: Falkner, 1881.

Burgess's Illustrated and Descriptive Guide to the International Health Exhibition. London: Burgess, 1884.

Cameron's Guide Through the International Exhibition of Industry, Science, and Art. Edinburgh: Cameron, 1886.

City of London Livery Companies' Commission. *Report and Appendix*, vol. 1. London: Eyre and Spottiswoode, 1884 [C.4073].

Colonial and Indian Exhibition. *Official Guide*. London: Clowes, 1886.

Dunlop, J. C. *Official Penny Guide to the Old Edinburgh Street*. Edinburgh: Constable, 1886.

Dunlop, J. C., and Alison Hay Dunlop. *The Book of Old Edinburgh, and Hand-Book to the Old Edinburgh Street*. Edinburgh: Constable, 1886.

Gillies, J. B. *Edinburgh Past and Present*. Edinburgh: Oliphant, Anderson and Ferrier, 1886.

Hake, Thomas St Edmund. *Ye Historical Sketch of Ye Olde London Streete*. London: Waterlow, 1884.

Hughes, Tom. 'June 8, 1881 – Ye Olde Fancy Fayre'. *Victorian Calendar*, at http://victoriancalendar.blogspot.co.uk/2011/06/june-8-1881-ye-olde-fancy-fayre.html, accessed 5 December 2012.

International Exhibition of Industry, Science, and Art. *Official Catalogue*. Edinburgh: Constable, 1886.

International Health Exhibition. 'Official Catalogue'. In *The Health Exhibition Literature*, vol. 18. London: Clowes, 1884.

International Inventions Exhibition. *Official Catalogue*. London: Clowes, 1885.

International Inventions Exhibition. *Official Guide*. London: Clowes, 1885.

International Inventions Exhibition. *Prospectus*. London, 1884.

Secondary Sources

Adams, Annmarie. 'The Healthy Victorian City: The Old London Street at the International Health Exhibition of 1884'. In *Streets: Critical Perspectives on Public Space*, edited by Zeynep Çelik, Diane Favro and Richard Ingersoll, 203–212. Berkeley: University of California Press, 1994.

Boyer, M. Christine. *The City of Collective Memory*. Cambridge, MA: The MIT Press, 1994.

Brown, Robert W. 'Albert Robida's Vieux Paris Exhibit'. *Yearbook of Interdisciplinary Studies in the Fine Arts* 2 (1991): 421–445.

Duncan, Ian. *Scott's Shadow: The Novel in Romantic Edinburgh*. Princeton: Princeton University Press, 2007.

Dziennik, Matthew P. 'Whig Tartan: Material Culture and Its Use in the Scottish Highlands, 1746–1815'. *Past and Present*, no. 217 (2012): 117–147.

Fahrmeir, Andreas. 'The London Lord Mayor's Show in the Nineteenth Century: Celebrating Achievement or Glorifying Tradition?' In *Political Rituals in Great Britain 1700–2000*, edited by Jörg Neuheiser and Michael Schaich, 55–73. Augsburg: Wissner, 2006.

Finlay, Richard J. 'National Identity, Union, and Empire, c.1850–c.1970'. In *Scotland and the British Empire*, edited by John M. MacKenzie and T. M. Devine, 280–316. Oxford: Oxford University Press, 2011.

Finlay, Richard J. 'Queen Victoria and the Cult of Scottish Monarchy'. In *Scottish History: The Power of the Past*, edited by Edward J. Cowan and Richard J. Finlay, 209–224. Edinburgh: Edinburgh University Press, 2002.

Geddes, Patrick. *Industrial Exhibitions and Modern Progress*. Edinburgh: Douglas, 1887.

Geddes, Patrick. *The Civic Survey of Edinburgh*. Edinburgh: Outlook Tower, 1911. Available at http://www.archive.org/details/civicsurveyofedi00gedduoft, accessed 13 August 2013.

Geppert, Alexander. *Fleeting Cities: Imperial Expositions in Fin-de-Siècle Europe*. Basingstoke: Palgrave Macmillan, 2010.

Greenhalgh, Paul. 'Education, Entertainment and Politics: Lessons from the Great International Exhibitions'. In *The New Museology*, edited by Peter Vergo, 74–98. London: Reaktion, 1989.

Greenhalgh, Paul. *Ephemeral Vistas: History of the Expositions Universelles, Great Exhibitions and World's Fairs*, Studies in Imperialism. Manchester: Manchester University Press, 1988.

Greenhalgh, Paul. 'The Art and Industry of Mammon: International Exhibitions, 1851–1901'. In *The Victorian Vision*, edited by John M. MacKenzie, 265–79. London: V&A, 2001.

Grenier, Katherine Haldane. *Tourism and Identity in Scotland, 1770–1914: Creating Caledonia*. Aldershot: Ashgate, 2005.

Gunn, Simon. *History and Cultural Theory*. Harlow: Pearson Longman, 2006.

Helland, Janice. 'Working Bodies, Celtic Textiles, and the Donegal Industrial Fund 1883–1890'. *Textile: The Journal of Cloth and Culture* 2, no. 2 (2004): 134–155.

Hobsbawm, E. J., and T. O. Ranger, eds. *The Invention of Tradition*. Cambridge: Cambridge University Press, 1983.

Home, Bruce J. 'Provisional List of Old Houses Remaining in the High Street and Canongate of Edinburgh'. *Book of the Old Edinburgh Club* 1 (1908): 1–30.

Kidd, Alan J. 'The Industrial City and Its Pre-Industrial Past: The Manchester Royal Jubilee Exhibition of 1887'. *Transactions of the Lancashire and Cheshire Antiquarian Society* 89 (1995): 54–73.

Kinchin, Perilla, and Julie Kinchin. *Glasgow's Great Exhibitions 1888, 1901, 1911, 1938, 1988*. Wendlebury: White Cockade, 1988.

Lowenthal, David. *The Past Is a Foreign Country*. Cambridge: Cambridge University Press, 1985.

Luckhurst, Kenneth. *The Story of Exhibitions*. London: Studio, 1951.

McCracken-Flesher, Caroline. *Possible Scotlands: Walter Scott and the Story of Tomorrow*. Oxford: Oxford University Press, 2005.

Mandler, Peter. '"The Wand of Fancy": The Historical Imagination of the Victorian Tourist'. In *Material Memories*, edited by Marius Kwint, Christopher Breward and Jeremy Aynsley, 125–141, Oxford: Berg, 1999.

Marshall, Rosalind K. 'Scotland's Westminster Abbey: The Memorials in St Giles' Cathedral, Edinburgh'. *Review of Scottish Culture* 23 (2011): 94–117.

Morris, Robert J. 'The Capitalist, the Professor and the Soldier: The Re-Making of Edinburgh Castle, 1850–1900'. *Planning Perspectives* 22 (2007): 55–78.

Morton, Graeme. 'The Most Efficacious Patriot: The Heritage of William Wallace in Nineteenth-Century Scotland'. *Scottish Historical Review* 77 (1998): 224–251.

Morton, Graeme. *Unionist-Nationalism: Governing Urban Scotland, 1830–1860*. East Linton: Tuckwell, 1999.

Morton, Graeme, and R. J. Morris. 'Civil Society, Governance and Nation, 1832–1914'. In *The New Penguin History of Scotland*, edited by R. A. Houston and William Knox, 355–416. London: Allen Lane, 2001.

Nauwelaerts, Mandy, ed. *De panoramische droom: Antwerpen en de wereldtentoonstellingen, 1885, 1894, 1930*. Antwerp: Antwerpen 93, 1993.

Nead, Lynda. *Victorian Babylon*. New Haven: Yale University Press, 2000.

Owen, David. *The Government of Victorian London 1855–1889*. Cambridge, MA: Belknap Press, 1982.

Paterson, Lindsay. *The Autonomy of Modern Scotland*. Edinburgh: Edinburgh University Press, 1994.

Pred, Allan. 'Spectacular Articulations of Modernity: The Stockholm Exhibition of 1897'. *Geografiska Annaler. Series B, Human Geography* 73 (1991): 45–84.

Richardson, Margaret. 'George Henry Birch (1842–1904)'. In *Oxford Dictionary of National Biography*, online edn, accessed 26 March 2013.

Rodger, Richard. 'The "Common Good" and Civic Promotion: Edinburgh 1860–1914'. In *Cities of Ideas*, edited by Robert Colls and Richard Rodger, 144–177. Aldershot: Ashgate, 2004.

Rodger, Richard. *The Transformation of Edinburgh*. Cambridge: Cambridge University Press, 2001.

Rosenburg, Lou, and Jim Johnson. '"Conservative Surgery" in Old Edinburgh, 1880–1940'. In *Edinburgh: The Making of a Capital City*, edited by Brian Edwards and Paul Jenkins, 131–149. Edinburgh: Edinburgh University Press, 2005.

Shaw, George. *Revived Guild Action, with a History of the Movement for the Registration of Plumbers*, 2nd edn. London: Simpkin, Marshall, 1889.

Smith, P. J. 'Slum Clearance as an Instrument of Sanitary Reform: The Flawed Vision of Edinburgh's First Slum Clearance Scheme'. *Planning Perspectives* 9 (1994): 1–27.

Spear, Jeffrey L. 'A South Kensington Gateway from Gwalior to Nowhere'. *SEL Studies in English Literature 1500–1900* 48 (2008): 911–921.

Thijs, Alfons. 'Old Antwerp 1894, a Bourgeois Dream World'. In *De panoramische droom*, edited by Mandy Nauwelaerts, 267–269. Antwerp: Antwerpen 93, 1993.

Tomlinson, Walter. *The Pictorial Record of the Royal Jubilee Exhibition, Manchester, 1887*. Manchester: Cornish, 1888.

Trevor-Roper, Hugh. 'The Invention of Tradition: The Highland Tradition of Scotland'. In *The Invention of Tradition*, edited by E. J. Hobsbawm and T. O. Ranger, 15–41. Cambridge: Cambridge University Press, 1992.

Welter, Volker. 'History, Biology and City Design – Patrick Geddes in Edinburgh'. *Architectural Heritage* 6 (1995): 60–82.

Wood, H. Trueman, and R. C. Denis. 'Sir Francis Philip Cunliffe-Owen (1828–1894)'. *Oxford Dictionary of National Biography*, online edn, accessed 26 March 2013.

Zelljadt, Katja. 'Presenting and Consuming the Past – Old Berlin at the Industrial Exhibition of 1896'. *Journal of Urban History* 31 (2005): 306–333.

ART AND DESIGN EXHIBITED

10

A Red-Letter Day: Evaluating Progress in New Zealand's Art at Dunedin's International Exhibitions, 1865 and 1889

Rebecca Rice

A red-letter day yearly marks the progress of every individual who treads life's path. With regret or satisfaction he then reviews the road already traversed, and with fresh hope and energy presses forward to the goal he would reach. At fixed times the careful trader tabulates his experience, and then draws the lines which in the future he can most safely follow. Nor are such wholesome and expedient customs confined to the individual. One of the prominent features of modern civilisation is that communities and nations have found it convenient to exhibit to the world at intervals the position they hold in the van of progress.[1]

In the nineteenth century, the international exhibition, *exposition universelle* or world's fair was the primary means by which countries competitively presented their progress to the world. New Zealand was no exception, and the New Zealand and South Seas Exhibition, Dunedin 1889–90, provided the vehicle by which the colony might 'recount the progress she has made during her short semi-centenary'.[2] This exhibition took place only fifty years after the establishment of New Zealand as a British colony, signalled by the signing of the Treaty of Waitangi in 1840.[3] Yet this was not the first time such an ambitious event was held in this young colony: in 1865 Dunedin hosted the first major international exhibition in Australasia, simply titled the 'New Zealand Exhibition'.

This chapter examines what these exhibitions tell us about the nature of progress made in New Zealand art between 1865 and 1889. For in spite of the country's marginal position and relative youth, both exhibitions boasted fine arts courts featuring works by both colonial and international artists. Further, the critical response to the New Zealand art displayed at the two exhibitions shows a distinct development. In 1865, the critical response to these exhibits was largely congratulatory. A reviewer writing under the pseudonym 'Rambler' suggested that the locally produced watercolours were 'quite equal to the works of some of the best artists at home and might be exhibited without fear of invidious comparison on the walls of old and new Water Color Societies of London'.[4]

However, by 1889, a more critical tone had emerged in reviews, challenging New Zealand artists to move beyond their reliance on the landscape as the predominant subject for their work. What had changed to produce this shift in tone from congratulation to criticism? This chapter argues that the 1865 exhibition had provided a boost for the development of arts infrastructure in New Zealand. By 1889, this infrastructure was well-developed and was having an impact on the artistic products of the colony. Yet this institutional progress ushered in more anxiety than triumphalism. The critics in 1889 questioned rather than celebrated New Zealand's artistic progress. They exhibited a new historical consciousness that the 'colonial' moment was past and an anxiety about the next, untested phase of national development in the fine arts. As Richard Jebb suggested following his survey of the British colonies in the first years of the twentieth century, the development of a national consciousness was never so progressively or easily achieved. Rather it was characterized by a 'process of internal friction, the old order slowly and painfully yielding to the new'.[5] In order to understand the complexities of this transitional stage I employ James Belich's concept of 'recolonization', which encourages a less progressive and more critical evaluation of New Zealand's history. In *Paradise Reforged* Belich describes how 'A sense of transition, of insecurity and uncertainty – indeed, something close to a collective identity crisis – can be detected in the New Zealand of the 1880s–1920s, partly masked by residues of the old ideology of progressive colonisation and, increasingly effectively, by the emergent new ideology of recolonisation.'[6]

While the ideology of progressive colonization celebrated the newness of the colony and the advance of industry and technology, Belich suggests that from the 1880s to the 1900s, New Zealand actively tightened its connection to Britain. In trade and economic relations, New Zealand effectively became a 'town-supply district of London', but ties to the British art tradition, upon which New Zealand's art practice was founded, were similarly reinforced during this period. A comparison of the art on show at the 1865 and 1889 exhibitions staged in Dunedin provides the opportunity to evaluate the shifting critical responses and to reflect upon the 'progress' of New Zealand art as it reached its sesquicentenary.

The International Context for Fine Art and the International Exhibition

The inclusion of art in Dunedin's exhibitions took as their model the precedents already set on the international circuit. By the second half of the nineteenth century, fine arts courts had become a standard feature of any great exhibition, yet their status within this context was not arrived at without debate. At the Great Exhibition in 1851, the British exhibits were classified into four categories: raw materials, machinery, manufactures and fine arts. The categories of machinery and manufactures dominated the exhibits, indicating the desire to promote the industrial developments over the manmade. A key factor leading to the conception of the Great Exhibition was the need for industrial education, directed at both workers, who were producing inexpensive but unattractive goods, and consumers, who lacked desire for the products. The inclusion of the fine arts category was part of an attempt to increase standards of taste and to encourage the makers of new mass-produced commodities to combine the finesse of the arts in their manufactured products. Despite these aims the organizers faced a problem concerning where the fine arts should be located in what was essentially an industrial exhibition. For this reason, paintings were excluded from the Great Exhibition: they were only admitted if they exhibited any improvements in colours, and further, were to be regarded 'not so much as examples of the skill of the artist, as that of the preparer of colours'.[7]

While the place of the fine arts was ambiguous at the 1851 exhibition in England, at the 1855 *exposition universelle* in France they took central place. By 1855 France felt that it had no rivals in the fine arts and therefore it was an integral element in the first exposition, with painting as the focus. What the British had perhaps not understood was that the fine arts provided the perfect counterpoint to industry because they brought status to an exposition.[8] The classification system of the exhibits, which progressed from raw materials through several stages culminating in the fine arts, reinforced their place at the peak of cultural development. As Paul Greenhalgh notes, there was no standard for the definition of fine arts at international exhibitions during this period.[9] Depending on the host country, the range of practices could be limited or extensive. For example, France restricted their definition to architecture, sculpture and painting, whereas the British had a wider-ranging definition which included lace, decorative designs and metal work, to name but a few.[10] The constant in all displays was the prominence of painting, which maintained its status as the pre-eminent art practice. Regardless of the range of objects included, the fine arts indicated the progress of a given culture. Even in this setting however, no one was under any pretensions that the display of fine arts at the Paris exhibition brought the crowds, for the fine art palace had only 100,000 visitors, compared with three million in the machinery hall.[11]

The fine arts at international exhibitions then, were markers of cultural status. Consequently, as the exhibition movement multiplied around the globe, the concept that a nation's fine art could serve as a tool of power in establishing international status gathered impetus.[12] This use of the fine arts aligns with one of the key theoretical approaches informing the discourse around international exhibitions, described by Rydell, Findling and Pelle as the 'cultural hegemony school' and which argues that 'world's fairs need to be understood as vehicles intended to win popular support for national imperial policies'.[13] So although the rhetoric of 'brotherly love and understanding between nations' dominated exhibition discourse, they quickly became sites where nations jostled for attention, sites of international competition. The development and proof through exhibition of a national style in art was but one arena in which these 'battles' took place.

This may have been true of great nations such as France and Great Britain but for a small British colony like New Zealand such an analytic framework does not hold up. At exhibitions held abroad, the colony's art was not shown in the extravagant fine art Galleries, but in the New Zealand Court, alongside raw materials, machinery and other manufactured items. Here it functioned primarily as illustrative material. New Zealand only entered the fine art galleries as one of Britain's colonies, first at the Colonial and Indian Exhibition in London, 1886, and at the British Empire Exhibition of 1924 at Wembley. Christine Boyanowski notes of this latter occasion that only Canada registered as having developed a national school by this time, one that was consciously constructed by the authorities to reinforce the notion of a collective national identity. Consequently the 1924 exhibition has achieved a status in Canada's national art history that is absent from those of the other colonies, New Zealand included.[14]

If showcasing a national school of art was not possible for New Zealand in 1865 or 1889, what then, was the intended function of the fine arts displays at Dunedin? Arguably, they were intended to provide evidence of the presence of culture and civilization. This was particularly important in the colonial context, for the inferiority of colonists, especially those born and raised in the colonies, was fiercely debated in the nineteenth century. A meeting of the Wellington Philosophical Society in 1876 read a report by Dr Newman, a recent arrival to the colony, that attempted to account for the inferiority of Europeans raised in Australia and New Zealand by scientific means, stating that a lack of phosphates and other minerals was the cause.[15] It was important therefore that colonists demonstrate their achievements in not only material, but also cultural wealth. As John Woolley stated at the inauguration of the Wollongong School of Art in Australia in 1861, 'Boundless pastures, bottomless depths of alluvial soil, inexhaustible mines are not the sole conditions of greatness; material without moral resources are of less than no avail'.[16] The illustration of cultural activities in the colonies at international exhibitions was therefore intended to feed the minds of the colonists. It was also hoped, however, that they might practically support the development of an infrastructure for art.

Dunedin 1865: Creating Culture

The Dunedin exhibition of 1865 both provided colonists with pleasing evidence of their success in bringing civilization to their part of the world and prompted the development of artistic infrastructure. A close analysis of the works in the fine arts display and the critical responses to it reveals that, just 25 years after the establishment of the colony, there was an art being produced that could be positively recognized as 'New Zealand'. Dunedin was the natural home for New Zealand's international exhibitions in the nineteenth century. The town's economy and population had grown exponentially from the 1860s, boosted first by a gold rush and then the beginnings of a successful export industry in the 1880s. Consequently, while Christchurch hosted an exhibition in 1882, Wellington in 1885, and Auckland in 1898, Dunedin, as host to the colony's two major international exhibitions of the nineteenth century, asserted itself as the cultural capital of New Zealand in the second half of the nineteenth century.[17]

On 12 January 1865 the New Zealand Exhibition opened to 'expressions of pleasure and surprise'.[18] Reports commented that although the exhibition was 'confessedly very incomplete ... the display was much better than could have been anticipated' and congratulations were voiced 'that so much should have been accomplished by a first effort in so young a colony'.[19] At the first planning meeting in 1863 the commissioners proposed that 'by exemplifying the great natural wealth and the adaptability of the soil and climate of New Zealand for almost every species of production', the exhibition 'would increase the confidence of its inhabitants in their resources'.[20] International exhibitors were welcomed – Tasmania reportedly put on a very good show and Britain sent local as well as imperial exhibits – but the greater emphasis was on fostering the development of local industry and internal relations than external trade and relations.

At the opening, Her Majesty's representative, the Superintendent of the Otago Province, John Hyde Harris, concluded:

> Through this Exhibition, New Zealand will be brought out from its obscurity into a face-to-face meeting with the outer world. Many of its great and varied resources will be exemplified – its vitality and progressive character will be attested – new industries will be created – existing ones revived – public taste and morals will be elevated, and a generous emulation between various provinces and colonies will be promoted.[21]

Such ambitions were shared by cities throughout the globe as they attempted to assert themselves on the world stage by hosting their own version of the Great Exhibition of the Arts and Industries of All Nations in London, 1851. In terms of Australasia, New Zealand was quick to take this step, for while Melbourne hosted an Intercolonial Exhibition in 1866, the first international exhibition did not take place until 1870 in Sydney, Australia. The 1865 exhibition took the 1851, as well as the more recent 1862 London Exhibitions as their model, ordering copies of 'all exhibition catalogues – illustrated and otherwise, 1862; all reports of Juries and committees, 1862', as well as 'all catalogues of the various collections in the South Kensington Museum' to aid their planning.[22]

James Hector, who went on to become the inaugural director of the Colonial Museum in Wellington, played a central role in determining the scope of the exhibition. In 1864 he toured New Zealand to promote the exhibition and gather information about possible exhibitors and exhibits from the various provinces. His report was intended to '... enable the commissioners to judge how far the Exhibition may be expected to indicate the value of the internal resources of the Colony, and its present advancement of the Arts of civilised life'.[23] Consequently Hector not only described the scientific exhibits that could be expected from the provinces, but also commented upon potential art exhibits. The town of Nelson, for example, would send a 'good show of works of art and especially of pictures illustrative of the romantic scenery of the Nelson Province by amateur and professional artists'.[24] The exhibition's secretary, Alfred Eccles, later recollected that while the commissioners had exercised 'every care to ensure that the Exhibition should be as comprehensive as they could make it, as regards its industry and scientific aspects', they had also 'wisely endeavoured to stimulate a love for Art'.[25] This was, however, as the critic writing under the pseudonym 'Rambler' noted in his review, an act of faith, for 'If there be a department in which failure might be predicted in a new colony, it would be that of the fine arts'.[26] Their faith was rewarded, however, and the display was reportedly most creditable to the 'artistic talent and taste of the colony'.[27] In opening his review, Rambler wrote:

> Three of the walls are covered with oil paintings – the fourth, with that species of art in which the English excel all other nations, watercolour. A great proportion of the former are paintings by English and other artists in the old country, and are lent for the occasion by the proprietors. It says much for the taste of the New Zealand community that so many good pictures should have been brought to the colony. The watercolours are almost exclusively the work of New Zealand artists and amateurs ... Many of them are quite equal to the works of some of the best artists at home and might be exhibited without fear of invidious comparison on the walls of old and new Water Color Societies of London.[28]

These words contain several points to reflect upon. By praising the 'taste' of colonists, Rambler draws attention to the relationship between the exhibition of art and the demonstration of class in colonial society. Rambler's identification of the role of the 'amateur' is also revealing of the conditions of artistic production in the colony. But most notably, Rambler's identification of a colonial body of works of art that could compare favourably to those at 'Home', particularly those British works on display, makes this the first review to group together the work of colonial artists as a distinctive category worthy of their own evaluation. Taken together, questions are raised around the classification of 'Art' in nineteenth-century New Zealand: firstly in terms of what might constitute 'New Zealand' art and secondly in terms of those classification systems that were articulated through catalogues, exhibitions and reviews.

The art exhibits were displayed in an annexe at the rear of the exhibition building alongside the refreshments area, and consisted of both works lent by colonists and works of art lent by colonial artists (Fig. 10.1). The loan works came primarily from the collections of Otago 'gentlemen' and were mostly by British artists. These were highly praised as providing evidence of colonial 'taste', and the prominence given to naming the lenders of works in the catalogue and in newspaper reports suggests that this public confirmation of cultural capital played a role in class definition in colonial New Zealand society.

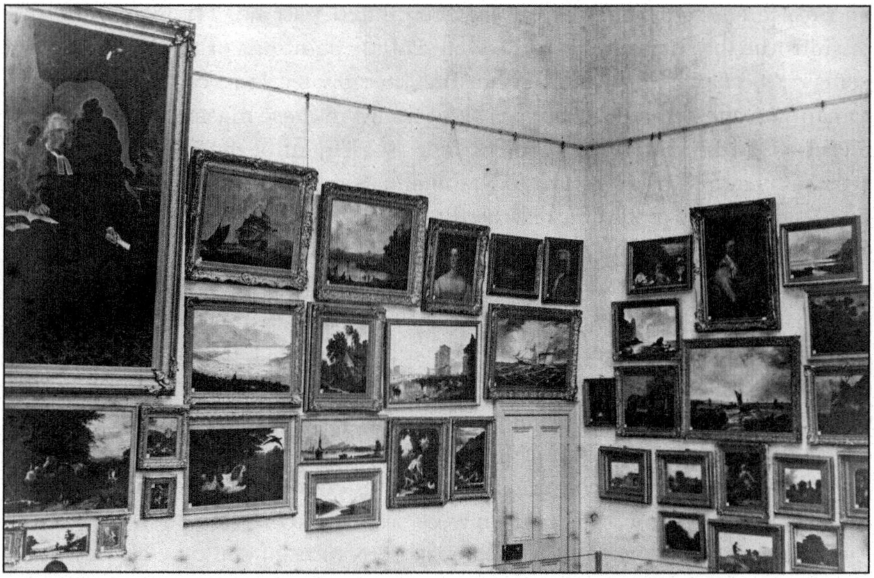

10.1 Photographer unknown, Dunedin Exhibition, 1865, photographic print. Alexander Turnbull Library (1/2-048820)

On the whole, however, it was the demonstration of 'native talent' that aroused the most impassioned reviews, tinged as they were with a sense of surprise and pride that such a fine group of works could be drawn together in such a young colony.[29] But how were these works identified as such? The catalogue lists all the works exhibited by title, artist and exhibitor, with no subcategories and little indication of medium or date of execution, meaning that Charles Decimus Barraud's watercolour landscape *Kai Kora Bay* [*sic*] was listed alongside John Martin's sepia drawing *Meeting of Adam and Eve in Paradise* and John Turnbull Thomson's *View of Dunedin, 1856* next to a version of Raphael's *Madonna del Gran Duca* by an Aliani artist, exhibited by F. Fulton. In contrast to the catalogue, the hang does appear to have been determined by medium. The report suggests that prints, engravings and chromolithographs were exhibited in the corridor, and 'Rambler' explains that the watercolours, mostly by New Zealand artists, covered 'the whole wall which faces the visitor as he enters'.[30] Was it only the reviewers and organizers who could readily identify the work of local artists, or were they visibly identified by name, or were they obviously 'New Zealand' in subject or style? Certainly, by the given titles, most colonial watercolours were, unsurprisingly, of New Zealand scenes. But it was certainly not through *inferiority* of subject choice or style that 'New Zealand' artists could be identified, for, as the report of the fine arts noted, the display 'was very creditable to New Zealand, not only that it contained so many good pictures, but also so few bad ones'.[31]

The reviewer for the *Otago Daily Times* gamely began a work-by-work review of the fine art exhibition following the catalogue order, but petered out having described only 119 of the 412 works on display. In this too, there was no clear distinction made between colonial and European artists. Unfortunately, this unfinished assessment meant that most works by colonial artists went unconsidered, but if the review had continued past no. 119, elaboration of the introductory boast that 'some watercolour paintings of the North are the gems of the Exhibition, and neither Italy nor Switzerland can boast of more romantic scenery than is brought under notice in these masterly productions' might have been provided.[32] For further assessment of the works of colonial artists it is necessary to return to Rambler's assessment of the fine arts which was less methodical, but more broadly analytical and descriptive.

The reviewer 'Rambler' is identified as William Fox in the annotated copy of the *Wellington Independent* in the Hocken Collections.[33] Fox was a prominent public figure in colonial New Zealand: his byline in the *Dictionary of New Zealand Biography* lists his attributes as explorer, politician, four-time Premier, social reformer and painter.[34] Fox himself was represented in the fine arts display by two works, *Wanganui* and *Oteriah Pah, Wellington*, both from the collection of the Supreme Court Judge, Henry Samuel Chapman. As Fox notes in his review, the watercolours of local production that he found compared favourably to those of British origin were 'almost exclusively the work of New Zealand artists and *amateurs*'[my italics].[35] Fox himself could be classed as an amateur, but this did not then carry the derogatory connotations implied by the twentieth-century use of the term. The 'gentleman' or amateur painter is typically generalized as one whose practice is largely confined to the private sphere, yet in the colonial setting, these spheres were less well defined, and the amateur practitioner regularly crossed into the professional and public sphere by exhibiting in contexts such as the 1865 Exhibition. Indeed, most New Zealand artists were necessarily 'amateurs', for as the report of the fine arts observed:

> Pioneer Colonists have rough, and generally stern, work before them. They may admire the beauties of nature with which they are surrounded; may, perhaps almost unconsciously, cultivate their taste for the picturesque and the grand as they explore the country they have come to, but their hands, even if skilled, are seldom at leisure to paint its beauties.[36]

In the case of the artist James Crowe Richmond, it was regretted by commentators that the realities of colonial life rendered art little more than a leisure activity, that his 'artistic activities should be smothered by the details of an official life'.[37] Richmond was one of the few New Zealand artists to exhibit in oil and his painting *Lake Rotoroa* was the work singled out by Fox. Alongside Richmond, John Gully and Barraud's works were greatly praised by reviewers. Fox described them as 'top of the ladder' in the field of watercolour and valued them for their competent blend of 'study in the open field and study in the atelier'.[38] However, as Gully was seen to achieve more of a union of the two, he was given the highest place by Fox.

Fox and Richmond were amateurs, but Gully and Barraud were defined by Fox as 'professionals' – despite the fact that both had 'day jobs': Barraud was a chemist and Gully worked as a draughtsman for the Survey Department until 1877. Obviously Gully, unlike other artists, was able to nimbly shift between the demands of topographical and 'artistic' practice, producing works of a different calibre for his surveying drawings as opposed to his exhibition pieces. James Turnbull Thomson, Chief Surveyor of the Otago Province, proved less able to refine his craft for exhibition purposes and was advised by Fox 'to throw all his colours but three or four, into the street' in the hope that 'His eye may in time get relieved of the jaundice with which a residence in a gold producing country has infected it'.[39] However, in the case of George O'Brien, it was the very qualities that were necessary to his trade as a draughtsman that rendered his work so appealing for Fox. While he also produced largely topographical views, Fox described them as the work of a 'man who has true artistic feeling and who has not condescended to any of those "picture manufacturing processes"'.[40] Such words were likely informed by the prevailing art criticism of the time, particularly that of John Ruskin, who wrote of truth in painting:

> ... the landscape painter must always have two great and distinct ends: the first, to induce in the spectator's mind the faithful conception of any natural objects whatsoever; the second, to guide the spectator's mind to those objects most worthy of its contemplation, and to inform him of the thoughts and feelings with which these were regarded by the artist himself.[41]

Consequently, Fox admired O'Brien's 'remarkable transparency of air ...' and for the 'great neatness with which the shipping is delineated' in his view of Port Chalmers, suggesting that O'Brien's charm lay in its meticulous realism, its 'truth to nature', matched with a sense of thoughtful contemplation.[42]

A consideration of O'Brien's work also raises a further point for discussion: the classification of 'art' in colonial New Zealand. For while only one work was attributed to O'Brien in the catalogue, Fox accurately diagnosed a view of the exhibition buildings as being by him, even though it was exhibited under the name of the architect's firm for which it was executed. Other artists, whose work was likewise executed in a truly professional capacity, but in the natural history and not the architectural field, were not singled out for comment in fine arts reviews. Architectural drawings were admitted in the fine arts display and classified as such in the catalogue, but natural history drawings, such as those by John Buchanan, were not. Buchanan's watercolour drawings were made from sketches executed by himself and Hector whilst exploring the West Coast and consisted of 17 'most interesting views, illustrative of the Geological features of Otago Scenery' as well as drawings of fossils and Alpine plants.[43] They were exhibited in 'Class XXIX, Educational Works and Appliances', and formed part of the Museum of the Geological Survey Department of Otago, a collection of specimens that visually converted the West Coast from a *terra incognita* into a known quantity, transforming what was supposed an 'inhospitable, desolate, and unfruitful region' into one that was 'endowed with rich resources'.[44] Buchanan received an honorary certificate for his drawings which were said to 'bring before us the botanical features and picturesque grandeur of this comparatively untrodden region', but it was clearly their integration into the display, alongside 'Rocks, Minerals, Fossils, Birds, Woods, and Plants' that rendered them meaningful.[45] In contrast, Charles Heaphy exhibited drawings that were not listed in the catalogue, but were described in the fine arts appendix as 'Drawings illustrative of the geology of Auckland'. These received a bronze medal, with the jurors commenting that 'independent of their scientific value, [they] were very beautiful sketches'.[46]

The Legacy of the 1865 Exhibition

As the report of the closing ceremony for the 1865 exhibition poetically noted, the exhibition was a short-lived event and a brave one for a young colony:

> As a mere spectacle it will fade away, as a museum, it will be rapidly scattered. But the impress it leaves on Industrial Art in New Zealand will be permanent, and to it will have to be attributed by the faithful historian much of the future progress made by the commerce of the Colony.[47]

Despite such contemporary assurances that they would leave their historical mark, the impact of such exhibitions is hard to measure. However, the positive reviews of the fine arts produced by colonial artists suggest the 1865 exhibition provided the confidence for developing an infrastructure for art in New Zealand. In 1869, Dunedin hosted the first major art exhibition in New Zealand organized by Alfred Eccles and William Mathew Hodgkins which featured a significant body of work by William Fox, as well as the esteemed Monrad collection of prints now housed in the Museum of New Zealand Te Papa Tongarewa. While it was hoped that any proceeds would enable the committee to purchase works that would form the nucleus of an art collection, numbers were disappointing and this did not eventuate. Nonetheless, while Auckland was the first city to form an art society in 1869, Dunedin followed closely, founding the Otago Society of Arts in 1876. New Zealand's first school of art opened in Dunedin in 1870 with David Con Hutton, who had recently arrived from Scotland, as the drawing master. In 1884 the first public art gallery was opened in Dunedin occupying a space in the Otago Museum. Consequently, it could be argued that Dunedin was the cultural capital of New Zealand in the mid-nineteenth century and that confidence in its cultural status was fostered by their hosting of New Zealand's first major exhibition in 1865.

The critical responses to the fine art on show at this exhibition provide insights into the nature of art produced in the colony up to 1865. They highlight the importance of the 'amateur' to colonial art histories, and through this draw attention to the permeable nature of classification in art. Both Buchanan's and Heaphy's works, for example, were illustrative of regional geology; that one set of these works was exhibited in the fine arts gallery, and the other in the Educational Works and Appliances section, speaks more of the weakness of classificatory boundaries rather than of any clear aesthetic differences between the works. Twenty-five years later, a similar blurring of boundaries was apparent, but in relation to the conception of the historical in art, rather than the scientific. The centrality of the amateur to local art production was less obvious in 1889, due to the gradual development of an infrastructure for art; however, this brought with it a more self-conscious and critical assessment of the artistic products of the colony. In contrast to the wholehearted celebration of the newness and potential of the colony in 1865, the 1889–90 Dunedin Exhibition both reflected and embodied a new historical awareness of New Zealand as a country that was creating history, as quickly as it was erasing it.

Creating History: Dunedin 1889–90

At the New Zealand and South Seas Exhibition in 1889, the Fine Arts Gallery constituted more than an annexe at the rear of the exhibition building. Instead, a purpose-built brick building housed some 1,500 works displayed in six galleries, within which New Zealand works were grouped together. The exhibition as a whole was recognized as a celebration of the jubilee of the colony. It was organized by a private company with input from several specialist committees that secured Government support for the buildings and for several of the exhibits, notably the Early History, Maori and South Seas Court. In particular, the Government was also asked to request the loan of the British pictures currently on show at the Melbourne Centennial Exhibition 1888–89.[48] In contrast to the 1865 exhibition, this one, as signalled by its title, was intended to 'promote closer economic and political relations with the Australian colonies and the Pacific Islands'.[49] Whether it achieved this is questionable, but generally the exhibition was a success – it made a modest profit and more than 600,000 people attended (more than the population of New Zealand at the time).[50] The 'red-letter day' represented by the exhibition provided the opportunity to survey the past in order to map the future, to negotiate the evolving relationship between the old and the new. In the case of the fine arts, for example, many of the colonial works that had, up to this moment, been received as contemporary, were contextualized in a manner that suggested they served more historical than aesthetic purposes. This was in part due to the presence of colonial art within the Early History, Maori and South Seas Court, which set out to visualize in text, picture and artefact, the history and pre-history of the colony to date, demonstrating an awareness of the need to preserve and record the past.

In the Fine Arts gallery, a consideration of the work of the star of the 1865 exhibition, the recently deceased John Gully, provided a turning point for the appreciation of colonial art. Eleven works by Gully were exhibited by their owners alongside works by living artists, but were not submitted for awards. Most of these were from local collections, but notably two views by

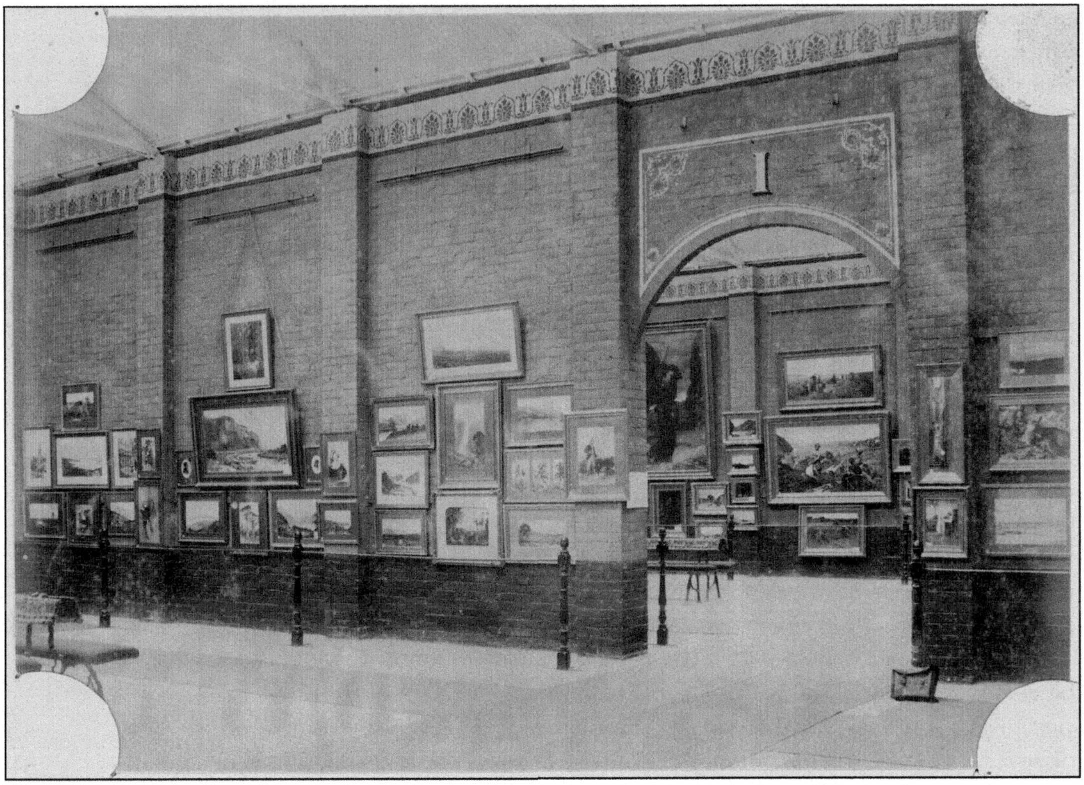

10.2 David Alexander De Maus, gallery interiors hung with paintings, at the New Zealand and South Seas Exhibition, Dunedin, 1889–90, albumen print. From Eric Hall McCormick's album. Alexander Turnbull Library (PA1-o-761-01)

Gully, *Waimea Valley* and *Mount Cook*, were exhibited by Melbourne art dealer Alexander Fletcher and the Melbourne businessman and philanthropist Alfred Felton respectively as part of the Victorian Loan Collection, indicating his status as an Australasian artist.[51] In addition, a number of Gully's drawings produced on a recent sketching trip with J. C. Richmond to Te Anau and Lake Manapouri in 1887 were lent by William Mathew Hodgkins and hung in the corridor leading to the Fine Arts galleries opposite a selection of old master engravings from the Monrad collection.[52] These constituted a kind of retrospective of Gully's work, allowing for comparison between his early and later styles, and their instructive example was noted by the primary reviewer of the fine arts, M. R.,[53] presumably Malcolm Ross, who thought they 'would provide of great advantage to students'.[54]

Ross also observed that New Zealand artists had a natural advantage with landscape painting, 'for our beautiful lake scenery, with towering snow-capped mountains, provides splendid material for pencil as well as pen', and predicted that 'before very long New Zealand is likely to have a distinct school of landscape painters, whose works are sure to be greatly sought after in the Australian colonies' (Fig. 10.2).[55] While this comment acknowledges the importance of inter-colonial traffic between Australia and New Zealand to the art world, Ross, as with other reviewers of the 1889 exhibition, was generally less congratulatory than those of 1865. While the usual praise was issued that 'New Zealand artists show a fair standard of excellence, considering the difficulties which have hitherto surrounded the development and progress of art in this country', the dependence on landscape as the dominant pictorial genre came under criticism from several sources. Ross noted that 'it is in *genre* painting that we are most behindhand'[56] and the judges concluded that 'if more attention were given to work in which the "figure" had place, it would vary the monotony of so much mere landscape work'.[57]

The introduction of the figure into painting required that some narrative element be integrated into New Zealand painting. The problem that faced New Zealand artists was where this narrative should be sought. In Australia, the 'Heidelberg' school, including artists such as Tom Roberts and Arthur Streeton, had initiated in the 1880s an impressionist mode of painting focusing on 'nationalistic themes that celebrate the white settlement of Australia and its economic foundations in extracting wealth from the cultivation of the land'.[58] Scenes of stockmen, shearers and bush rangers at work in a bleached dry landscape became synonymous with emergent Australian nationalism.[59] Several New Zealand artists had been producing and exhibiting works that dealt with historical or genre subjects, but no definite school comparable to that emerging in Australia was noticeable by the time of the Dunedin Exhibition. A number of artists had sent work of rural or agricultural themes to the Colonial and Indian Exhibition in 1886, such as Peter Power's painting *A New Zealand Homestead*,[60] or Benjamin A. Branfill's two works *After a Long Day on the Sheep Run* and *Sheep-Shearing on the Waimea Plains*. However, rather than demonstrating a direct response to the local, these examples may better represent a transplantation of the 'rural idyll' genre, popular in British nineteenth-century landscapes, to the New Zealand context.[61]

As far as historical subjects were concerned, William Mathew Hodgkins had exhibited watercolours at the 1888 Melbourne Centennial Exhibition which re-presented moments of initial contact by explorers, such as *Tasman's Encounter with the Natives of Massacre Bay, N. Z., on the 19th of December 1642* (1886; Hocken Library) as well as a scene from Cook's voyages. The contemporary art historian Leonard Bell argues that history painting only emerged as a prevalent genre in New Zealand art in the 1890s, a delay that was undoubtedly linked to the state of art education and patronage.[62] History painting, as the pinnacle of traditional academic painting, required a high level of study, particularly from the figure, and there was limited access to such training in colonial New Zealand. Further, there was little state support for such painting in New Zealand, for the colonial government tended to foster and promote works of art of a primarily illustrative nature for promotional purposes.

Historical or genre subjects in colonial art were consequently scarce, and works dealing with such themes were largely absent from the Dunedin Exhibition, a fact lamented by judges and reviewers alike. This meant that the historical representation of New Zealand was left to the Early History, Maori and South Seas Court. This section closely reflected the composition of Thomas

Morland Hocken's own collection, which would form the basis of the Hocken Library.[63] Included in the display were documents relating to the early history of New Zealand's settlement, maps, plans, drawings, photographs, artefacts and paintings. Works that had been displayed in the Fine Art section of the 1865 Dunedin Exhibition, such as Charles Heaphy's watercolour sketches of Rangitoto Island, were now displayed for their historical interest. Gilbert Mair exhibited 'five portraits in oil, illustrative of the Maori race' in this court, all of which were by his wife, Eleanor Mair (née Sperrey), some of which had recently been exhibited at the Melbourne Centennial Exhibition 1888–89 in New Zealand's Fine Art display.[64] Portraits of Maori by Gottfried Lindauer also found their place in this court, including the large painting *The Maori at Home* (1885; Wanganui, Whanganui Regional Museum), which had formed the centrepiece of Walter Buller's collection at the Colonial and Indian Exhibition in London 1886. Consequently, the framing of works of art could swing from aesthetic to historical and/or ethnographic depending on the chronology and the context of display. The placement of these paintings in the Early History, Maori and South Seas Court encouraged a reading of them as historical or ethnographic records, rather than as works of art.

Overall then, the critical responses to the 1889–90 exhibition differed from the positive reception New Zealand art had received at previous exhibitions, particularly those held abroad which were more concerned with illustrating the country for an international audience. Abroad, the landscape could act as a point of difference for an international jury, but in Dunedin, a largely colonial jury used their feedback to encourage New Zealand artists to participate in a more international artistic dialogue. By making these comments they also emphasized the importance of these exhibitions in disseminating ideas about and approaches to art and art-making.

Conclusion

Between 1865 and 1889 the colony of New Zealand developed at an unprecedented rate. Libraries, museums, galleries and art societies were formed and an infrastructure for art in the colony was slowly being created. However, it seems that the development of a more professional framework within which art was practised and received brought with it greater anxiety about New Zealand's artistic progress. The critics of the fine art at the Dunedin Exhibition in 1889 demonstrated a new historical consciousness that the 'colonial' moment was past, accompanied by apprehension about the next, untested phase of national development in the fine arts. The new historical awareness signalled by the displays at 1889 Exhibition, particularly the Early History, Maori and South Seas Court, no doubt coincides with the emergence of a new 'national' consciousness. The example of Gully consequently illustrates a key shift that had occurred between 1865 and 1890. In 1865, his work was celebrated as the best work on display, while in 1889–90, its role was markedly different. In contrast to other works of art on display, Gully's works were neither for sale, nor open to the competitive aspect of the exhibition. Their function was a new one for the colonial context. For the embryonic New Zealand art world they provided the artistic equivalent of an individual's 'red-letter' day. They represented the realization that the *colonial* was on the verge of no longer being contemporary but something of the past that was perhaps worth collecting. For artists, they stood as an example of what had been achieved to date, and their exhibition allowed this achievement to be evaluated and mapped against future expectations. The art world was aware of the need for art to continue to progress, but the path it should follow to do so, as with the road towards nationalism, was not an easy one.

Notes

1 'The New Zealand and South Seas Exhibition', *Otago Witness*, 28 November 1889, 13.

2 'New Zealand', 13.

3 The Treaty of Waitangi is New Zealand's founding document, first signed at Waitangi, in the Bay of Islands on 6 February 1840. It is an agreement, in Māori and English, which was made between the British Crown and about 540 Māori rangatira (chiefs).

4 'Rambler', *Wellington Independent*, 16 March 1865, 3.

5 Richard Jebb, *Studies in Colonial Nationalism*, London, 1905, ix.

6 James Belich, *Paradise Reforged: A History of the New Zealanders from the 1880s to the Year 2000* (Auckland, 2001), 76.

7 Jeffrey Auerbach, *The Great Exhibition of 1851: A Nation on Display* (New Haven and London, 1999), 115.

8 See Paul Greenhalgh, *Ephemeral Vistas: The 'Expositions Universelles', Great Exhibitions and World's Fairs, 1951–1939* (Manchester, 1988), 198–199; and Greenhalgh, *Fair World: A History of World's Fairs and Expositions from London to Shanghai 1851–2010* (Winterbourne, 2011), 237.

9 Greenhalgh, *Ephemeral Vistas*, 199.

10 Greenhalgh, *Ephemeral Vistas*, 199.

11 Greenhalgh, *Ephemeral Vistas*, 199.

12 David Park Curry, 'American Art, American Power at the Paris Expositions Universelles', in *Americans in Paris 1860–1900*, ed. Erica E. Hirshler and Helena Barbara Weinberg (London, 2006), 192.

13 Robert W. Rydell, John E. Findling and Kimberly D. Pelle, *Fair America: World's Fairs in the United States* (Washington, DC and London, 2000), 5.

14 Christine Boyanowski, 'Selective Memory: The *British Empire Exhibition* and National Histories of Art', in *Rethinking Settler Colonialism: History and Memory in Australia, Canada, Aotearoa New Zealand and South Africa*, ed. Annie E. Coombes (Manchester, 2006), 156.

15 *New Zealand Mail*, 22 July 1876, 6.

16 John Woolley, *A Lecture delivered at the Inauguration of the Wollongong School of Arts, 28 May 1861* (Sydney, 1861), 5.

17 These are the only nineteenth-century New Zealand exhibitions to be listed in John E. Findling and Kimberly D. Pelle, eds, *Encyclopedia of World's Fairs and Expositions* (Jefferson, NC, 2008).

18 'Opening of the New Zealand Exhibition', *Otago Daily Times*, 13 January 1865, 4.

19 'Opening', 4.

20 Minutes of meeting held 21 September 1863, Minute Book of the Commissioners, Hocken Library: MS-0329, 10.

21 'Opening of the New Zealand Exhibition', *Otago Witness*, 14 January 1865, 9.

22 Minute Book of the Commissioners, Hocken Library: MS-0329, 61.

23 Minutes of meeting held 27 July 1864, Minute Book of the Commissioners, Hocken Library: MS-0329, 184.

24 Hector to Eccles, 30 July 1864, Hocken Library: MS-0366.

25 Alfred Eccles, *The First New Zealand Exhibition and Dunedin in 1865* (Dunedin, 1925), 10.

26 'Rambler', *Wellington Independent*, 16 March 1865, 3.

27 'Rambler', 3.

28 'Rambler', 3.

29 'Native' here refers to the colonial, rather than indigenous occupants of New Zealand. 'The New Zealand Exhibition', *Supplement to the Lyttleton Times*, 14 February 1865.

30 'Rambler', 3.

31 *New Zealand Exhibition, 1865: Reports and Awards of the Jurors and Appendix* (Dunedin, 1866), 499.

32 *New Zealand Exhibition*, 499.

33 R. D. J. Collins and Wendy-Anne Curtis, *The New Zealand Exhibition 1865: Documents on the Fine Arts* (Dunedin, 1988), 33.

34 Keith Sinclair and Raewyn Dalziel, 'Fox, William – Biography', *Dictionary of New Zealand Biography*, at http://www.TeAra.govt.nz/en/biographies/1f15/1, accessed 27 December 2012.

35 Roger Blackley addresses the role of the amateur within the developing culture of 'fine art' in nineteenth-century New Zealand. See Roger Blackley, 'John Kinder and New Zealand Art', in *John Kinder's New Zealand*, ed. Ron Brownson (Auckland, 2004).

36 *New Zealand Exhibition*, 497.

37 'Rambler', 3. See also *New Zealand Exhibition*, 498.

38 'Rambler', 3.

39 'Rambler', 3.

40 'Rambler', 3.

41 John Ruskin, *Modern Painters*, ed. and abridged by David Barrie (New York, 1980), 24.

42 'Rambler', 3.

43 *New Zealand Exhibition*, 245.

44 *New Zealand Exhibition*, 261.

45 *New Zealand Exhibition*, 261.

46 *New Zealand Exhibition*, 499.

47 *New Zealand Exhibition*, 499.

48 'The New Zealand and South Seas Exhibition', *Otago Witness*, 28 November 1889, 15.

49 Conal McCarthy, 'Dunedin 1889–90', in *Encyclopedia of World's Fairs and Expositions*, ed. Findling and Pelle (Jefferson, NC, 2008), 108.

50 McCarthy, 108.

51 See no. 78 and 120 of the 'Victorian Loan Collection', in *Official Guide to the Picture Galleries: And Catalogue of the Fine Arts Department*, ed. William Hodgkins (Dunedin, 1889), 58–66.

52 See 'The Exhibition Art Gallery', in *New Zealand and South Seas Exhibition, 1889: Official Catalogue of the Exhibits*, 246–251.

53 The reviewer was presumably Malcolm Ross, journalist for the *Otago Daily Times* and exhibitor with the Otago Art Society in the 1890s. See Una Platts, *Nineteenth-Century New Zealand Artists: A Guide and Handbook* (Christchurch, 1980), 209.

54 Malcolm Ross, 'The Art Gallery, No. X: New Zealand Pictures', *Otago Witness*, 27 February 1890, 17.

55 Ross, 17.

56 Ross, 17.

57 John H. Scott, Chairman of Jurors, Class XVIII, in *Official Record of the New Zealand and South Seas Exhibition Held at Dunedin, 1889–90*, ed. D. Harris Hastings (Wellington, 1891), 293.

58 Anne-Marie Willis, *Illusions of Identity: The Art of Nation* (Sydney), 1992, 73.

59 See also Richard White, *Inventing Australia: Images and Identity 1688–1980* (Sydney, London and Boston, 1981).

60 *New Zealand Herald*, 3 August 1886, 5.

61 See Rosemary Treble, 'The Victorian Picture of the Country', in *The Rural Idyll*, ed. G. E. Mingay (London, 1989).

62 See Leonard Bell, *Colonial Constructs: European Images of Maori 1840–1914* (Auckland, 1992).

63 The Hocken Library opened on 31 March 1910, six weeks before Hocken's death.

64 *New Zealand and South Seas Exhibition, 1889: Official Catalogue of the Exhibits* (Dunedin, 1889), 235.

Bibliography

Primary Printed Sources

Hocken Library, Dunedin, Minute Book of the Commissioners, MS-0329.

Hocken Library, Dunedin, MS-0366.

Hodgkins, William. *Official Guide to the Picture Galleries: And Catalogue of the Fine Arts Department*. Dunedin: Evening Star Job Printing Works, 1889.

Lyttelton Times.

New Zealand and South Seas Exhibition, 1889: Official Catalogue of the Exhibits. Dunedin: Exhibition Commissioners, 1889.

New Zealand Exhibition, 1865: Reports and Awards of the Jurors and Appendix. Dunedin: Mills, Dick and Co., 1866.

New Zealand Herald.

New Zealand Mail.

Otago Daily Times.

Otago Witness.

Wellington Independent.

Secondary Sources

Auerbach, Jeffrey. *The Great Exhibition of 1851: A Nation on Display*. New Haven and London: Yale University Press, 1999.

Belich, James. *Paradise Reforged: A History of the New Zealanders from the 1880s to the Year 2000*. Auckland: Penguin, 2001.

Bell, Leonard. *Colonial Constructs: European Images of Maori 1840–1914*. Auckland: Auckland University Press, 1992.

Blackley, Roger. 'John Kinder and New Zealand Art'. In *John Kinder's New Zealand*, edited by Ron Brownson, 122–129. Auckland: Random House New Zealand and Auckland Art Gallery, 2004.

Boyanowski, Christine. 'Selective Memory: The *British Empire Exhibition* and National Histories of Art'. In *Rethinking Settler Colonialism: History and Memory in Australia, Canada, Aotearoa New Zealand and South Africa*, edited by Annie E. Coombes, 156–170. Manchester: Manchester University Press, 2006.

Collins, R. D. J., and Wendy-Anne Curtis. *The New Zealand Exhibition 1865: Documents on the Fine Arts*. Dunedin: University of Otago, 1988.

Curry, David Park. 'American Art, American Power at the Paris Expositions Universelles'. In *Americans in Paris 1860–1900*, edited by Erica E. Hirshler and Helena Barbara Weinberg, 191–205. London: National Gallery, 2006.

Eccles, Alfred. *The First New Zealand Exhibition and Dunedin in 1865*. Dunedin: Crown Print, 1925.

Findling, John E., and Kimberly D. Pelle, eds. *Encyclopedia of World's Fairs and Expositions*. Jefferson, NC: McFarland, 2008.

Greenhalgh, Paul. *Fair World: A History of World's Fairs and Expositions from London to Shanghai 1851–2010*. Winterbourne: Papadakis, 2011.

Greenhalgh, Paul. *Ephemeral Vistas: The 'Expositions Universelles', Great Exhibitions and World's Fairs, 1951–1939*. Manchester: Manchester University Press, 1988.

Jebb, Richard. *Studies in Colonial Nationalism*. London: Edward Arnold, 1905.

McCarthy, Conal. 'Dunedin 1889–90'. In *Encyclopedia of World's Fairs and Expositions*, ed. John Findling and Kimberly D. Pelle, 108–109 (Jefferson, NC: McFarland, 2008).

Platts, Una. *Nineteenth-Century New Zealand Artists: A Guide and Handbook*. Christchurch: Avon Fine Prints, 1980.

Ross, Malcolm. 'The Art Gallery, No. X: New Zealand Pictures'. *Otago Witness*, 27 February 1890, 17.

Ruskin, John. *Modern Painters*. Edited and abridged by David Barrie. New York: Knopf, 1980.

Rydell, Robert W., John E. Findling and Kimberly D. Pelle. *Fair America: World's Fairs in the United States*. Washington, DC: Smithsonian Institution Press, 2000.

Sinclair, Keith, and Raewyn Dalziel. 'Fox, William – Biography'. In *Dictionary of New Zealand Biography*, at http://www.TeAra.govt.nz/en/biographies/1f15/1, accessed 27 December 2012.

Treble, Rosemary. 'The Victorian Picture of the Country'. In *The Rural Idyll*, edited by G. E. Mingay, 50–56. London: Routledge, 1989.

White, Richard. *Inventing Australia: Images and Identity 1688–1980*. Sydney, London and Boston: George Allen and Unwin, 1981.

Willis, Anne-Marie. *Illusions of Identity: The Art of Nation*. Sydney: Hale and Iremonger, 1992.

Woolley, John. *A Lecture delivered at the Inauguration of the Wollongong School of Arts, 28 May 1861*. Sydney: Reading and Wellbank, 1861.

11

International Exhibitions and Urban Aspirations: Launceston, Tasmania, in the Nineteenth Century

Anne Neale

Introduction

As separate British colonies (prior to Federation in 1901), the Australian states, including Tasmania, participated in most of the large international exhibitions of the nineteenth century, beginning with London in 1851.[1] The Australian contributions to the Great Exhibition were, however, assembled in rather an *ad hoc* fashion, with many submissions made by individuals without much reference to an overall organization. Recognizing the time and work involved in obtaining and preparing suitable exhibits, shipping them safely and displaying them to best advantage, some colonies decided to organize local or inter-colonial exhibitions, to take place in the year preceding a major international exhibition, both as an aid to preparation and for the benefit of the local community. Melbourne, for example, hosted such exhibitions in 1854, 1861, 1866, 1872 and 1875, prior to the Paris 1855, London 1862, Paris 1867, Vienna 1873 and Philadelphia 1876 events.

Beginning in the 1870s, the prosperous and confident Australian colonies decided to hold their own international exhibitions, starting at Sydney, New South Wales, in 1878–79. Melbourne, Victoria (1880–81; 1888–89) and Adelaide, South Australia (1887) soon followed.[2] The international exhibition held in Launceston, Tasmania, in 1891–92 was the fifth to be held in the Australian colonies and the first outside a capital city. Viewed from the perspective of Europe, North America or even the Australian mainland, the small city of Launceston, situated on the island of Tasmania, was remarkably 'marginal': how and why did it come to host an international exhibition?

Tasmania is approximately the size of Ireland and, situated at latitude 42° south, enjoys a climate similar to the south of France. European settlement began in the first decade of the nineteenth century and by 1891 the island's population numbered approximately 147,000, with Hobart in the south functioning as the political and administrative capital, while Launceston in the north (population 17,000) was the financial and commercial hub. Settlement occurred principally along the north-west coast, through Launceston and south through the Midlands, to Hobart and the south-east. Tasmania's chief industries and exports in the late nineteenth century were connected with wool, tin, gold and silver.

Accounts in the local press suggest that nineteenth-century Tasmanians did not regard the island's remote location, or its small population, as necessarily presenting significant barriers to development. As a colony financed largely by the British government until the early 1850s, when the island became self-governing and the transportation of convicts ceased, Tasmania was seen by its ethnically European inhabitants as a closely connected off-shoot of Great Britain. Cultural, economic, social and political developments in Britain were closely followed in the Tasmanian press.

Launceston was a prosperous community, which had the advantage of a substantial group of civic leaders who were far-sighted, public-spirited and, in a broad sense, liberal. Particularly in the development of public amenities, such as the reliable supply of clean water, reticulated sewerage, street lighting and public parks, Launceston in the nineteenth century was frequently in advance of substantial British cities, as well as larger settlements in the Australian colonies, and knew itself to be so. Justifiable civic pride apparently led to a desire for greater recognition of Launceston's achievements and this, coupled with a determination to remain *au fait* with the latest developments around the world, led to the idea of hosting an international industrial exhibition. While others might ask 'why?' Launceston's response was evidently 'why not?'.

The Tasmanian International Exhibition, Launceston, 1891–92

Plans were made to hold an Exhibition of Art and Industry in Launceston as early as 1857,[3] but the notion of holding an *international* industrial exhibition first emerged in 1883.[4] After several major setbacks, largely due to a lack of financial support from the Tasmanian government, the committee's plans finally came to fruition in 1891–92. In the interim, the organizers retained their high aspirations, but pragmatically planned for a more modest affair.

Stimulated by the 1888 Centennial International Exhibition in Melbourne, plans were made in that year for a Juvenile and Industrial Exhibition to be held in Launceston in 1890–91 and work on a permanent exhibition building commenced in 1889. The New Zealand government's decision to participate in the exhibition widened its proposed scope and in July 1890 led the Tasmanian Premier to officially invite all the Australian colonies to contribute, thus making the event officially both Tasmanian and inter-colonial.

It was initially planned to open the exhibition in December 1890, but 'in consequence of the paralysing effect upon commerce and transport of the Australian shipping strike and the labour troubles in England', it was decided to defer the opening for 12 months.[5] This delay came to be seen as the proverbial 'blessing in disguise'. It enabled Tasmania's Agent-General in London to form an exhibition committee in that city and secure the cooperation of numerous contributors from Britain and the Continent. The enterprise had now, finally, become international in scope and also considerably larger than the organizers had anticipated: realizing that they needed professional assistance, they obtained the services of Jules Joubert (1824–1907) as general manager.[6] French by birth, Joubert has been rightly described as 'Australia's nineteenth-century exhibition impresario': he was involved in the organization of numerous large exhibitions, including the overall management of the Calcutta International Exhibition (1883) and the New Zealand and South Seas Exhibition in Dunedin (1889–90), immediately before coming to Tasmania.[7]

The claim of the 1891–92 Tasmanian Exhibition at Launceston to be truly 'international' has been challenged in recent times,[8] but the exhibition did have official representatives from Great Britain, France, Germany, Austria and New Zealand,[9] with exhibits coming not only from those countries but also from Italy, Belgium, Switzerland, the US, Canada and Japan.[10] This was in addition to representatives and exhibits from the Australian colonies of Victoria, New South Wales, Queensland, South Australia and, of course, Tasmania. Reference to the floor plan of the exhibition gives some indication of the relative contributions of the different regions.[11] Tasmania and its closest neighbour, Victoria, received about equal space, with New South Wales somewhat less, while the areas set aside for New Zealand and South Australia were very much smaller. The 'British Court' and the collective 'Foreign Courts', facing each other across the 'Avenue of Nations', were similar in size to the Tasmanian and Victorian courts.[12] A total of 1,372 exhibitors displayed 6,826 exhibits, spread across 59 classes.[13] While the exhibition did not pretend to rival the really large international exhibitions, its use of the term 'international' does appear to be justifiable.

At the time, the main challenge to Launceston's aspirations to hold such an exhibition came from within the island colony itself. A spirited rivalry has existed between Launceston and Hobart since the earliest days of European settlement.[14] While Launceston was undeniably the commercial and industrial hub of the island, the exhibition organizers, acknowledging opposition in the south to any 'Tasmanian' exhibition being held outside the capital, marshalled their arguments in Launceston's favour.

The US was cited as a precedent, in that its great exhibitions had been held in San Francisco and Philadelphia and the next one was to be in Chicago, rather than New York. The recent New Zealand exhibition being held in Dunedin, rather than the capital, Wellington, was also mentioned. It was asserted that most inter-colonial and international visitors to the exhibition would travel to Tasmania via Melbourne and as the voyage from Melbourne to Launceston would be considerably shorter than that to Hobart, this made the

Launceston site a much more attractive proposition for visitors. Additionally, it was argued, Launceston was centrally placed on the island, permitting all Tasmanians to more readily visit the exhibition, while allowing inter-colonial and international visitors to see more of the island's industries and attractions than would be the case if they were based in Hobart. Should anyone question the validity of holding an industrial exhibition at all, the usual arguments were advanced: the potential advances to be made in commerce, trade and investment; and the educative role of the exhibition, in broadening the horizons and raising the aspirations of artisans and manufacturers and, indeed, of the whole community.[15]

Two key results of the exhibition indicate that the confidence of Launceston's leaders was not misplaced. Approximately a quarter of a million people passed through the turnstiles in the six months the exhibition was open.[16] The Victorian obsession with statistics, always on display in association with the industrial exhibitions, led the organizers to proudly declare that: 'The attendance from the day of opening to the closing numbers 243,000, which, taken as compared with the total population of this city (17,248), or that of Tasmania (146,667), compares more than favourably with the records of previous Exhibitions *in any part of the globe*.'[17]

Perhaps even more significantly, the Launceston exhibition achieved a rare thing in the annals of nineteenth-century industrial exhibitions: it made a profit![18]

The Launceston Exhibition Buildings

Most structures erected in connection with the international exhibition movement were designed to be temporary. Nevertheless, some provided a significant but unforeseen legacy as icons of either the exhibition itself, or its location. Thus the distinctive and innovatory 'Crystal Palace', London [destroyed], is effectively synonymous with the Great Exhibition of 1851, while the Eiffel Tower, created for the *Exposition Universelle* of 1889, is internationally recognized as a symbol of Paris. However, intentionally permanent exhibition buildings were rare in the nineteenth century and their survival into the twenty-first century even more so.

The buildings designed to house the first Tasmanian international exhibition consisted of four linked structures situated in the north-west corner of Launceston's City Park. The buildings comprised the 'Albert Hall', a substantial permanent building capable of seating 2,500 people; temporary annexes covering about three acres (1.2 hectares) of the twelve-acre (about 5-hectare) park; a large fernery which served as the vestibule to the exhibition; and a 'Mineral Arch' which provided the main entrance to the exhibition complex via the fernery (Fig. 11.1).

The Mineral Arch was designed by Launceston architect E. A. Luttrell (1865–1924), the official architect to the exhibition. It was a plain classical

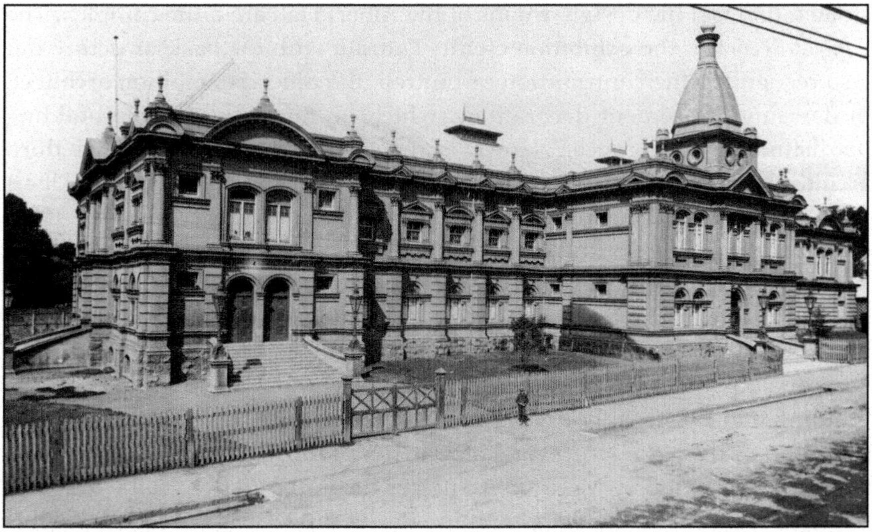

11.1 The Albert Hall, Launceston, Tasmania, erected for the Tasmanian International Exhibition of 1891–92. Spurling's Studios, Roger Holdsworth (compiler), n.d. (1890s?). State Library of Victoria, H2012.90/10

Along with the Royal Exhibition Building in Melbourne, Victoria (1879–80) this is a very rare survivor amongst 19th century international exhibition halls. It is still regularly used for large events.

design, constructed from a range of Tasmanian bricks and building stones, incorporating small samples of the island's more impressive minerals. Built to last, it remained as the fernery entrance until 1910 and then survived in splendid isolation in the park's work yard until it was demolished, probably in the early 1960s.[19]

The Fernery was a lightweight cruciform structure previously known as 'the pavilion', which had been built to the designs of Launceston architect/builder Peter Mills (1828–86) in 1873 and used as a venue for large events.[20] With a new pavilion provided in the form of the Albert Hall, it was decided to convert the old pavilion into a fernery, complete with waterfalls, rills and classical statuary, to form an impressive entrance vestibule to the exhibition. The widely acclaimed work was carried out under the direction of Scottish landscape gardener William McGowan (1858–1939), who had oversight of Launceston's public parks and gardens.[21] By 1910, the fernery had fallen into disrepair and was demolished.[22]

The temporary annexes designed by Luttrell for the exhibition were unpretentious structures of timber and corrugated iron. In addition to the courts devoted to the various colonies and countries, about a third of the annexe space was set aside for the exhibition of heavy machinery, in a court cleverly connected to the local railway by a branch line for the duration of the exhibition. The main axis of the annexes ran east–west and centred on an existing fountain (see below). The annexes appear to have been demolished shortly after the exhibition.

The Albert Hall (1889–91), designed for the exhibition but always intended as a permanent structure, was originally referred to as the 'New Pavilion' to distinguish it from the 'Old Pavilion', which became the fernery. Credit for its design is usually given to the Launceston amateur architect, John Duncan (1870–1936), which certainly accords with the inscription on the foundation stone but in fact the design origins of the Albert Hall are quite complex. The *Official Record* of the exhibition credits Duncan with the basic structure but also recognizes the contribution of Luttrell, the official exhibition architect, in devising the cement decorations, which do much to give the building its distinctive appearance.[23] However, new research indicates that a third architect deserves at least as much recognition for the design of the Albert Hall as Duncan and Luttrell and probably more: Gordon McKinnon (?–1922) of Parramatta, New South Wales.

McKinnon won the design competition for the New Pavilion in 1889 but the council declared that they would only use the front elevation of his design and denied him the promised right to supervise the works.[24] Duncan subsequently emerged as the council's approved 'architect' for the building and he strongly denied that the final structure owed anything to McKinnon's design.[25] However, comparison of the Albert Hall with similar structures designed by McKinnon, such as the Forbes Town Hall (1891) and the Albury Town Hall (1908), both in New South Wales, demonstrates this claim to be false. It also indicates that Luttrell referred to McKinnon's design in preparing his cement embellishment of the public facades of the building. Drawings of McKinnon's design have not been found, but visitors to the exhibition were able to see for themselves the design origins of the building. In Class IV, 'Architectural Drawings and Models, Elevations and Plans of Buildings', McKinnon's 'front elevation design of the Albert Hall' received a first award, while Luttrell's 'isometrical perspective drawing of Tasmanian Exhibition and surroundings' received a second.[26]

The Albert Hall is built in the 'Free Classical' manner of architecture popular in the late nineteenth and early twentieth centuries. Unhindered by any concern for the grammatically correct use of the classical language of architecture, this design approach utilized classically derived elements in a loose fashion that would have appalled more academically inclined practitioners.[27] The overall dimensions of the building were approximately 75 × 18 metres, while the main interior space measured 45 × 18 metres and was believed to be 'the tenth largest hall in the British dominion' at the time of its erection.[28]

The claims of the Albert Hall to cultural significance, above and beyond the Tasmanian context, depend upon its role in the international exhibition movement of the nineteenth century. When the Melbourne Exhibition Building of 1880 was accorded UNESCO 'world heritage' status in 2004, it was primarily because 'The Exhibition Building is the only nineteenth-century Great Hall to survive largely intact, still in its original landscape setting and still used as an exhibition and event venue.'[29] Seen in this context, the Albert Hall of Launceston must also be seen as possessing considerable cultural significance, at least at a national level, for despite its smaller size, it too is a nineteenth-century Great Hall, surviving largely intact, still in its original landscape setting and still used as an exhibition and event venue.

Fountains at the International Exhibitions

For visitors to the international exhibitions, as well as for those further afield whose knowledge was dependent upon published illustrations, fountains were frequently the most striking items on display. This was due to their substantial size, aesthetic appeal and prominent placement.

Osler's famous 'crystal fountain', which formed the centrepiece of the so-called 'Crystal Palace' in Hyde Park, London in 1851, set a remarkable precedent in the arrangement of major exhibition buildings. In the Dublin International Exhibition of 1853, a terracotta fountain by Ferguson & Miller of Glasgow provided a focal point in the central hall. At the Paris Exposition of 1855, a cast-iron fountain by Barbezat & Co. took pride of place ... and so it went on. Two of Britain's most famous fountains, the cast-iron Ross Fountain in Edinburgh (1872) and the terracotta Doulton Fountain in Glasgow (1888), were legacies of the London 1862 and Glasgow 1888 international exhibitions respectively.

It therefore comes as no great surprise to find that the Tasmanian International Exhibition in Launceston in 1891–92 also centred on a substantial fountain. However, rather than being a new exhibit, this fountain had been in place for thirty years and the temporary annexes were constructed around it. Despite its age, the fountain had justifiable claims to its honourable position, for it was a replica of the terracotta fountain at the centre of the 1853 Dublin Exhibition. Yet, it was not the only, or even the first fountain to be erected in Launceston, subsequent to featuring in international exhibitions in Europe.

Civic pride in the establishment of a reliable water supply network in the late 1850s, combined with the development of significant urban parks, led the Launceston municipal authorities to import large ornamental fountains for two of these parks. It was characteristic of their view of Launceston as thoroughly up-to-date that they would look to Europe for the latest developments. And where better to find the best fountains that Europe could produce than at the international exhibitions?

The Prince's Square Fountain, Launceston

The first 'exhibition fountain' to be erected in Launceston was an overt celebration of the town's achievements in obtaining a reliable clean water supply (Fig. 11.2). To commemorate the opening of the town's waterworks in 1857, it was decided to landscape an entire town block as a public square or 'promenade', with a large ornamental fountain as its centrepiece.[30] The undeveloped ground, then known as 'St John's Square' or 'Church Green', was located on the southern border of the town centre, between the business and residential areas and flanked by churches. It acquired the name 'Prince's Square' (unofficially at first) at the time of its public opening in

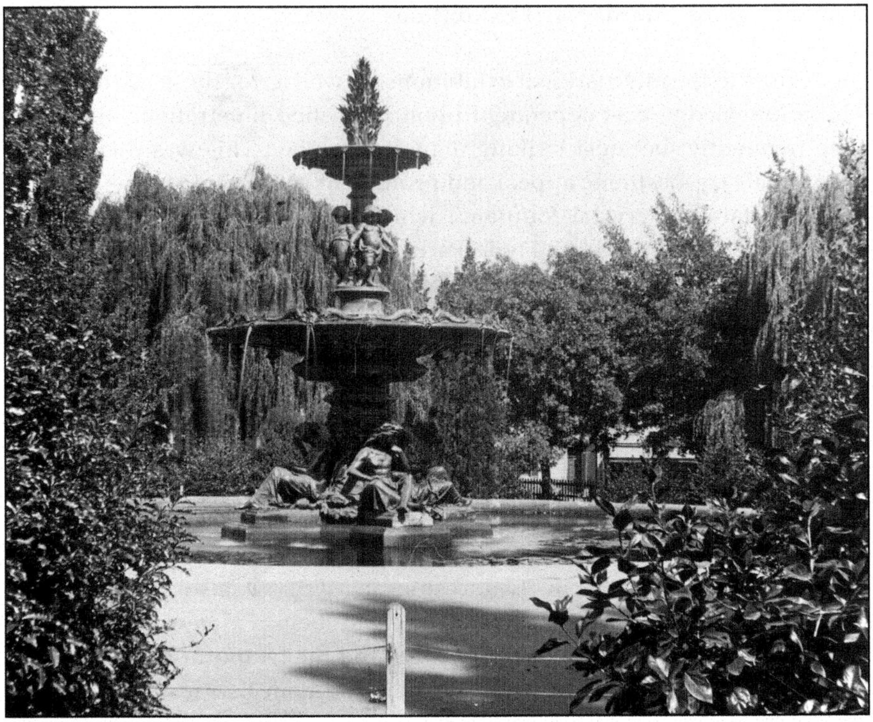

11.2 Fountain, Prince's Square, Launceston, Tasmania, erected in 1859. Anon., Roger Holdsworth (compiler), n.d. (1890s?). State Library of Victoria, H2012.90/59

The prototype of this cast iron fountain was manufactured by the Val d'Osne Foundry near Paris and displayed at the Paris Exposition Universelle *of 1855. The Launceston fountain, ordered in 1857, incorporated finial of rushes and reeds in place of the fish-mouth spouted urn of the original design. The fountain, believed to be the oldest surviving public fountain in Australia, remains on its original site.*

November 1859, in recognition of the simultaneous coming of age of the heir–apparent to the British throne.[31]

The fountain selected for Prince's Square in 1857 was manufactured by Barbezat & Co., of the Val d'Osne Foundries near Paris. Designed by Michel Lienard and incorporating substantial statuary by the sculptor Mathurin Moreau, it was constructed of cast iron with a special bronze-effect finish. The prototype of the design had been exhibited at the Paris *Exposition Universelle* of 1855, where it was given pride of place, occupying the central position in the main hall.[32]

In its original exhibited form, the fountain consisted of a cruciform base of four classical semi-nude figures, grouped in pairs. The Val d'Osne catalogue identifies them as couples from classical mythology: *Neptune & Amphitrite* and *Acis & Galatea,* all of whom were connected with water.[33] Above the figures and supported by a substantial pedestal, was a large richly decorated octagonal basin with an ornamented rim and pendants at each angle. At the centre of the large basin was a plain round column, surrounded by four standing half-clothed children, holding items such as ropes, a fishing net and the rudder of a small boat, again with suitably watery associations. Above the children was a smaller circular basin and above this was a finial in the form of an urn with spouts placed at right angles. In this standard form, with only minor variations in decoration, at least 19 examples are known world-wide (see Table 11.1 opposite).

Table 11.1 Fountains based on that first exhibited at the *Exposition Universelle*, Paris, 1855, by Barbezat & Co., Val d'Osne Foundries, near Paris

Country	City	Location/name	Date and state	Finial	Notes
AUSTRALIA					
Australia	Launceston	Prince's Square [St John's Square]	1859 Extant	Reeds & rushes	Ordered 1857
EUROPE					
France	Paris	Exhibited Paris Exposition 1855	1854–55 Untraced	Spouted urn	The 'original model' 554T as per catalogue
France	Angers	Jardin du Mail	1855 Extant	Spouted urn	NOT the actual example from the 1855 Expo. Gardens c. 1859
France	Bordeaux	Allées de Tourny. Fontaines de Tourny [2]	1857 Removed c. 1960s	Spouted urn	Originally two. Re-erected at Soulac-sur-Mer [qv] & Quebec [qv]
France	Soulac-sur-Mer	Place Georges-Clemenceau [ex Bordeaux]	1960s? Extant	Spouted urn	One of the 1857 fountains from Bordeaux, re-erected
France	Lyon	Place Guichard	By 1859 ?	Spouted urn	Postcard, c. 1900 Destroyed?
France	Saint Quentin	Place Dufour-Denelle	? ?	Spouted urn	Postcard, c. 1900 Destroyed?
France	Hyères	Ave Alexis Godillot, Fontaine Godillot	3rd qtr. of 19th cent. Extant	Spouted urn + lamp above	
France	Troyes	Fontaine Argence	1897 Extant	Spouted urn	
Switzerland	Geneva	English Garden, Four Seasons Ftn	1862 Extant	Spouted urn	Garden created 1854
Spain	Valencia	Paseo de la Alameda, Four Seasons Ftn	1863 Extant	Figure	

Table 11.1 *continued*

Country	City	Location/name	Date and state	Finial	Notes
Portugal	Lisbon	Praça Dom Pedro IV [a.k.a. 'Rossio'] [2]	c. 1874? Extant	Spouted urn	Two [N&S], with outlying fountain elements, flanking a column erected c. 1874
SOUTH AMERICA					
Argentina	Buenos Aires	Avenida 9 de Julio y Avenida Cordoba [2]	c. 1868 Extant Current locations from 1960s?	Spouted urn	Originally located in Plaza de Victoria, in c. 1868, E & W of the 'Pyramide'. Relocated at various times from 1880s
Peru	Tacna	Plaza de Armas	1868? Extant Relocated 1885?	Spouted urn	
Chile	Valparaiso	Plaza de la Victoria, Four Seasons Ftn	1877? Extant	Spouted urn	
Argentina	Córdoba	Plaza Cólon	1890 Extant	Spouted urn	Exhibited Paris Expo., 1889
Argentina	Mendoza	?	Late 1800s Removed 1920s	Spouted urn + basin above	Relocated to Maipu 1927
Argentina	Maipú, near Mendoza	Plaza de Maipú Fuente de Maipú [ex Mendoza]	1927 Extant	Spouted urn + basin above	Relocated from Mendoza
NORTH AMERICA					
US	Boston	Boston Common Brewer Fountain	1868 Extant	Spouted urn	Purchased by Brewer, Paris Expo., 1867; originally located in front of his Boston house
Canada	Quebec	Fontaine de Tourny [ex Bordeaux]	2007 Extant	Spouted urn	One of the 1857 fountains from Bordeaux, restored & re-erected

Table 11.1 was compiled by reference to the website Fontes d'art en France et dans le monde: un réseau international *(http://www.fontesdart.org), supplemented by extensive online visual and word searches. The dating of the fountains is usually derived from secondary sources, and may not be reliable.*

Secondary sources mention similar fountains in Dijon and in Egypt, but no visual or descriptive records of these have been found. An example at Salvador de Bahia, Brazil, is thought by this author to differ too markedly from the other examples to be considered truly comparable. The English-manufactured Steble Fountain in Liverpool appears to be a somewhat clumsy assemblage of castings copied from various Val d'Osne designs.

The Prince's Square fountain in Launceston differs from the original Paris exhibition and catalogue example in only one obvious respect: it has a markedly different finial. In December 1857, apparently referring to some drawings supplied by the manufacturer, the town councillors agreed to 'send to Paris for a fountain as per drawing produced, marked T but with the jet of the drawing marked A. B.'[34]

A persistent local myth holds that this unusual finial, taking the form of a dense cluster of rushes, was an afterthought, made locally to replace a figurative finial supplied by the French manufacturer, which was considered to be an affront to public decency. However, no nineteenth-century records of this have been found and recent research has revealed that the Val d'Osne Foundries produced fountain spouts in an identical form to that found in Launceston.[35] Thus there can be little doubt that the finial seen on the fountain today was part of the original order.

Nevertheless, the question arises as to why those placing the order for the Launceston fountain felt it necessary to specify an alternative finial design. A close inspection of the standard design reveals the probable cause. What appear at a distance to be simple spouts issuing from the urn-like finial actually take the form of fish mouths, rather like medieval gargoyles. In a fountain commissioned to celebrate the establishment of a reliable supply of pure drinking water, this imagery could have undesirable associations: would anyone want to be reminded that the water they were drinking had previously passed through, or been inhabited by, fish? How much more wholesome and desirable were the associations of a dense cluster of reeds, not only as vegetable rather than animal matter, but also as a filter for impurities!

It is difficult to establish who was responsible for drawing the councillors' attention to the French design. It may have been the consulting engineer to the new waterworks, W. R. Falconer (1818–69), who took a very close interest in the Paris Exposition of 1855 and could have visited it while he was in Britain, *en route* from Montreal to Tasmania.[36] Another possibility is Henry Reed (1806–80), formerly of Launceston and principal in the London firm of Reed & Hawley, Launceston's European agents in connection with the waterworks project.[37] Reed took a personal interest in manufactured fountains, subsequently commissioning the original 'Dunorlan' fountain, exhibited by the Pulham Company at the London Exhibition of 1862, for his English estate at Tunbridge Wells.[38] However, the *Art Journal Illustrated Catalogue* to the Paris Exposition of 1855 was widely distributed and direct knowledge of the fountain from the illustration in this alone may have prompted the councillors to approach the French firm.

Research conducted by the author in 2009 (summarized in Table 11.1) on the occasion of the Launceston fountain's 150th anniversary indicated that this design was popular around the world. The continuing influence of the international expositions in the wide dissemination of the design is indicated by the dates of the fountains erected outside Europe. In some cases, such as the Brewer Fountain in Boston in the US and the Plaza Colon fountain in Cordova, Argentina, the fountain's purchase has been directly connected to the Paris expositions of 1867 (Boston) and 1889 (Cordova).

The fountain in Prince's Square, Launceston, is believed to be of international significance. Of the 19 known models built to the Barbezat & Co. design 554T, first exhibited at the Paris Exposition of 1855, it is one of only four extant examples dating from the 1850s and the second oldest example in the world remaining on its original site. It was the first example of the internationally popular French design to be erected in an English-speaking country and is unique as the only example to incorporate the distinctive bull-rush finial. The Prince's Square fountain is also believed to be the oldest extant public fountain in Australia, as well as one of the finest.[39]

The City Park Fountain, Launceston

In December 1857, the same month the Launceston municipal councillors ordered the fountain from France for Prince's Square, Mayor Henry Dowling ordered another substantial fountain, this time from Scotland, with the intention that it would be presented by the mayor and alderman to the Launceston Horticultural Gardens, now known as City Park.[40]

Dowling wrote to Messrs Ferguson, Miller & Co. of the Heathfield Works, near Glasgow, to obtain a copy of the terracotta fountain that had been exhibited at the Dublin Exhibition of 1853. The reply, which came from the Garnkirk Coal Co.'s 'Terracotta Works, Glasgow', said they would be happy to comply, but that it would take some time, as the original moulds had been destroyed in a fire and would need to be remade (Fig. 11.3).[41] Their initial quotation of £200 nearly lost them the order, but when this was halved the decision was made to go ahead.[42] It was two years before the large fountain arrived in Launceston,[43] and another two years before it was finally erected in the Horticultural Gardens, when it was described in detail:

> The fountain is very chaste in design and is a great addition to the attractions of the gardens. A jet of water rises in the air from the flower of an Ethiopian lily and flowing down its expanded leaves, falls into the upper bason [*sic*] which is supported by three cranes. Below the pedestal on which these birds stand is the middle cup borne up by

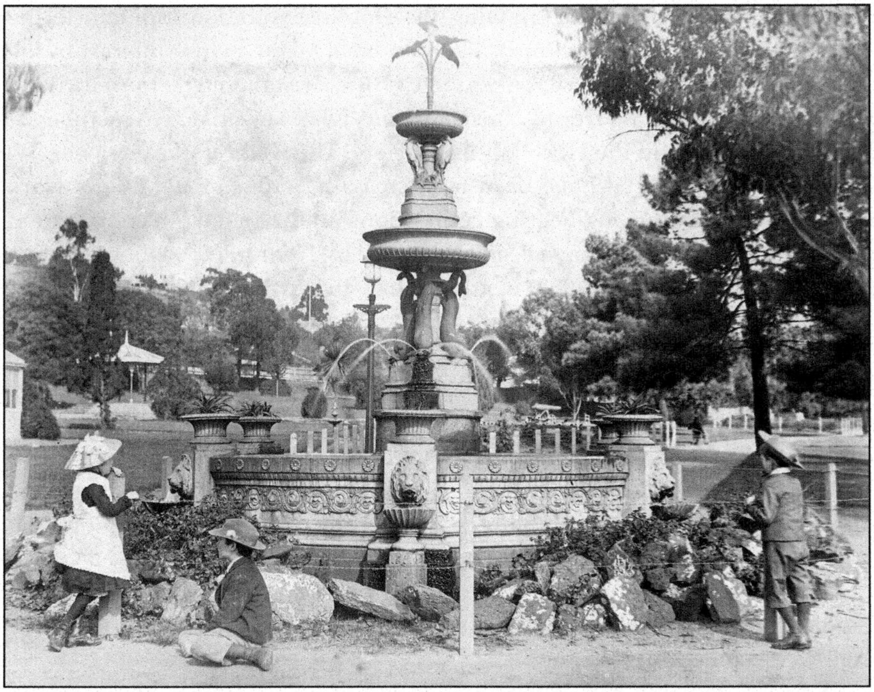

11.3 Fountain, City Park, Launceston, Tasmania, erected in 1861. Anon., H. Grattan-Guinness (compiler), n.d. (1880s?). National Library of Australia, PIC/8510/12 LOC Album1040

The prototype of this fountain, manufactured by Ferguson, Miller & Co., near Glasgow, was displayed at the Dublin Exhibition of 1853. The Launceston example was ordered in 1857 and received in 1859. In its original location, it formed the centre-piece of the 'Avenue of Nations' in the Tasmanian International Exhibition of 1891–92. It is believed to be the second-oldest surviving public fountain in Australia.

three dolphins, out of whose heads spout forth eight [*sic*] silvery streams which fall into the lower bason [*sic*] and the water thus collected gushes out of the mouths of six lions into shells, placed immediately under the same number of elegantly shaped vases. … The outer portion of the basons [*sic*] and the pedestal are embellished with design in bas-relief.[44]

The fountain had been illustrated in the *Art Journal Illustrated Catalogue* to the Dublin exhibition and even more lavishly in Volume II of the *Illustrated Magazine of Art*, both published in 1853. The *Art Journal* described it as 'remarkably good in character; its proportions are just and graceful; the manipulation is excellent; and, taken altogether, it is one of the best efforts of British Art in the collection'.[45] The *Journal* went on to state that the parts of the fountain could 'be constructed separately, each part forming a distinct and practically useful fountain'.[46]

Comparison with descriptions and illustrations of the 1853 Dublin fountain show that the Launceston example differed in only one respect: it omitted the largest of the monolithic terracotta basins, together with its supporting atlantes. This may have been done to reduce the cost, or it may have been because that particular component had proved extremely difficult to reproduce successfully. In the original Dublin example, the size of the largest monolithic basin (1800mm in diameter) tested the limits of terracotta technology. In consequence of repeated problems encountered in manufacturing this basin, the fountain arrived embarrassingly late at the exhibition.[47]

Who was responsible for drawing the Dublin Exhibition fountain to the mayor's attention? The man most likely to have taken a close interest in the Dublin Exhibition was John Crookes (1805–70) an Irishman who ran a thriving import-export business from Launceston and who was an active member of both the Council and the Horticultural Society. The waterworks engineer, W. R. Falconer, could well have been familiar with Ferguson & Miller's work through his Glasgow engineering connections and may have recommended them. Dowling may also have been influenced by Henry Reed, whose famous 'Dunorlan Fountain' of 1862 would utilize the artificial stone known as 'Pulhamite' for its large basins, but terracotta for all its finer details.[48]

Fountains based upon Ferguson & Miller's Dublin exhibition design are extremely rare internationally: in fact the Launceston example is one of only two examples known to have survived more or less intact. Several factors could account for this. The design was effectively 'out of production' for a period in the 1850s, after the moulds were destroyed in a fire between 1853 and 1857 (see above). Susceptibility to cold weather could also account for the fountain's rarity. Unglazed terracotta is not impervious to water and so when placed outdoors in a frosty climate, it is subject to spalling and potentially rapid deterioration. The manufacturers recommended that the fountains be kept well-painted for this reason.

Extensive research has so far revealed only four fountains comparable to that exhibited in Dublin in 1853 and erected in Launceston in 1861. Of these, two date from the 1870s in Scotland and two from the 1880s in Ireland (Table 11.2 opposite).

In his detailed 1878 account of *The Ceramic Art of Great Britain*, Llewellyn Jewitt not only clarified the relationship between the Ferguson & Miller's Heathfield works and the Garnkirk Fire-Clay company (itself an offshoot of the Garnkirk Coal Co.), but also reviewed the recent output of the Garnkirk factory including 'fountains, notably an example of five tiers, supported by figures of dolphins and cranes, with basin twenty four-feet in height and sixteen feet across, erected in the public park at Aberdeen'.[49]

Jewitt was evidently referring to the Victoria Park in Aberdeen, founded in 1870. The Aberdeen Council ordered the fountain from the Garnkirk company in 1872 at a cost of £105, and it was erected at the intersection of the park's two main pathways in 1873. In 1880, after it had fallen into disrepair, it was replaced by a fountain of Aberdeen granite, donated by the Aberdeen Stone Masons' Association.[50] No trace has been found of Aberdeen's terracotta fountain after 1880.

A fountain of broadly similar appearance was formerly located on the Isle of Skye. It is thought unlikely, though not impossible, that this fountain had been relocated from Aberdeen. Described as a Garnkirk fountain, c. 1870, from Waternish House (formerly Fasach House), it realized the astonishing sum of £106,000 when offered for sale by Sotheby's South (Sussex, UK) in 1999.[51] In terms of size, composition and intactness this Scottish example falls far short of the Launceston fountain, but the price paid indicates the remarkable value

Table 11.2 Fountains relating to that first exhibited at the International Exhibition, Dublin, 1853, by Ferguson & Miller (later Garnkirk), Glasgow

Example, original location	Date	Relation to original	Current status	Notes
Industrial Exhibition, Dublin, Ireland	1853	Original model	Unknown	Offered for sale by auction in Dublin, November 1853; untraced
Horticultural Gardens, Launceston, Tasmania	1861	Lacked atlantes Now lacks lily finial	Extant in City Park, Launceston	Made in Glasgow from new moulds, 1858–59
Fasach House, Isle of Skye, Scotland	c. 1870	Lacks atlantes, dolphins, lily finial Different urns on outer basin	Believed extant at unknown location	Sold at auction by Sotheby's UK in 1999
Victoria Park, Aberdeen, Scotland	1873	Lacked atlantes	Unknown	Replaced by granite fountain 1880
Baggotrath House, Sandymount, Dublin, Ireland	By 1880	Lacks fine decorative mouldings on basin rims, central pedestal stages, lily finial	Extant at Bewley's Hotel, Ballsbridge, Dublin	Bequeathed by Richard Sawyer to Masonic Female Orphan School, Ballsbridge, 1882 (now Bewley's Hotel)
People's Park, Waterford, Ireland	1883	As original	Destroyed. One atlante survives in council storage	Demolished c. 1990 after major vandalism c. 1982

and interest attached to the work by collectors.[52] It thereby indicates the even greater value that should attach to the earlier, larger and significantly more intact example in Tasmania, which to date has lacked recognition.

Two terracotta fountains, built according to the Dublin Exhibition model, have been identified in Ireland, one of which is still extant. Standing in the grounds of what is now Bewley's Hotel, Ballsbridge, Dublin, is a fountain erected in 1882, when the buildings functioned as the Masonic Female Orphan School.[53] The fountain had been bequeathed to the institution by Richard Sawyer of the nearby Baggotrath House, Sandymount, Dublin. Sawyer died in 1881: the codicil bequeathing the fountain dates from 1880 and is the earliest identified documentation known for it.[54] Given its location, the question naturally arises: was this the very fountain displayed at the Dublin Exhibition of 1853? The fountain was advertised for sale at the end of the exhibition, but no record has been found of any similar fountain in Ireland between then and 1880.[55] The Baggotrath example could conceivably be the exhibition original, but certain variations in detail raise doubts that can only be resolved by further research.[56]

The second Irish example of the Dublin Exhibition fountain was erected in the People's Park, Waterford, in 1883. As a popular subject for photographs and postcards, its appearance is well recorded, showing it to be an accurate and full reproduction of the original 1853 design.[57] Sadly, after almost a century of varying fortunes it was subject to serious vandalism c. 1982 and demolished c. 1990. All that remains is one of the atlantes, now kept in storage.[58]

The terracotta fountain in City Park, Launceston, is believed to be of international significance, as one of only two known surviving copies of the elaborate fountain first exhibited by Ferguson and Miller of Glasgow at the Dublin Exhibition of 1853 and probably the earlier of the two. In addition, it is thought to be the second-oldest extant public fountain in Australia, after that in Prince's Square.[59]

Conclusion

The international exhibition movement of the nineteenth and early twentieth centuries served international commerce in a similar way to the Internet of today. An extraordinary range of materials and manufactures from around the world could be viewed together in one place. There they were assessed against their competition and rated accordingly. Awards and commendations by independent assessors provided a useful buying guide for consumers and were avidly seized upon by advertisers hoping to boost sales. Through widely distributed illustrated exhibition catalogues and journal articles, manufacturers and consumers in remote parts of the world were able to keep abreast of the latest international developments. Geography was no longer a significant barrier to material consumption: anyone anywhere was able to purchase and import the world's newest and best products, provided they had the financial means.

Thus it was that Launceston in Tasmania acquired an elaborate French cast-iron fountain, first displayed at the centre of the main hall of the Paris *Exposition Universelle* of 1855. Ordered in 1857 and erected in 1859, the Prince's Square fountain is believed to be the second oldest example of its type, still on its original site, in the world and the oldest surviving public fountain in Australia. Also in 1857, Launceston councillors ordered a copy of an impressive Scottish terracotta fountain, first displayed on the central axis of the main building at the Dublin Exhibition of 1853. Received in 1859 and erected in 1861, the City Park fountain is believed to be one of only two substantially intact examples of its type in the world and the second oldest extant public fountain in Australia. Finally, despite its small population and relatively remote location, Launceston hosted its own 'Tasmanian International Exhibition' in 1891–92, which was regarded at the time as remarkably successful, attracting approximately a quarter of a million visitors and finishing in profit. The Albert Hall, Launceston (1889–91), built especially for this event, is a rare surviving nineteenth-century exhibition hall on the international stage and in the Australian context is only surpassed in significance by the world-heritage listed Royal Exhibition Building in

Melbourne of 1880. These three Launceston structures together provide a remarkable testimony to the extraordinary influence of the international exhibition movement in the nineteenth century, showing it to be as strong 'at the margins' as it was in the centre of a fundamentally Eurocentric world.

Notes

1 For an overview see Louise Douglas, 'Representing Colonial Australia at British, American and European International Exhibitions', *reCollections* 3, no. 1 (March 2008): 13–32.

2 For an overview see Linda Young, '"How Like England We Can Be": The Australian International Exhibitions in the Nineteenth Century', in *Seize the Day: Exhibitions, Australia and the World*, ed. Kate Darian-Smith et al. (Melbourne, 2008), 12.1–12.19.

3 *Launceston Examiner*, 20 October 1857, 2.

4 *Launceston Examiner*, 15 May 1883, 3.

5 *Official Record of the Tasmanian International Exhibition Held at Launceston 1891–2* (Launceston 1893), 15B.

6 *Official Record*, 15C.

7 Louise Douglas, 'Joules Joubert: Australia's Nineteenth-Century Exhibition Impresario', *National Library Magazine* (December 2012): 27–30. For the Dunedin exhibitions, see Rebecca Rice's text 'A Red-Letter Day …' in this volume.

8 Young, 1.

9 *Official Record*, viii.

10 *Official Record*, 17, 27 and 118.

11 *Launceston Examiner*, 26 November 1891, 'Exhibition Supplement'.

12 The Avenue of Nations was first introduced at the Paris Exposition of 1867 and first used in Australia in Melbourne in 1888. See Young, 9.

13 *Official Record*, 17.

14 The intense rivalry between the cities accounts for Hobart's decision to host its own 'Tasmanian International Exhibition' in 1894–95, despite the woeful economic situation. See Young, 11.

15 *Launceston Examiner*, 25 September 1891, 4.

16 The attendance at the exhibition was variously recorded as 262,059 and 243,000: *Official Record*, 17, 117.

17 *Official Record*, 117. Present author's emphasis.

18 *Official Record*, 137. Admittedly, this only amounted to £180.

19 Mary Knaggs et al., 'Launceston City Park Conservation Management Plan' (Launceston 2006), 1: 68.

20 *Launceston Examiner*, 23 January 1873, 1S.

21 *Official Record*, 49.

22 *Mercury* (Hobart), 23 August 1910, 3.

23 *Official Record*, 15B.

24 *Launceston Examiner*, 28 May 1889, 3; *Launceston Examiner*, 3 June 1889, 2.

25 *Launceston Examiner*, 15 November 1890, 7.

26 *Official Record*, 64. McKinnon evidently never forgot his treatment by the Launceston council: thirty years later, after a full career extending over four decades, his winning design for the hall at Launceston was the first thing mentioned in his obituary (*Sydney Morning Herald*, 5 July 1922, 9).

27 For more on 'Free Classicism' in Australia see Richard Apperly, Rorbert Irving and Peter Reynolds, *A Pictorial Guide to Identifying Australian Architecture* (Sydney, 1994), 104–107.

28 *Mercury* (Hobart), 26 November 1891, 3.

29 'World Heritage Site: Royal Exhibition Building', at http://museumvictoria.com.au/reb/history/heritage-listing. During the nomination process, the author was approached by the Victorian nominators and asked if it was true that the Albert Hall was constructed for an international exhibition and was substantially intact. They were somewhat dismayed to have this confirmed and seem to have done their best to overlook it!

30 *Launceston Examiner*, 22 August 1857, 2; *Launceston Examiner*, 3 November 1857, 2.

31 *Launceston Examiner*, 3 November 1860, 3.

32 *Art Journal Illustrated Catalogue: The Exhibition of Art-Industry in Paris, 1855* (London, 1855), 8. The Lienard design of c. 1854 was almost certainly inspired by the pair of fountains erected in the Place de la Concorde, Paris in 1836–40, to the designs of Jakob Ignaz Hittorf. Flanking an Egyptian obelisk, these three-tier fountains symbolized 'Rivers' and 'Seas' respectively.

33 *Catalogue de la fonderie*, vol. 2, pl. 532.

34 'Launceston Municipal Council Minute Book 1853–1862', 280 (Archives Office of Tasmania), cited by Eric Ratcliff, 'A Permanent Advantage: The French Fountain in Princes Square, Launceston', *Tasmanian Historical Research Association Papers* 44, no. 2 (1997).

35 *Catalogue de la fonderie*, vol. 2, pl. 530.

36 Falconer possessed a copy of the *Exposition Universelle de 1855: Rapports du jury mixte international* (Paris: Imprimerie impériale, 1856) which he presented to the Royal Society of Tasmania: *Courier* (Hobart), 22 September 1858, 3.

37 *Launceston Examiner*, 15 April 1856, 2.

38 'The Pulham Legacy', at http://pulham.org.uk/cuttings/dunorlan-park/pulhamite-and-terracotta-fountain-at-dunorlan-park.

39 While there are records of two other ornamental fountains being erected in 1859, one in central Melbourne and one in the Botanic Gardens at Adelaide, these are no longer extant. The oldest identified extant examples outside Tasmania are all in Victoria: the 'River God' fountain (1862) and the Grey Street fountain (1863), both in the Fitzroy Gardens, East Melbourne; the Burke and Wills fountain in Ballarat (1867); and the Stanford fountain in East Melbourne (1867–1870).

40 Until 1863, the Horticultural Gardens were technically a privately funded institution, although they effectively functioned as a public park. Leading figures in the Launceston Horticultural Society were also leading figures in local politics and the boundaries between institutions were apt to become blurred.

41 Letter to Dowling from the Garnkirk Coal Co., 11 February 1858, Launceston City Council Archives, Queen Victoria Museum and Art Gallery, Launceston (hereafter LCC Archives, QVMAG).

42 Letter to Dowling from Reed & Hawley, 11 March 1858, LCC Archives, QVMAG. The price reduction was possibly achieved by omitting the largest tier of the fountain (see below).

43 *Launceston Examiner*, 22 November 1859, 2.

44 *Launceston Examiner*, 23 November 1861, 5.

45 *Art Journal Illustrated Catalogue: The Exhibition of Art-Industry in Dublin, 1853* (London, 1853), 55.

46 *Art Journal ... Dublin*, 55.

47 *Freeman's Journal and Daily Commercial Advertiser* (Dublin), 19 May 1853, 3.

48 'Pulham Legacy'.

49 Llewellyn Jewitt, *The Ceramic Art of Great Britain* (London, 1878), 2: 508.

50 *Aberdeen Journal*, 8 May 1872; 3 July 1872; 5 November 1873. *Aberdeen Weekly Journal*, 22 July 1879; 3 August 1880.

51 Illustrated at http://www.scottishpotterysociety.co.uk/imageuploads/garnkirk fountain.

52 The Skye example lacked not only the atlantes but also the distinctive 'dolphin' tier, which was replaced by a rather awkward-looking disjunctive pier, decorated with lion heads, while the perimeter urns of the original design were supplanted by different, disproportionately large models. It is possible that these discordant elements were later ill-informed attempts at 'restoration', rather than original variations. Another example was offered for sale by Sotheby's in 2004 (Lot 206: Garnkirk stoneware fountain, c. 1870; Sotheby's UK, 2004; illustrated at http://www.artfact.com/auction-lot/a-garnkirk-stoneware-fountain,-circa-1870). Sotheby's noted that it had been subject to restoration and some elements replaced. This example lacked the atlante and dolphin tiers, as well as the pedestals, urns, lion heads and shells of the original pool surround. According to Sotheby's, a similar fountain was illustrated in the 'Garnkirk Fireclay Co. specimen sheets no. 3a', but the author has not been able to access this document.

53 'History of Bewley's Hotel, Ballsbridge, Dublin', at www.bewleyshotels.com/ballsbridge/history.

54 *Freeman's Journal*, 15 March 1882.

55 *Freeman's Journal*, 13 November 1853.

56 From available illustrations the Baggotrath example, in comparison with the original and Launceston models, appears to lack the fine decorative mouldings on the basin rims, some central pedestal stages and the lily finial (always the most vulnerable element of the design).

57 For one example among many, see Robert French, 'People's Park, Waterford City' [L ROY 03646] in the Lawrence Collection, National Library of Ireland, at http://catalogue.nli.ie/Record/vtls000322783.

58 A modern fountain, known as the 'Big Ball', was erected on the site c. 2003. For information regarding the history of the Waterford fountain, I am indebted to Daniel Moore, Waterford City Archivist, pers. comm., April 2013.

59 See note 37 in relation to the Prince's Square fountain.

Bibliography

Printed Primary Sources

Art Journal Illustrated Catalogue: The Exhibition of Art-Industry in Dublin, 1853. London: George Virtue, 1853.

Art Journal Illustrated Catalogue: The Exhibition of Art-Industry in Paris, 1855. London: George Virtue, 1855.

Art Journal Illustrated Catalogue: The Industry of All Nations, 1851. London: George Virtue, 1851.

Illustrated Magazine of Art, vol. II. London: Cassell, 1853.

Jewitt, Llewellyn. *The Ceramic Art of Great Britain*. 2 vols. London: George Virtue, 1878.

Official Catalogue of the Great Industrial Exhibition, 1853. 3rd edn. Dublin: John Falconer, 1853.

Official Descriptive and Illustrated Catalogue of the Great Exhibition of the Works of Industry of All Nations, 1851. 3 vols. London: Spicer, 1851.

Official Record of the Tasmanian International Exhibition Held at Launceston 1891–2. Launceston: Examiner, 1893.

Secondary Sources

Apperly, Richard, Robert Irving and Peter Reynolds. *A Pictorial Guide to Identifying Australian Architecture*. Sydney: Angus & Robertson, 1994.

Australian newspapers accessed through the National Library of Australia's search engine, *Trove* at http://trove.nla.gov.au.

Catalogue de la fonderie du Val d'Osne, vol. II, c. 1900, having been compiled cumulatively from c. 1850 onwards. Accessed at www.fontesdart.org.

Douglas, Louise. 'Representing Colonial Australia at British, American and European International Exhibitions'. *reCollections* 3, no. 1 (March 2008): 13–32.

Douglas, Louise. 'Joules Joubert: Australia's Nineteenth-Century Exhibition Impresario'. *The National Library Magazine* (December 2012): 27–30.

'Réseau International de la Fontes d'Art'. Accessed at www.fontesdart.org.

Knaggs, Mary, et al. 'Launceston City Park Conservation Management Plan'. 2 vols. Launceston: unpublished report, 2006.

Neale, Anne. 'Prince's Square Fountain, Launceston, Tasmania: Assessment of Cultural Significance'. Launceston: unpublished research paper, 2009.

Ratcliff, Eric. 'A Permanent Advantage: The French Fountain in Princes Square, Launceston'. *Tasmanian Historical Research Association Papers* 44, no. 2 (1997): 74–90.

Young, Linda. '"How Like England We Can Be": The Australian International Exhibitions in the Nineteenth Century'. In *Seize the Day: Exhibitions, Australia and the World*, edited by Kate Darian-Smith, 12.1–12.19. Melbourne: Monash University ePress, 2008.

12

'Urbi et Orbi': Decentralization and Design in Nancy's International Exposition of Eastern France, 1909

Claire O'Mahony

The Parc Sainte Marie, a green space specially created in Nancy to house the International Exposition of Eastern France in 1909, afforded the strolling visitor glimpses of a palimpsest of this contested margin between France and Germany: industry and craft; modernity and historicism; international and regional constituencies (Fig. 12.1). Pierre Roger Claudin's official poster for the event captures these rival discourses within the spectacle. A distinctive local form of 'art nouveau' typography declares 'Nancy 1909', intimating the French tricolour in its dark blue and red outlined white lettering, whilst the golden highlights suggest the famous gilt-edged rococo gates designed by Jean Lamour for Nancy's place Stanislas at the heart of the poster's composition. The exposition title – International, East and France – signalled the mediation of the national and the international by the regional. The vibrant colour separations, recently pioneered by poster makers such as Jules Chéret and Toulouse-Lautrec, define the dynamic force of dawn sun rays. A young woman in traditional Lorrain costume in the foreground echoes the solar orb and the axial effects of the sunrays in the encircling gesture of her arms as she releases pigeons, announcing the festivities to 'urbi et orbi'.[1] The black edge of the figure's coiffe and the remembrance of carrier pigeons' duty in the dark days of the Franco-Prussian War vouchsafe the poster and the Exposition from any charge of ignoring the troubled recent history of the annexation. The architectural elements which frame the scene along the right and bottom edges of the composition define silhouettes of landmarks in old and new Nancy. The elegant features of the ornamental urns of the rococo Hôtel de Ville on the central place Stanislas and the medieval fortified gate of la Craffe in the 'ville vielle' are juxtaposed with the graceful dome of the 1909 exposition's Palais de Fêtes amidst elegant unobtrusively smoking chimneys in this more

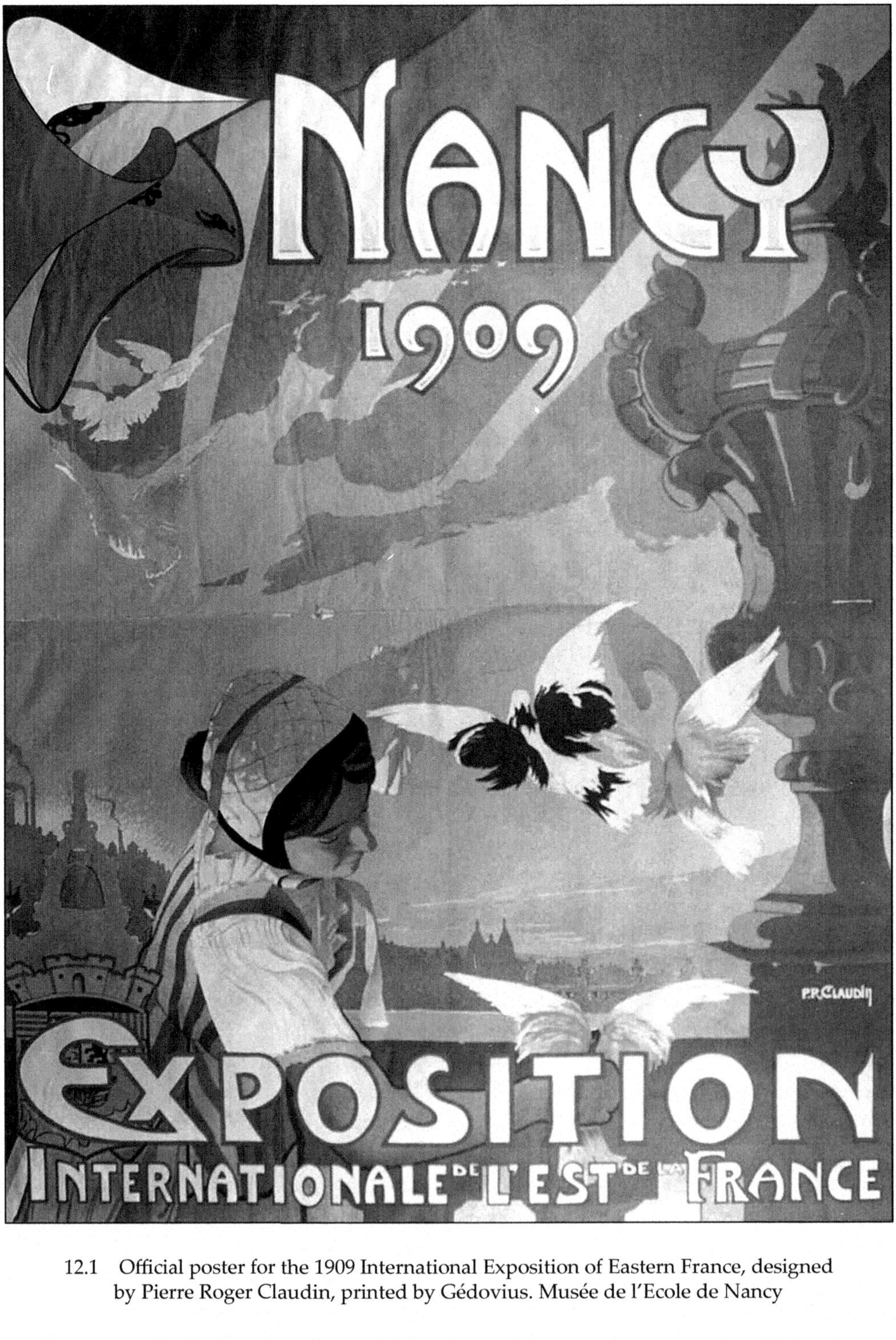

12.1 Official poster for the 1909 International Exposition of Eastern France, designed by Pierre Roger Claudin, printed by Gédovius. Musée de l'Ecole de Nancy

industrial suburb of Nancy. Claudin's poster was part of a wide publicity campaign which attracted over 50 million visitors to this display of the new dawn of Lorraine as a region uniquely poised between heroic, tragic histories and hopeful futures, between France and the world. The 1909 exposition would prove to be an important model for decentralization. Its strategy of integrating art, commerce and industry to express regional vigour and coherence was resonant in Paul Vidal de la Blache's famous article on 'Régions françaises', a cornerstone of the emergent disciplines of cultural and economic geography.[2] The 1909 exposition funded and conceived by Nancy's Chamber of Commerce proposed a conduit for regional cultural and industrial production to international audiences and markets which prefigures the model of economic regions embraced in the 1920s.

Scholarship examining international expositions has concentrated particularly upon events held in capital cities such as London or Paris or in urban centres selected for important anniversaries such as Chicago for the 1893 Columbian exposition. The commissioning process and the architecture of the pavilions is often attended to more than the contents of the displays housed within them.[3] Displays of art and design have figured most prominently in analyses of both the five great *expositions universelles* held in Paris in each decade from 1855, as well as the specifically decorative and industrial arts expositions of 1925 and 1937.[4] However, local international expositions, such as Bordeaux's seven international expositions held between 1850 and 1907 or the hundred or more analogous undertakings in France's regional cities documented in a 1912 Archives Nationales survey await close analysis.[5] Each of these provincial initiatives offers fascinating insights into how 'la petite patrie' of the regions wrestled with their position within the nation, its empire and the wider world. The case study of the 1909 International Exposition of Eastern France speaks to these tensions with particular acuity. Its host city of Nancy held the iconic status of the last outpost of France on the borders of the new 'Alsace-Lorraine' region, annexed to the victorious new nation of Germany after the Treaty of Frankfurt signed on 10 May 1871. The present argument outlines the political and economic climate which underpinned the organization of the exposition before discussing how the multiple displays of decorative objects in 1909 constructed Lorraine not as a troubled borderland, but rather as a centre for international economic and cultural exchange. The positioning of the Ecole de Nancy Industrial Alliance of designers within the topography of the exposition declared this collective's importance as the lynch pin of a new reconciliation of traditional craft and industrial methods. Rather than the paradigmatic solipsistic retreat and 'revanche' usually ascribed Alsace-Lorraine and Belle Epoque decorative arts, the displays in this marginal borderland exposition envisioned a marriage of art and industry in which regionalism offered a vehicle for new dialogues between nations, producers and consumers.

Annexation, Design and Industry: Displaying the Economic and Aesthetic Miracle of Lorraine in 1909

Nancy was in many ways an unlikely site upon which to forge the delicate balance of national pride and international relations which lies at the heart of planning an international exposition. This contested borderland has often been described as the epicentre of the fin-de-siècle climate of bellicose 'revanchism' in Third Republic France.[6] Undoubtedly defeat and annexation scarred the French cultural imagination profoundly, prompting national revenge fantasies such as those voiced in prose by Maurice Barrès or paint by Edouard Detaille.[7] The 1877 children's classic *The tour of France by two children* forged a conception of Republicanism rooted in the rupture of the annexation for generations; by 1901 six million copies had been disseminated.[8] However, unlike the lost cities of Phalsbourg, Metz or Strasbourg, Nancy embodied a distinct dynamic role as the 'last city of France'.[9] In many ways the city benefited from the large influx of families who 'opted' for France, from members of elite professions such as university professors and industrial magnates to the skilled workforces in their employ.[10]

In 1906 the Nancy architect Emile Jacquemin captured the complex mixture of local and international focus at the heart of the exposition project: 'It must be an industrial, commercial and artistic event of great appeal, of a noble and prestigious character, like the great expositions in Paris, and not one such as other cities have given us in recent times – banal and fastidious spectacles.'[11] However, this plea for education and excellence in the regions was not prompted by obsequiousness towards Paris:

> our compatriots and fellow citizens must not be supplanted by foreigners. Our architects, our entrepreneurs, our tradesmen must unite; they must agree, they must form an active and powerful consortium so as not to lose even the tiniest benefit of this happy enterprise which is in their home, and therefore theirs. Firstly the exposition belongs to us, and only to us ....[12]

The exposition would dispel the shame of defeat and annexation by demonstrating the industrial might and cultural coherence of a borderland regional constellation of Eastern France. Exhibitors were invited from the 12 eastern French departments: Ardennes; Marne; Haute Marne; Meuse; Aube; Haute-Saône; Doubs; Territoire de Belfort; Côte-d'Or; Jura; Meurthe-et-Moselle (formed in 1870 from fragments of Moselle and Meurthe) and the Vosges. The annexed zone played a central, but sublimated, role – present in the demonstration of successful migrations of vast numbers of Alsatian firms and a peace-offering invitation extended to 'those who remained behind' to share in the collective imaginary of the Alsatian village display.

The 32 members of the Comité de l'Exposition were largely former Alsatian entrepreneurs, architects and municipal officials who had 'opted' for Lorraine. The four protagonists behind the inception of the 1909 Exposition embody a triumvirate of municipal government, local business and creative practitioners which fostered the flowering of Lorrain commerce and craft in the decades before the First World War. Two local entrepreneurs represented the views and primary funding sources for the exposition which were provided by Lorrain businesses. Louis Vilgrain (1856–1929) left his business in Metz to create the largest flour mill concern in Lorraine and served as president of Nancy's Chamber of Commerce from 1903 and 1919. Eugène Corbin (1867–1952), Director of the Magasin Réunis department store chain, key patron of the Ecole de Nancy designers and founding editor of the journal *Art et industrie,* commissioned a special art nouveau pavilion to represent his stores in the exposition. A Dreyfusard and founder member of the Solidarist Université Populaire, Victor Prouvé (1858–1943), represented arts and creative industries of Nancy. His directorship of the Ecole des Beaux-Arts coincided with the school's new building within the grounds of the Parc Sainte Marie housing the one of the four key art and design displays of the exposition. As President of the Ecole de Nancy industrial art alliance, Prouvé helped to designed the group's 1909 pavilion intended as a permanent design museum for the city, a hope sadly unrealized until the 1950s. Ludovic Beauchet (1855–1914), a law professor elected as the progressive Republican mayor of Nancy from 1904 to 1912, oversaw the municipal contribution to realization of the exposition, second only to the Chamber of Commerce. Once Lami from the Parisian Ministry of Public Works, which had granted vital pump-priming funds at the start of the project, had resigned from the Exposition committee, responsibility for the general oversight of the project resided solely with Louis Laffitte (1873–1914), appointed in 1908 as general secretary of the Chamber of Commerce and impresario and peace broker for the whole project. A formal contract granted all the pavilion designs to local Lorrain architects. With the exception of the wooden framework underpinning the elaborate dome of the Palais des Fêtes, a local firm, Chenut, was contracted to construct identical iron substructures for all the exposition's pavilions. The very framework of the Exposition was both ideologically and materially the embodiment of the alliance of art, commerce and industry.

The exposition was conceived in four main zones (Fig. 12.2). The monumental entrance led to an opening section juxtaposing education, region and empire through the new Ecole des Beaux-Arts, the Alsatian and Senegalese villages. A central 'English' garden, planted by local industrial horticulturalists, housed the Ecole de Nancy alliance's pavilion as well as smaller pavilions for the newly established University Forestry Faculty, local firms and the amusements. The densest and grandest constellation of displays was housed in the main Palaces which flanked a 'French' garden. Finally, a brief train ride provided access to the outlying aeronautical displays where the 'City of Nancy' dirigible was moored. The main palaces celebrated every aspect of Lorraine life and commerce, from lavish displays of commodities to a working farm; from the latest scientific discoveries to manufacturing techniques. Lucien Bentz's Palace of Textiles demonstrated the importance of the region's cloth production, both as industrial manufacture and traditional

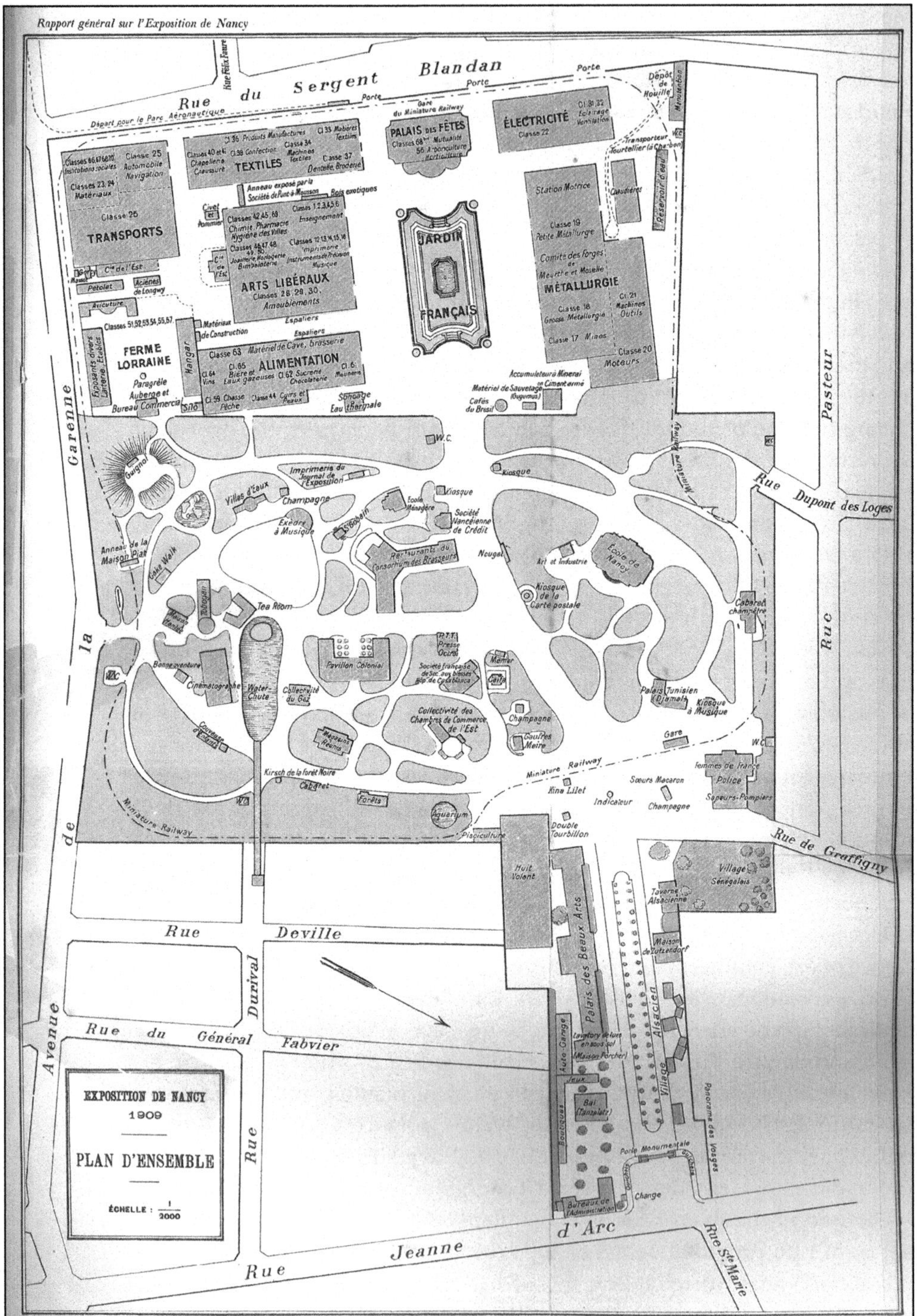

12.2 Plan of the 1909 International Exposition of Eastern France, Nancy. Musée de l'Ecole de Nancy

handwork. Vilgrain's mills, which provided 50 per cent of the nation's bread flour, had centre stage in the Food Palace. Significantly, the 'Liberal Arts' palace juxtaposed displays devoted to salt mining, chemistry and financial services with installations by creative industries. A specially created exposition newspaper, *Revue Générale de l'Exposition de Nancy 1909*, included illustrated articles devoted to the expositors be they long-established or new companies. The Société anonyme de la Cristallerie Baccarat, for example, announced in its pages that its over-2,230 staff produced more than seven million francs worth of crystal ware in 1909. The hat designer Aimé Delchard whose patented designs had been worn by the military and school children since 1640 had an advertisement alongside new Ecole de Nancy design firms such as those directed by the glassmaker Antonin Daum and the furniture designer Louis Majorelle. The Belgian entrepreneur Ernst Solvay, renowned for his enlightened philanthropic support for free health care for mothers and children, funded displays both for his soda firm located in Dombasle-sur-Meurthe and University departments and scientific institutes founded with his support.

Despite this staunch commitment to local firms and talents, the completed site of the 1909 Nancy exposition nonetheless also welcomed international association. The date of 1909 was favoured as it nestled between the 1908 London and 1910 Brussels expositions. The displays and festivals of the Nancy exposition specially celebrated the region's trade and diplomatic links to these two nations, most notably through invitations to international delegations for highlighted week-long visits. Echoes of the décor of the Paris 1889 and 1900 displays are discernible in the 1909 site, but with a new distinctive conjunction of artistry and industry embodying Eastern French aesthetics and identity. The Palais des Fêtes, designed by the architects Georges Biet, Henri Marchal and Emile C. Toussaint, was the only building to utilize a wooden substructure to support an elaborate neo-rococo plaster facing reminiscent of the highly ornamented Palaces of Electricity and Forests, Hunting and Fishing at the Paris 1900 Universal Exposition. The fresco by Louis Guingot which ornamented its entrance nonetheless asserted Nancy's primacy not only through a personification of the city, but also with a panorama of contemporary successful sons of Nancy paying homage to her.

The Exposition opened on 1 May and ran until 31 October 1909, housing over 1,500 expositors in a site of over 20,000 square metres. Open from 10 in the morning to 11 at night, the exposition operated variable entrance fees, punctuated by a panoply of symbolic festivities, celebrating local identity as well as national and international visitors and relations.[13] The inauguration boasted the presence of the Minister of Public works as well as 'tout-Nancy' suggesting a détente with the capital and its desirable tourists. Tram lines were extended from the railway station to the exposition grounds in the Parc Sainte-Marie to facilitate construction and visitor transport; main line rail services to Nancy were also tripled during the five-month run. One of the highlights of the exposition was a model railway to move visitors around the displays commissioned from the English firm Bassett-Lowke, whose director's Northampton home, 78 Derngate, was commissioned from Charles Rennie and Margaret Mackintosh.[14] Indeed the idea of the amusements and the 'English' gardens at the heart of the exposition found their origin in the re-enforcement of the 1904 *Entente cordiale* treaties through an exchange between the entrepreneurial and municipal authorities at the heart of the 1908 London White City and 1909 Nancy Expositions. It would seem Vilgrain and Laffitte had won the affection of the 1908 London committee, who arranged a week of 'Nancy' events in London. During a brief additional visit to Manchester, Laffitte was so impressed by English fairgrounds that he commissioned a set of English rides which were a highlight of the exposition. The week from 1–6 June 1909 was 'English' week when Nancy returned the 1908 compliment to the Mayor of London's delegation. A wreath laid at the feet of Jeanne d'Arc's monument marked the close of old rivalries and a garden party the consolidation of the Franco-British alliance in the shadow of escalating German militarism. The cycling race Nancy-Strasbourg-Nancy suggested a counter-impulse reinforcing confidence in the profundity of Alsace-Lorraine identity, deeper than national boundaries, whilst also highlighting the rupture of the annexation which imposed two sets of border controls requiring pauses within the race.

The exposition's monumental entrance celebrated both Lorraine's rich artistic past and its vibrant industrial future (Fig. 12.3). The iconic metal gate designed by Paul Charbonnier and manufactured by the local Pompey firm envisioned a multivalent iconography of Lorraine's economic and artisanal riches. Despite the assiduous precision with which the Treaty of Frankfurt had been devised to ensure the richest mining resources and the 11 principal mining firms and foundries were allocated to the annexed region of Alsace-Lorraine, neither the diplomats nor their geographers could have predicted the economic miracle of the Briey Valley discoveries in 1884 in French Lorraine or the advent of the Thomas-Gilchrist process which would render these deposits, high in phosphorous, utilizable in the founding of high-grade steel. The exploitation of these new resources and techniques by Meurthe-et-Moselle's 72 of France's 117 foundries increased production from 3 million tons in 1895 to 13 million tons by 1910; 77 per cent of the Thomas Process steel and 55 per cent of the total French national steel production.[15] The gate imaged the alliances between culture and industry which lay at the heart of Lorraine's fin-de-siècle vitality. The gate was produced by the Pompey firm which had long been at the forefront of high specification metal components production, be it for the Eiffel Tower at the 1889 Paris Exposition, the substructure for the Grand and Petit Palais for the 1900 Paris Exposition or a fleet of hulls for the Compagnie Générale Transatlantique liners.[16] The 1909 Exposition demonstrated a new greater integration of aesthetics and functionality. Rather than producing bespoke components as had been done to assemble the Eiffel Tower project in the capital, these gates weave standard products (wheels and braces) to create the lacework of the monumental gate: industrial production

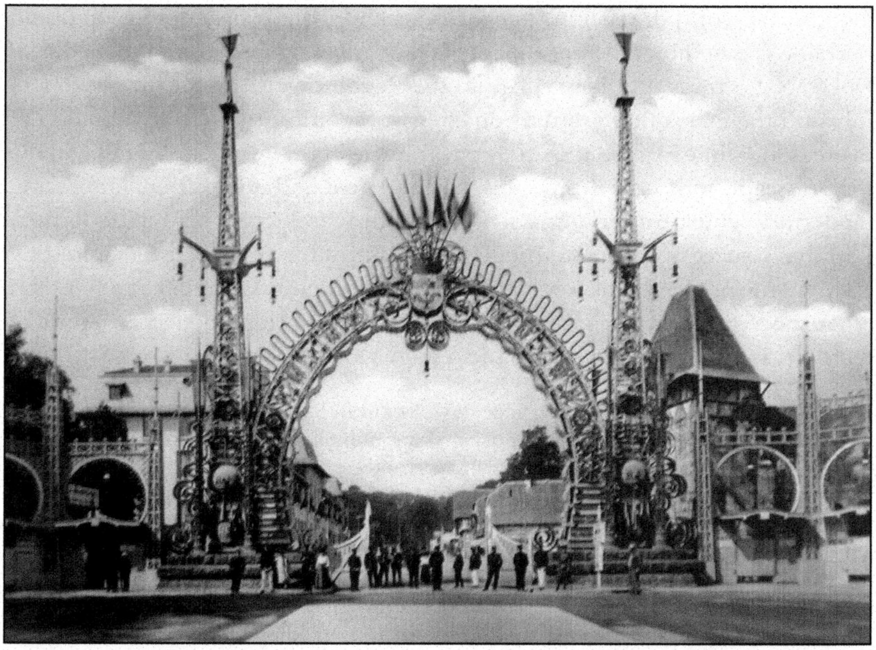

12.3
Monumental gate to the 1909 Exposition, designed by Paul Charbonnier and manufactured by Pompey firm, photograph. Musée de l'Ecole de Nancy

is shown to have its own intrinsic aesthetic beauty. The intricate pattern effects of the metal structure thus interweave the region's traditional handicrafts of embroidery and lacemaking with metallurgical manufacturing, prefiguring the fusion of handicraft and industrial decorative art in the museum displays of the Alsatian Village and the contemporary installations in the Ecole des Beaux-Arts, the Ecole de Nancy Pavilion and the Palace of Liberal Arts.

The rational construction and ideologically rich ornament of the pavilions encapsulated the core principles of the Chamber of Commerce and the Ecole de Nancy Industrial Alliance designers who strove to provide decorative industrial art to a wider clientele without sacrificing the quality of the product or that of the lives of their makers. The Palace of Mines and Metallurgy designed by the architect Louis Lanternier typifies the intermingling of pragmatic modernity and cultural aspiration. The striking factory methods used, cement over an iron girder substructure fenestrated with vast sheet glass and girder frameworks, prefigures with the early work of Peter Behrens for AEG or Bruno Taut's Deutscher Werkbund pavilion at the Cologne 1914 exposition.

However, Eugène Vallin's designs for the façade sculptural ornament punctuating this austere rigour is rooted in the organicism of Ecole de Nancy decoration. The corner pylons representing blast furnaces glowed red with electric light at night, but their fanlike flames emerge from tree trunk-like forms naturalizing industry. Furthermore the decorative fresco panels painted by his son, Antoine Vallin, and representing the stages of iron production humanized the factory: Lorrain labourers are celebrated as the life force of industry.[17] At the 1909 exposition, Eastern French identity hovered between the futurist world of machine age factories and dirigibles and a vision of old Lorraine typified by the pageants of 600 historical figures, from the medieval Dukes of Lorraine and Jean d'Arc to the Revolution, through the city on 11 July and the exposition grounds on 25 July. Significantly in his evaluation of the region Lafitte located the economic miracle of fin-de-siècle Lorraine in the region's capacity to rise from the ashes of defeat and annexation, embracing modernity whilst preserving a sense of history and artistry.[18] Lafitte located Lorraine's success in this marriage of art and industry as being inspired by the dynamism of a dispossessed people triumphing in commerce rather than bloodshed. Importantly he hastens to align the entrepreneurialist spirit of these firms 'des *self-made men*' to the region's artistic vitality:

> Since the war, the qualities of our race have found in Lorraine many opportunities for self-expression; but one of the most striking manifestations of our intellectual vitality was the appearance of the artistic methods which have won the Ecole de Nancy such renown. In a country so harshly tested by war and where thought, it would seem, might never find the lasting calm and rest without which arts remain sterile, one saw burgeoning graceful masterworks where the balance, the freshness, the variety of colour only rivals the purity of form ... Thus an external cause of sorts became one of the main causes underpinning the rise of our province, the reinforcement of regional energy with the arrival of the displaced.[19]

The design ethos and cultural production of the Ecole de Nancy formulated an integration of art and industry which lay at the heart of the region and the nation's regeneration, as such displays of art, craft and design informed every sector of the 1909 Exposition.

Lorraine Past and Present: Crafting Modern Design Practice

The visitor to the 1909 Exposition found art, craft and designed objects everywhere in the Parc Sainte-Marie. Three display areas and the assessments of them in the Exposition report capture of the conception of this cultural production with particular acuity: the historic farmhouse in the Alsatian village, the Ecole des Beaux-Arts and the Ecole de Nancy Pavilion. These three displays declare the core tenets of the decentralizing ethos at the heart of the Exposition which advocated regionalism, education and commerce.

Placed immediately after the triumph of engineering and artistry of the Pompey firm's monumental entrance gate, the Alsatian village suspended the visitor's sense of time and place through immersion in a synesthetic fantasy rooted in the material cultures of the lost provinces. The tavern and kiosks offered the tastes and sounds of an Alsatian childhood whilst Auguste Ramet's painted panorama created a subsuming vista of the forests and the now broken protective natural border of the 'blue line of the Vosges mountains'.[20] Perhaps the most oft commented-upon festivity of the Exposition was the festival of Alsace held on 27 June, a pageant of 200 peasants composed of deputations from specific villages wearing distinct local dress, varying from

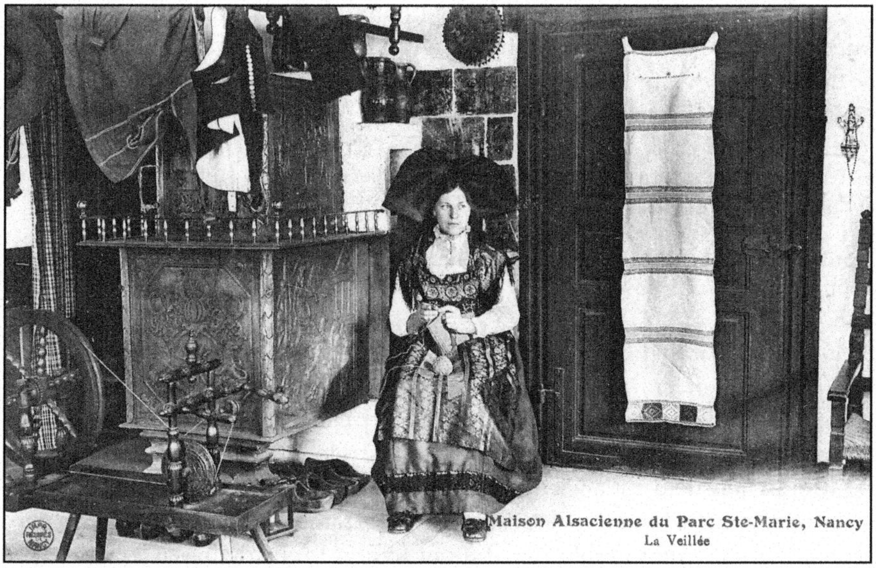

12.4 Interior displays in the transplanted Zutzendorf farmhouse in the Alsatian Village, 1909, postcard. Author's collection

village to village in the details of headgear or fabric design. The pageant reinforced the fiction of living history created by the focal point of the Alsatian village: an actual 1798 farmhouse, each timber and stone of which had been dismantled and transported in 14 wagons from Zutzendorf in the annexed regions to be reassembled for the Nancy 1909 Exposition. Léon Dollinger, curator of the Musée Alsacien in Strasbourg identified the farmhouse as a middle-class protestant farmhouse type known as a 'stub', referring to its central communal kitchen room (Fig. 12.4).[21] Passing by the functioning press demonstrated by old men in regional costume in the courtyard, one entered a living exhibition of artisanal female craft workers, children and elders. The first floor housed a loan exhibition of two rooms, a display organized by Dollinger comprised of objects principally deriving from Catholic Alsatian traditions: sumptuous women's jewellery and ornamental textiles from the seventeenth and eighteenth centuries such as the 'Schlaufkappe' (coiffe) alongside ecclesiastical objects (Christian and Jewish). A second room housed a display selected by Charles Sadoul, curator of the Musée Historique Lorrain and editor of the preeminent Nancy cultural journals the *Pays Lorrain* and *Revue Lorraine Illustrée,* demonstrating the distinct traditions of Lorraine: secular objects for use, plain-carved wooden furniture, ceramics, metal utensils, simple peasant clothing.

The many postcard sets of the Alsatian Village only portrayed women, children and old men (none of fighting age). Coupled with the material culture exhibits, this village was in a form of suspended animation more often deployed in the service of colonialist projects, such as the Senegalese Village nearby where weavers and the chief jeweller Mamadou Seck performed traditional craft as if outside of time.[22] However, the two curators who organized these displays contributed articles to the exposition report both asserting not only the importance of traditional crafts, but also the continuing importance of such everyday objects within the cultural and economic life of the modern regions. Dollinger proposed that the authenticity of Alsatian crafts offered a lesson for modern urbanites:

> Naiveté, which is a form of sincerity, and utility, i.e. the perfect suitability for use and place is proposed by [these objects]. The attentiveness to ornament which can be found in these humble objects of everyday life bespeak a permanent striving after beauty. We believe that even the highly cultured man can find a lesson to meditate upon, and even sometimes follow in these works of the humble village artist.[23]

Sadoul first asserts the distinctness of the two regions of Alsace and Lorraine typified by their craft traditions, emphasizing the austerity of the Lorrain context embodied in its craft. Its popular art is characterized by 'simplicity of line and sobriety of decoration' and a discrete ornament 'inspired by the flowers of the field or ancient symbols'. Whereas Alsatian dress favoured elaborate embellishment, peasant dress in Lorraine, 'a place of hardship and economy', was 'comfortable, one did not wear it for its richness'.[24]

Sadoul recognized the encounter of handwork and modernity at the heart of the Exposition, admonishing modern manufacturers in Lorraine not to destroy all craftwork and to emulate its fulfilling simplicity:

> Let us hope that this disappearance will not be definitive and that after the Nancy Exposition, where so much effort has gone into the linkage of the past and present, a new flowering of gracious and practical conceptions such as our fathers loved will embellish and make happy the life of the Lorrain people.[25]

These texts reinforce a core tenet of the Ecole de Nancy alliance which, whilst embracing modern industrial methods, still strove to ensure the production processes through which these objects are made remain fulfilling for the creative workers who make them by deploying artistry as well as modern techniques and knowledges. The syllabus of the new art school and the designers' alliance housed nearby would be the mentors and Maecenas of this art for all.

The report's chapters devoted to the displays in these two spaces, the Ecole des Beaux Arts and the Ecole de Nancy Pavilion, by René d'Avril and Emile Nicholas champion a new alliance of modernity, science and artistry, locating Lorraine as a distinct design tradition which reconciled industry and craft in ways rarely ascribed to art nouveau:

> Does not decoration have close ties to industry itself and to the sciences from whence it comes? If the botany which was cultivated with such love by Emile Galle, the chemistry pursued to such acclaim by Nancy's institutes, did not both these researches render great service to glassmaking and all the arts of fire? Certainly the unintelligent use of that which industry places at our disposition could produce and has produced in other contexts practical results which annihilated the work of art. It will be the glory and originality of Lorrains that they understand that one need not regard modern progress as the enemy, but rather as an ally, or indeed if one so wishes an auxiliary. Just as

floodwaters which at first devastate our fields, once controlled, fulfil a nourishing role. Similarly the practical conclusions of science, if canalized in some way can be made to serve the goals of art, far from destroying, they feed and give new youthfulness. This has been the achievement of the Ecole de Nancy.[26]

The article directly aligns the number of new enterprises in Meurthe-et-Moselle with the creative dynamism of the region, expressed in decorative art. D'Avril and Nicholas locate the Chamber of Commerce and the Ecole de Nancy alliance as the wellspring of a cultural and economic vitalism in Lorraine: 'Wealth engenders the taste for wellbeing and comfort. In all ages the Lorrain temperament, at once delicate and contemplative has been committed to the introduction of art into the public and private life of the city'.[27] This review interweaves its discussion of the displays in the Ecole des Beaux-Arts, the Ecole de Nancy pavilion and the Palace of Liberal Arts and champions the determined choice to install 'fine' art, decorative art and craft alongside industrially produced objects for the home. The elite handmade work of educators and practitioners such as Victor Prouvé and the stained glass maker Jacques Gruber and Lucien Weiner the bookbinder rub shoulders not only with the production of entrepreneurs within the Ecole de Nancy alliance such as the glass manufacturer Antonin Daum, but also with the production of manufacturers beyond the fold, such as the industrial embroidery firm, Heymann. Prouvé pioneered competitions in which student designers and workers responded to specifications defined by the industrial manufacturers, the successful example entering serial production with a stamp 'E de N'. The goal is a pedagogic model where creative industries and art education collaborate to forge a generation of designers who combine imagination and pragmatism: 'The hope is to create a general education in decorative art which will ensure young people can find lucrative and practical vocations in the world of industrial art and that this formation will be an artistic laboratory for our regional manufactures.'[28] This education serves a social mission and a practical approach to garnering employment as well as self-fulfilment in a mode rarely recognized as a key force within French decoration, and stymied within Anglo-Saxon Morrisian inspired strategies which fail to recognize the potential of a policy of détente and collaboration amongst art, industry and science:

This is the work of the Ecole de Nancy, elegant and new in its inspiration, practical and precise in its application, generous and nobly social in its aims ... It is the honour of our region to have understood and to have known how to realise for the first time in France the tripartite union of the beautiful, the truthful and the useful.[29]

Importantly, the displays themselves embodied this integration of the arts, industry and social conscience. In three conjoining rooms, shaded in red, yellow and green, in the Palace of Liberal Arts every form of visual and material culture intermingled. The acid-patterned velvets of Charles Fridrich are declared to be artworks as much as the paintings of Emile Friant and Aimé

Morot. In the 'Arts of Women' display, pieces executed by women industrial embroiderers were juxtaposed side by side with the bespoke book bindings of Mlles Julie and Marie Thiriet. D'Avril and Nicholas celebrate the social significance of this integrated society of creative industry:

> For the first time in an art display, one sees mixed together and welcomed with equal status all categories of artists and artisans, thus creating the effect of a family banquet where there are no poor parents. All hierarchy, all privilege has been banished in this section, and every exhibitor will find their work shown there where they would have chosen to see it placed.[30]

Significantly, a number of displays integrated furnishings, decorative objects and textiles creating whole interiors: 'a free and supple display strategy which creates an equivalent to the real adaptation of the elements which contribute to making the decoration of an interior'.[31] Whilst still celebrating the individualism of Gallé and Majorelle furnishings, in their assessment of these ensembles of furniture and interior design De Roche du Teilloy and Weissenberger echoed the praises accorded to textiles and decorative objects by D'Avril and Nicholas:

> [The techniques of furniture making …] also apply themselves to aligning art and industry, that is to say to find that balance between the exigencies of taste and the love of progress, the need for the ideal and the necessities of life which are reconciled for the greatest joy for the eye and the spirit.[32]

The Gauthier Poinsignon firm was selected out for particular praise, their affordable products having embraced the simplified forms of modern art 'without vulgarity or banality'.[33] Durability was achieved without sacrificing distinctive individual expression; the use of local timber of 'antediluvian' oak trees and ornamental motifs infused the furniture of this firm which catered to the 'hotel' market with an aura of calm: 'One imagines an easy and gentle life in such interiors'.[34]

The displays of decorative and industrial art in Nancy's 1909 Exposition Internationale de l'Est de France proposed a productive and peaceful integration of rival forces; commerce, education and creativity coalesced in the ambition to create a new construct of a region within national and international networks of identity. The Exposition attracted attention throughout the provincial and Parisian French and German press, but despite the 'English' week the exposition seems to have merited only a descriptive sentence or two in Anglo-Saxon periodicals, hence perhaps its subsequent neglect in the scholarship. These articles mostly position the decentralizing agenda underpinning the project as an exemplary model of collaboration, not a dangerous separatism.[35] The fortunes and national flags of this troubled region would rise and fall many more times in the world wars of the ensuing century.

Perhaps the last word should be from Louis Lafitte's closing remarks about his adopted city and the colleagues with whom he created the 1909 exposition:

> The decorations will all fall down, but behind what will appear is Nancy capital of the East, modernised and triumphant, inheritor of Strasbourg and Metz, centre of art and industry, militarized city, summing up the living ideals of France ... Lorraine is an admirable reserve of energy and riches, in which industrialists, artisans and workers of the lost provinces all contributed to this validation, a place of iron, the soil as well as the men.[36]

Lafitte, like so many of the visitors and expositors, would die on the battlefields of the First World War. This tragedy has in many ways led to this extraordinary moment of modernity and the design of the Ecole de Nancy at its heart to be positioned as a 'swan song' for the idyll of the Belle Epoque. The present argument proposes it might be construed as a gateway where the utopian social aspirations and practical partnerships of an age of high industrialization, social utopianism and nascent decentralization became productive bedfellows. The truncated and bloodied region of Lorraine, deluged with displaced professionals and workers, seemed an unlikely site of regional vibrancy in 1909. This economic and aesthetic miracle would be repeated again when in the midst of crushing losses on 25 August 1917 Etienne Clémentel, Minister of Commerce and Industry, circulated a letter to the nation's Chambers of Commerce proposing that 20 economic regions be formed through which to decentralize and to regenerate France's industries both in the present crisis and in a hoped-for bright future. Clémentel chose Nancy and its Chamber of Commerce as the exemplar upon which to model the 17 regional economic alliances which would be constituted on 5 April 1919; the thistle of Lorraine was to rise phoenix-like from the ashes yet again.[37]

Notes

1 'A Lorraine peasant girl in traditional costume releas[ing] pigeons, urbi et orbi, from the balcony of the Hotel de Ville in Nancy to tell the world the news [of the Exposition]'. Emile Baudel cited in Béatrice Damamme-Gilbert, 'The 1909 Nancy International Exhibition: Showcase for a Vibrant Region and Swansong of the Ecole de Nancy', *Art on the Line* (special issue: Emile Gallé and la Lorraine Artiste) 1 no. 5 (2007): 10, at artontheline.bris.ac.uk, accessed 1 October 2007. I am indebted to Dr Damamme-Gilbert for introducing me to the exposition, her generous extended loan of her grandfather Louis Laffitte's report on his organization of the exposition and her contribution to the original conference from which her article ensued.

2 Paul Vidal de la Blache, 'Régions françaises', *Revue de Paris* (1910): 821–849. For his central role in the formation of the discipline of cultural and economic geography see Mona Ozouf-Marignier, 'Le *Tableau* et la division régionale: de la tradition à la modernité', in *Le tableau de la géographie de la France de Paul Vidal de la Blache dans le labyrinthe des formes*, ed. Marie-Claire Robic (Paris, 2000), 151–181; and Annie Sevin, 'Les acteurs économiques et le regionalism lorrain de la Belle Epoque', *Annales de géographie* 2, no. 648 (2006): 174–196, JSTOR, accessed 25 June 2012.

3 Exposition studies have been typified by comparative surveys, from the groundbreaking work of Paul Greenhalgh, *Ephemeral Vistas: The Exposition Universelles, Great Exhibitions and World's Fairs 1851–1939* (Manchester, 1988) to most recently Alexander C. T. Geppert, *Fleeting Cities: Imperial Expositions in Fin-de-Siècle Europe* (Basingstoke, 2011) and the examination of landmark expositions in isolation, such as Richard Mandell, *Paris 1900: The Great World's Fair* (Toronto, 1967) and Arnold Lewis, *An Early Encounter with Tomorrow: Europeans, Chicago's Loop and the World Columbian Exposition* (Urbana, 1997).

4 Landmark studies in the discussion of art and design displays at the 1855, 1867, 1878, 1889 and 1900 Paris International Expositions are Patricia Mainardi, *Art and Politics of the Second Empire: The Universal Expositions of 1855 and 1867* (New Haven, 1987); and Deborah Silverman, *Art Nouveau and Fin-de-Siècle France: Politics, Psychology, Style* (Berkeley, 1989). For decorative and industrial arts expositions see Peter Clericuzio, 'Modernity, Regionalism, and Art Nouveau at the Exposition Internationale de l'Est de la France, 1909', *Nineteenth Century Art Worldwide* 10, no. 1 (Spring 2011): n.p., at 19thc-artworldwide.org, accessed 11 June 2014; Tag Gronberg, *Designs on Modernity: Exhibiting the City in 1920s Paris* (Manchester, 1998); Shanny Peer, *France on Display: Peasants, Provincials and Folklore in the 1937 Paris World's Fair* (New York, 1998); and James Herbert, *Paris 1937: Worlds on Exhibition* (Ithaca, 1998).

5 Christiane Demeulenaere-Douyére, 'Expositions universelles et internationals (hors les expositions universelles présentées à Paris) (1844–1920) – Expositions nationales, foires et salons français (1895–1921) – Expositions régionales françaises (1852–1913)', Archives nationales Paris, SOUS-SERIE F/12 (Commerce et industrie) (Paris, 2008), at http://www.archivesnationales.culture.gouv.fr/chan/chan/fonds/edi/sm/sm_pdf/7-Expositions%20diverses.pdf, accessed 5 February 2013; doctoral research ('Etudes coloniales', Asociation Etudes Coloniales), at http://etudescoloniales.canalblog.com/archives/2006, accessed 5 February 2013; and local history projects ('La mémoire maritime', Bordeaux Maritime), at http://bordeauxmaritime.free.fr/index.html, accessed 5 February 2013, have begun to open up research into field of the regional and international expositions in France.

6 Richard Thomson, *The Troubled Republic: Visual, Cultural and Social Debate in France 1889–1900* (London, 2004). For the insistence upon elite craft and retreat see Silverman, *Art Nouveau and Fin-de-Siècle France*; for the more multivalent conception of French decoration allowing for engaged politics and alliances between craft and industry, see Roger Marx, *L'Art Social* (Paris, 1913); Nancy Galeries Poirel, *L'Ecole de Nancy, 1889–1909: Art nouveau et industries d'art* (Nancy, 1999); Rossella Froissart Pezone, *L'art dans tout les arts décoratifs en France et l'utopie d'un Art nouveau* (Paris, 2004); Blandine Chavanne et al., *Roger Marx, un critique aux côtés de Gallé, Monet, Rodin, Gauguin* (Nancy, 2006).

7 See for example Maurice Barrès's notorious *Les Deracinés: le roman de l'énergie nationale* (Paris, 1897), the first part of his trilogy of 'novels of national energy'; and Edouard Détaille, *The Dream*, 1888, oil on canvas, h. 300 cm, w. 400 cm, Musée D'Orsay.

8 G. Bruno (Augustine Fouillée), *Le Tour de france par deux enfants* (Paris, 1877); for the dissemination of the book and its role in national education see Jacques and Mona Ozouf, '*Le Tour de France par deux enfants*: The Little Red Book of the Republic', in *Realms of Memory: The Construction of the French Past, Volume 2: Traditions*, ed. Pierre Nora, trans. Arthur Goldhammer (Chichester, 1997), 125–150.

9 For the distinct construction of annexed Alsace-Lorraine, see Jean-Marie Mayeur, 'A Frontier Memory', in *Rethinking France: Les Lieux de Mémoire, Volume 2: Space*, ed. Pierre Nora, trans. David Jordan (London, 2006), 409–442.

10 See Nancy Galeries Poirel; and Hélène Sicard-Lenattier, *Les Alsaciens Lorrains à Nancy 1870–1914* (Haroué, 2002).

11 This 1906 article appeared in the journal *L'Immeuble et la Construction dans l'Est* which Jacquemin edited, and is cited in Frédéric Descouturelle et al., *Nancy 1909: Centenaire de l'Exposition Internationale de l'Est de France* (Nancy, 2008), 25–26.

12 Descouturelle et al., 26.

13 1F for a day entry; 50 centimes in the evening; 2 francs on gala days with options for season tickets priced at 20F for the heads of households, 15F for wives and 10F for each child over the age of 5.

14 See Bassett Lowke letter illustrated in Descouturelle et al., 37; and Perilla Kinchin, *78 Derngate Northampton* (Northampton, 2005).

15 Louis Lafitte, 'L'Evolution économique de la Lorraine', *Annales de géographie* 21, no. 120 (1912): 393–417, at 411–413, JSTOR, accessed 25 June 2012.

16 The entrance arch to the displays of the firm Pont-à-Mousson which produced the vast numbers of standardized structural rings which formed the tunnels of the Parisian metro deployed a similar juxtaposition of function and aesthetics, integrating art nouveau flourishes into a standard metro tunnel support ring.

17 The Electricity Pavilion, housing the separate generator from which the exposition drew all its power, also bespeaks the persistence of an organicist conception of this energy force, 'la fée éléctricité', so easily forgotten from a twenty-first-century standpoint.

18 'One notes that most of these innovations were achieved to the detriment of German industry, helping to free us from the reparations we continue to pay even now. This has been another form of reconquering.' Lafitte, 'L'Evolution économique', 406.

19 Lafitte, 'L'Evolution économique', 406–407.

20 A phrase coined by Jules Ferry in 1890 which became a rallying cry throughout the ensuing century of border conflicts; see Vincent Morinaux, 'Le mythe de la ligne bleue de Vosges at la vision nationaliste de la Sapinière', in *Les Sapins Enjeux anciens, enjeux actuels*, ed. Andrée Corvol (Paris, 2001), 373.

21 Jean René Trochet, *Maisons paysannes en France et leur environment XVe–XXe siècles* (Paris, 2006), 495.

22 The scholarship on colonial displays is extensive; see for example Sadiah Qureshi, *Peoples on Parade: Exhibitions, Empire, and Anthropology in Nineteenth-Century Britain* (London, 2011).

23 Léon Dollinger, 'La Maison Alsacienne de Zutzendorf', in *Exposition Internationale de l'Est de la France*, ed. Louis Lafitte (Nancy, 1910), 20. Lafitte's report of over 800 pages anthologized every aspect of the exposition from financial matters to the critical reception of every element of the displays in the local Nancy, Paris and foreign press.

24 Charles Sadoul, 'La Chambre Lorraine de la maison Alsacienne', in *Exposition Internationale de l'Est de la France*, ed. Louis Lafitte, 21.

25 Sadoul, 'La Chambre Lorraine', 24.

26 René D'Avril and Emile Nicholas, 'Les Beaux Arts', in *Exposition Internationale de l'Est de la France*, ed. Louis Lafitte, 35–36.

27 D'Avril and Nicholas, 36.

28 D'Avril and Nicholas, 41.

29 D'Avril and Nicholas, 44.

30 D'Avril and Nicholas, 44.

31 De Roche du Teilloy and Weissenberger, 'La construction et l'ammeublement', in *Exposition Internationale de l'Est de la France*, ed. Louis Lafitte, 476.

32 De Roche du Teilloy and Weissenberger, 475.

33 De Roche du Teilloy and Weissenberger, 477.

34 De Roche du Teilloy and Weissenberger, 477.

35 See Louis Daussat, *Petite Gironde*; Pascal Fortuny *Le Matin*; A Froemer, *L'Evenement*; E. Martin, *La Croix*; and Munich's *Neueste Nachrichten* and the *Dresdner Volkzeitung*; all cited in *Exposition Internationale de l'Est de la France*, ed. Louis Lafitte, 823–848.

36 Cited in Louis Lafitte, ed., *Exposition Internationale de l'Est de la France*, 780.

37 Sevin.

Bibliography

Printed Primary Sources

Barrès, Maurice. *Les Deracinés: le roman de l'énergie nationale*. Paris: Charpentier, 1897.

Bruno, G. (Mme Fouillée). *Le Tour de France par deux enfants*. Paris: Librarie classique d'Eugène Belin, 1877.

Lafitte, Louis. 'L'Evolution économique de la Lorraine', *Annales de géographie* 21, no. 120 (1912): 393–417, JSTOR, accessed 25 June 2012.

Lafitte, Louis, ed. *Exposition Internationale de l'Est de la France*. Nancy: Imprimerie de l'Est, 1910.

Marx, Roger. *L'Art Social*. Paris: Charpentier, 1913.

Secondary Sources

Chavanne, Blandine, et al. *Roger Marx, un critique aux côtés de Gallé, Monet, Rodin, Gauguin*. Nancy: Editions Artlys, 2006.

Clericuzio, Peter. 'Modernity, Regionalism, and Art Nouveau at the Exposition Internationale de l'Est de la France, 1909'. *Nineteenth-Century Art Worldwide* 10, no. 1 (Spring 2011): n.p., at 19thc-artworldwide.org, accessed 11 June 2014.

Corvol, Andrée, ed. *Les Sapins Enjeux anciens, enjeux actuels*. Paris: L'Harmattan, 2001.

Damamme-Gilbert, Béatrice. 'The 1909 Nancy International Exhibition: Showcase for a Vibrant Region and Swansong of the Ecole de Nancy'. *Art on the Line* (special issue: Emile Gallé and la Lorraine Artiste) 1, no. 5 (2007): 1–13, at artontheline.bris.ac.uk, accessed 1 October 2007.

Descouturelle, Frédéric, et al. *Nancy 1909: Centenaire de l'Exposition Internationale de l'Est de France*. Nancy: Editions Place Stanislas, 2008.

Geppert, Alexander C. T. *Fleeting Cities: Imperial Expositions in Fin-de-Siècle Europe*. Basingstoke: Palgrave Macmillan, 2011.

Greenhalgh, Paul. *Ephemeral Vistas: The Exposition Universelles, Great Exhibitions and World's Fairs 1851–1939*. Manchester: Manchester University Press, 1988.

Gronberg, Tag. *Designs on Modernity: Exhibiting the City in 1920s Paris*. Manchester: Manchester University Press, 1998.

Herbert, James. *Paris 1937: Worlds on Exhibition*. Ithaca, NY: Cornell University Press, 1998.

Kinchin, Perilla. *78 Derngate Northampton*. Northampton: 78 Derngate Trust, 2005.

Lewis, Arnold. *An Early Encounter with Tomorrow: Europeans, Chicago's Loop and the World Columbian Exposition*. Urbana: University of Illinois Press, 1997.

Mainardi, Patricia. *Art and Politics of the Second Empire: The Universal Expositions of 1855 and 1867*. New Haven: Yale University Press, 1987.

Mandell, Richard. *Paris 1900: The Great World's Fair*. Toronto: University of Toronto Press, 1967.

Mayeur, Jean-Marie. 'A Frontier Memory'. In *Rethinking France: Les Lieux de Mémoire Volume 2: Space*, edited by Pierre Nora, trans. David Jordan, 409–442. London: University of Chicago Press, 2006.

Nancy Galeries Poirel. *L'Ecole de Nancy, 1889–1909: Art nouveau et industries d'art*. Nancy: Réunion des Musées Nationaux, 1999.

Ozouf, Jacques, and Mona Ozouf. '*Le Tour de France par deux enfants*: The Little Red Book of the Republic'. In *Realms of Memory: The Construction of the French Past, Volume 2: Traditions*, edited by Pierre Nora, trans. Arthur Goldhammer, 125–150. Chichester: Columbia University Press, 1997.

Ozouf-Marignier, Mona. 'Le *Tableau* et la division régionale: de la tradition à la modernité'. In *Le tableau de la géographie de la France de Paul Vidal de la Blache dans le labyrinthe des formes*, edited by Marie-Claire Robic, 151–181. Paris: Comité des Travaux Historiques et Scientifiques, 2000.

Peer, Shanny. *France on Display: Peasants, Provincials and Folklore in the 1937 Paris World's Fair*. New York: State University of New York Press, 1998.

Pezone, Rossella Froissart. *L'art dans tout les arts décoratifs en France et l'utopie d'un Art nouveau*. Paris: CNRS Editions, 2004.

Qureshi, Sadiah. *Peoples on Parade: Exhibitions, Empire, and Anthropology in Nineteenth-Century Britain*. London: University of Chicago Press 2011.

Sevin, Annie. 'Les acteurs économiques et le regionalism lorrain de la Belle Epoque'. *Annales de géographie* 2, no. 648 (2006): 174–196, JSTOR, accessed 25 June 2012.

Silverman, Deborah. *Art Nouveau and Fin-de-Siècle France: Politics, Psychology, Style*. Berkeley: University of California Press, 1989.

Thomson, Richard. *The Troubled Republic: Visual, Cultural and Social Debate in France 1889–1900*. London: Yale University Press, 2004.

Trochet, Jean René. *Maisons paysannes en France et leur environment XVe–XXe siècles.* Paris: Editions Creafis, 2006.

INTERNATIONAL AMBITIONS

13

Merging Peripheries and Centres: The Transnational Interconnectedness of the Helsinki National Exhibition of 1876

Taina Syrjämaa

In July 1876, a quarter of a century after the Crystal Palace exhibition in London, the entrance gates to the first Finnish industrial and art exhibition opened in Helsinki. Whilst the attention of the wider world was focused on the contemporaneous world's fair in Philadelphia, which celebrated the American centennial, the national exhibition was at the hub of things in the Finnish context. The exhibition attracted some 90,000 visitors during the two-and-a-half months that it was open to the public and gained a great deal of attention in the printed media throughout the country.[1] The event was unprecedented in size and popularity in Finland, although it was tiny in comparison to the great exhibitions that were seen by millions of visitors.

At the time, exhibitions had already consolidated their position as one of the leading – and most spectacular – forms of modern mass medium. Indeed, the largest exhibitions became globally renowned. Yet these well-known events were merely the tip of the iceberg, representing the most visible and imposing epitomes of the medium.[2] The great majority of exhibitions were smaller-scale events, whether at a regional, national or local level. The Helsinki exhibition was one of these countless embodiments of the exhibition medium, which was much more multifaceted and complex than the chronological chain of single great exhibitions might suggest (Fig. 13.1).

The colossi have undoubtedly functioned as prestigious models for others, including the Finnish national exhibition. However, the inter-exhibition contacts were not simple and straightforward. As Alexander C. T. Geppert has recently shown in his study on five imperial exhibitions, arranged in Berlin, Paris and London, the exhibition medium was 'dominated by far-reaching internal references and formative transnational and inter-urban connections'.[3] In this chapter I examine the complex interrelatedness at

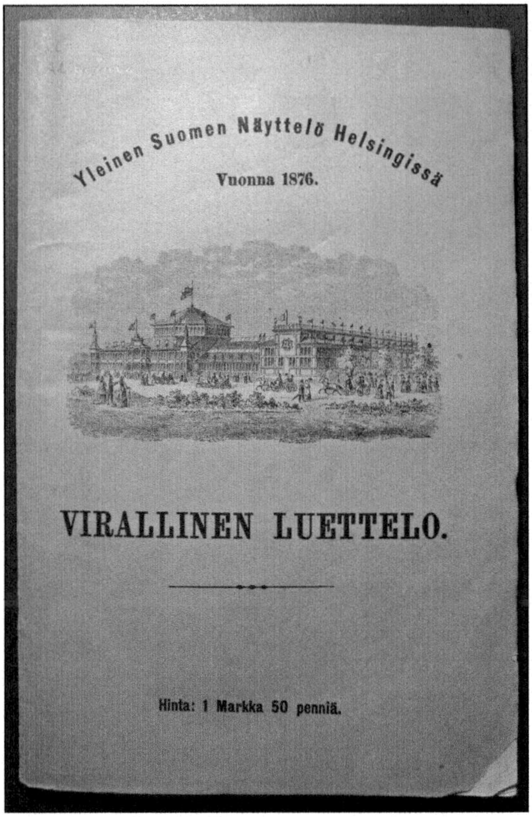

13.1 Helsinki exhibition pavilion, from *Yleinen Suomen Näyttelö Helsingissä Vuonna 1876. Wirallinen luettelo*, front cover, Helsinki, 1876

yet another level by focusing on the medium from the perspective of a minor, national and seemingly peripheral case. I explore the interconnectedness and interdependence of exhibitions by examining the links of the Helsinki exhibition to a variety of previous grand and more modest expositions. I also make reference to the subsequent events to bring out the plurality embedded in the interaction and to show that minor exhibitions could also contribute to the medium.

Putting on an exhibition posed a myriad of logistical, ideological, commercial and even epistemological challenges for organizers. Planners looked to existing models and earlier adopted practices, either after meticulous consideration or accepting many things as self-evident requirements. Knowledge of a range of exhibition technologies and practices was conveyed via printed media, photographs, correspondence, or by individual, first-hand experiences. Explicit commentaries on how various practices had been chosen are of the utmost rarity, and in general temporary exhibitions have left only fragmentary and lacunose sources. However, a lot can be deduced if one undertakes a careful reading and combines a variety of sources.

The analysis in this chapter is based on the transnational paradigm, which seeks to make visible contacts and flows that transcend national borders and surpass or bypass the agency of the nation state. In the nineteenth-century context exhibitions were described in terms of 'inter-national' operations, but from the current research perspective it is possible to discern other, more varied layers, in the exhibition medium. The transnational approach, which lays emphasis on hybridity, bricolage and cultural translation, also assists in bringing to the fore the dynamic interaction of various exhibitions.[4] Therefore, to trace and explore these fluid forms of connections, I focus on two intertwined dimensions: practices and persons. My intention is by no means to track the exact genealogy of a set of practices or to honour a certain individual for having invented them; rather it is to make visible the plurality and dynamicity of connections and the inherent internationality of national exhibitions. In order to locate the Helsinki exhibition within the international sphere of such events, I will trace inter-exhibition flows and manifold entanglements. A particular focus will be placed on the exhibition pavilion, which was a key characteristic of temporary exhibitions, as well as what were still the rather novel qualities of the ethnographic display.

Exhibitions Far and Near

The first national exhibition in Finland, which at the time was an autonomous grand duchy within the Russian Empire, was a much-acclaimed and long-awaited event.[5] A solitary voice had proposed an exhibition – based on the model of national exhibitions in Europe – as early as the 1840s, but the idea had gained momentum in the 1860s. This decade not only witnessed the consolidation of great exhibitions, hosted in London and Paris, but many other kinds of expositions pullulated. From the Finnish perspective it is important to note that a number of exhibitions were arranged in neighbouring countries: a Nordic industry and art exhibition was staged in Stockholm and Russian industrial and agricultural expositions were held in St Petersburg and Moscow.[6] Furthermore, the decade had opened with major economic and social reforms in Finland, such as the introduction of the Finnish mark and the launching of a new municipal and secular school system. The national exhibition had been proposed in this progressive and optimistic atmosphere, and in parallel with the gradual intensification of nationalist sentiment. Also some experimentation had been undertaken, most notably in 1862, when an agricultural and industrial exhibition was held in Viipuri, the biggest city in Eastern Finland and in close vicinity to St Petersburg.[7] However, successive severe crop failures and the consequent large-scale famine defeated these hopes and postponed the realization of a major national exhibition until the mid-1870s.

Finally, in the spring of 1875 progress was made on the planning of a Finnish national industrial and art exhibition. A group of businessmen, tired of waiting for the realization of more official initiatives, succeeded in securing partial funding from the Finnish Senate and also arranged a nationwide fundraising drive.[8]

Among the key persons and members of the exhibition committee was Leonard Borgström, for example, who represented a notable cluster of family businesses that included the textile company Forssa Ab. This firm was not only one of the biggest in the country, but had actively participated in a number of exhibitions. It had won medals, for example, in the great exhibitions in London in 1862, Paris in 1867 and Vienna in 1873. In a similar manner, Oskar Osberg and August Bade were awarded a medal at the Moscow exhibitions in 1864 and 1865 for their fire-extinguishing equipment.[9] The chairman of the exhibition committee, Baron Johan August von Born, had gained a great deal of experience attending agricultural meetings and exhibitions both at home and abroad.[10]

The head commissioner of the Helsinki exhibition, Axel Wilhelm Wahren, provided considerable first-hand experience for the preparatory committee. The Swedish-born businessman, who was in his sixties at the time and the managing owner of Forssa Ab, had played an active role in exhibitions ever since the Crystal Palace exhibition, where his firm had been awarded a medal.[11] He had also served as a juror in the Nordic exhibition in Stockholm in 1866.[12]

Even the young engineer Robert Runeberg, who was nominated as commissioner of the Helsinki exhibition, had previously gained some experience. Although he was only 30 years of age at the time, he had already visited two great exhibitions – in Paris 1867 and Vienna – as well as the Nordic exhibition in Stockholm. Runeberg had also entertained the idea of travelling to the Philadelphia World's Fair. His role in Paris had been merely that of a curious spectator; while in Vienna he was enlisted as the technical correspondent of a Finnish newspaper. Furthermore, he had gained some insight into the exhibition medium from the point of view of an exhibitor when he had assisted his brother, a sculptor living in Rome, in having his art works exhibited at these expositions.[13]

Although the Helsinki exhibition was the first larger industrial and art exhibition arranged in Finland, it was not created without an expertise based on manifold experience and insights of the exhibition medium. From the very outset, when the idea of a national exhibition was still in its embryonic stage, it was connected in multiple ways to previous events and deeply embedded in the varied landscape of large-scale international and national exhibitions.

Looking Like an Exhibition: An International Genre in a Helsinki Park

The Helsinki exhibition had all the basic characteristics of expositions of the time: a separate exhibition area and pavilion for paying visitors; grandiose opening ceremonies in the presence of dignitaries, speeches and choruses; exhibits which had been classified in a number of classes and endowed with explanatory labels; printed catalogues and prizes awarded by a jury of invited specialists. The focal point of the exhibition was a temporary, purpose-built wooden pavilion in Kaivopuisto Park. It hosted the display of industrial products, handicrafts, the Finnish school system, fine arts, ethnographic display and machinery, including the 'clou' of the exhibition: the first locomotive ever constructed in Finland. Furthermore, a separate display of foreign agricultural machines and an agricultural exhibition of shorter duration were arranged in the same area. The site was also furnished with restaurants and bars.[14]

The pavilion, designed by the young architect Theodor Höijer, consisted of a rotunda that spanned the main entrance and that was crowned by a polygonal elevated lantern. A longish gallery for machines was annexed to this in the rear, and there was also a two-storeyed hall on the right-hand side for the school section and fine arts display. On the left-hand side of the temporary pavilion was to be found an older, permanent edifice, which had been built to serve the spa that had been founded in the park in the 1830s.[15]

It was out of the question that Helsinki could accommodate the sheer scale and grandiosity of the pavilions of the great exhibitions, but notwithstanding this the basic requirements were similar. It was necessary for the exhibition premises to be easily constructed and dismantled. They also had to be luminous and spacious in order to accommodate the exhibits and the flow of visitors, who were supposed to closely scrutinize the displays whilst also savouring wide panoramas. Some sort of spectacularity was also an important factor. Indeed, some commentators explicitly drew parallels between the Helsinki pavilion and the Crystal Palace, which must have been due to the genre of the buildings, rather than to any specific similarity in their appearance.[16] Höijer was successful in creating an edifice which was easily recognized as an exhibition pavilion.

The most striking visual feature of the pavilion was its central rotunda, placed in a rectangular hall, which bore resemblance to the pavilion of the Vienna great exhibition. The Helsinki pavilion lacked the numerous lateral 'fishbone' galleries of Vienna, but the rotunda was in a similarly dominating position. The Vienna exhibition had been rather closely followed in Finland: a Finnish commissioner had been nominated to take care of displaying Finnish items and of purchasing interesting objects for the Finnish design collection.[17] The architect Höijer himself also travelled to Vienna with the help of a state grant.[18] Thus, the structural solution of the Vienna pavilion and its central rotunda were certainly familiar to the organizers of the Helsinki exhibition. The Vienna experience was also fresh in the memory when preparations commenced in Finland.

At first sight, the scenario seems to be straightforward: the architect and organizers of a small national exhibition sought to emulate the model utilized in the most recent great exhibition, thereby effectively constructing a simplified miniature version. However, when we broaden the scope of the examination to other international exhibitions, another connection becomes evident: there are some striking similarities between the Helsinki exhibition and the international Nordic exposition held in Stockholm a decade earlier. Not only did they both maximize the position of a central rotunda, but they also made use of the same construction material, wood, which was uncommon during the heyday of glass and iron architectural design. Ulf Sörenson has claimed that a specific Nordic style of wooden pavilion architecture developed on the basis of the Stockholm pavilion (Fig. 13.2). Thus, he argues that this style consequently re-appeared in Helsinki in 1876, Copenhagen in 1888 and again in Stockholm in 1897.[19] In the Nordic countries, wooden architecture had historical and cultural resonances, as well as being logistically and financially amenable, even though the material lacked the glamour of glass and iron. Although wood could link together various Nordic pavilions – including a number of constructions from the region constructed in the great exhibitions – only the pavilions of the Stockholm and Helsinki exhibitions of 1866 and 1876 respectively shared a similar form.

The resemblance was not only in the external appearance of the architecture, but also in the way the pavilions were depicted. The nameplate of the exhibition newspaper that showed the façade of the Stockholm pavilion

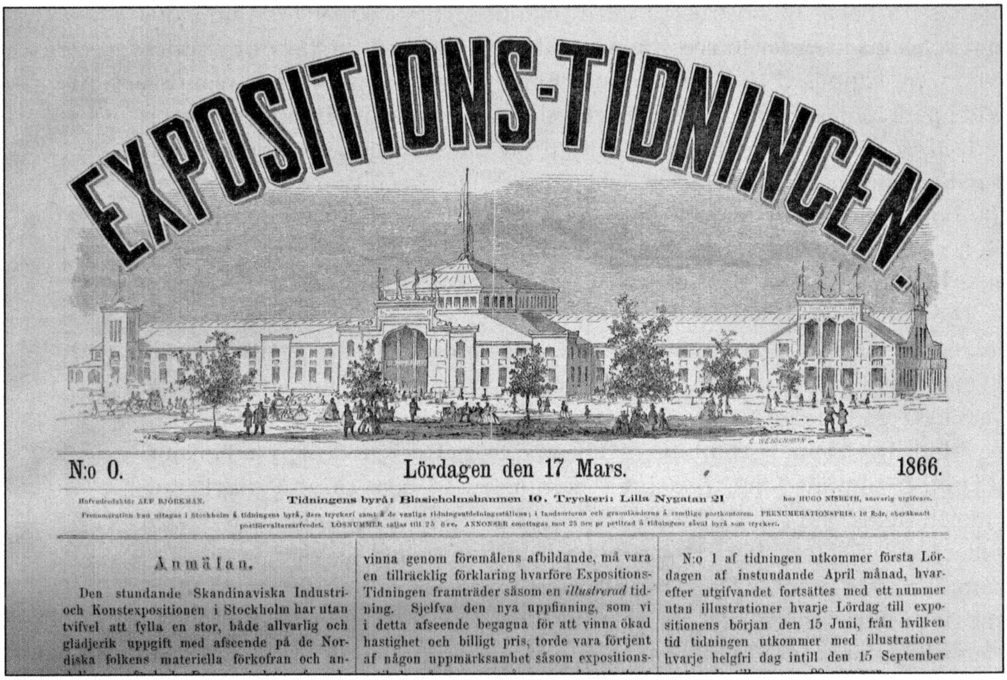

EXPOSITIONS-TIDNINGEN.

N:o 0. Lördagen den 17 Mars. 1866.

Hufvudredaktör ALF BJÖRKMAN. **Tidningens byrå: Blasieholmshamnen 10. Tryckeri: Lilla Nygatan 21** hos HUGO NISBETH, ansvarig utgifvare.

Prenumeration kan uttagas i Stockholm å tidningens byrå, dess tryckeri samt å de vanliga tidningsutdelningsställena; i landsorterna och grannländerna å samtliga postkontoren. PRENUMERATIONSPRIS: 10 R:dr, oberäknadt postförvaltarearfvodet. LÖSNUMMER säljas till 25 öre. ANNONSER emottagas mot 25 öre pr petitrad å tidningens såväl byrå som tryckeri.

Anmälan.

Den stundande Skandinaviska Industri- och Konstexpositionen i Stockholm har utan tvifvel att fylla en stor, både allvarlig och glädjerik uppgift med afseende på de Nordiska folkens materiella förkofran och an-

vinna genom föremålens afbildande, må vara en tillräcklig förklaring hvarföre Expositions-Tidningen framträder såsom en *illustrerad* tidning. Sjelfva den nya uppfinning, som vi i detta afseende begagna för att vinna ökad hastighet och billigt pris, torde vara förtjent af någon uppmärksamhet såsom expositions-

N:o 1 af tidningen utkommer första Lördagen af instundande April månad, hvarefter utgifvandet fortsättes med ett nummer utan illustrationer hvarje Lördag till expositionens början den 15 Juni, från hvilken tid tidningen utkommer med illustrationer hvarje helgfri dag intill den 15 September

13.2 The Stockholm Pavilion, from *Expositions-Tidningen*, 1866

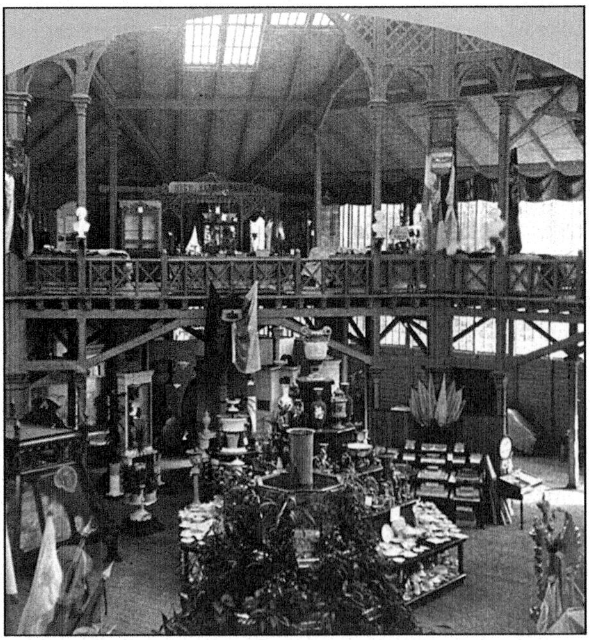

13.3 Charles Riis, interior of the Helsinki exhibition with a fountain and a display of ceramics in the centre of the rotunda, photograph. City Museum of Helsinki

and the drawing on the Helsinki exhibition catalogue, for example, share the same angle, a similar use of cropping and the presence of human figures in the foreground.[20]

When examining the interior of the pavilion, other similarities appear. Not only were the forms of the rotunda similar in Helsinki and Stockholm, but also in both cases a fountain acted as a dominant centrepiece of the rotundas. Fountains have an illustrious history, being utilized as imposing pieces of urban décor, as well as being used in exhibitions, including at Crystal Palace. In Stockholm, however, the fountain, which was created by the Swedish sculptor Johan Peter Molin, was tremendously popular and became the symbol of the entire exhibition. It was even cast in bronze in 1873 and subsequently placed in the same central location in the city in which the temporary exhibition pavilion had been situated.[21] The fountain was also enthusiastically described in the Finnish media at the time.[22]

In Helsinki, the fountain was more prosaic in its form and lacked depictions of Scandinavian mythology (Fig. 13.3). However, its location in the topography of the exhibition corresponded to that of the Swedish fountain. There also was an eye-catching similarity in the flamboyant displays of decorated vases next to both fountains. In Helsinki the porcelain on show was produced by Arabia which mainly manufactured household ware at the time. In 1876, Arabia, which was later to grow notably both in size and in the versatility of its goods and to gain international fame, was still quite a small company and only known in Finland. It was quite new, having been founded in 1873 by the well-established Swedish firm Rörstrand, whose porcelain vases had also been proudly displayed next to the Molin fountain in Stockholm. Although the decision on where to locate various object classes in the pavilion was certainly not made by a single exhibiting company, it is evident that the Arabia display followed the arrangement of its Swedish predecessor. It is also well-known that Rörstrand transferred plenty of knowhow to Arabia, in terms of its workforce and porcelain models and patterns.[23] Some commentators at the Helsinki exhibition even suggested that Arabia was altogether more Swedish than Finnish.[24]

The similarities between the Stockholm and Helsinki exhibitions suggest that the Finnish organizers were not only closely acquainted with the Nordic exhibition but also willing to act in a similar manner. Indeed, all the principal organizers belonged to the Swedish-speaking high society, whether as businessmen, engineers or architects, and enjoyed many connections to Sweden. Actually, the head commissioner Wahren was himself born in Sweden and had moved to Finland as a young industrialist in the 1830s. Many of the organizers, if not all, had personally visited the Nordic exhibition in Stockholm. An important factor may well have been the architect Höijer's years-long sojourn in Stockholm as a student of the Royal Swedish Academy of Arts in the 1860s.

The key features of the Helsinki exhibition premises and display notably resembled those of the preceding Nordic exhibition in Stockholm. The Finnish manner of conceiving how a pavilion should look and how it should function coincided with the Stockholm pavilion too much to be put down to mere chance. In fact, these analogies and references make visible the interconnectedness of these exhibitions in a similar manner pointed out by Geppert between the imperial exhibitions.[25] Thus, even though the organizers of the Helsinki exhibition were aware and even had first-hand knowledge of more recent great exhibitions, they sought inspiration and followed the example set by the relatively small international exhibition of Stockholm arranged already a decade earlier. This shows that the great exhibitions did not dominate the exhibition medium alone, but it evolved in a more polyphonic interplay.

Yet despite the evident similarities between the settings of the Helsinki and Stockholm expositions, the first Finnish national exhibition was not merely a replica of any previous exhibition. A replication of the Vienna fishbone galleries would have been out of the question for many practical reasons. Moreover, neither was the Stockholm model adopted as such. Whilst in Stockholm the pavilion was located in a central urban area, for example, the Helsinki pavilion was situated in a park. In placing the exposition here, the Finnish organizers followed the method adopted in many other exhibitions. The selection of Kaivopuisto Park as an appropriate location was justified at the initial planning stage by referring to previous 'great exhibitions' which had utilized sites that already possessed gardens and flower arrangements. These were considered to be necessary for any exhibition, but constructing them was expensive. In other words, a park would be a more suitable place in economical terms than the city centre.[26] The explicit reference to great exhibitions is interesting in this context, although it must be borne in mind that different circumstances could also affect the choices. In fact, Forssa Ab's other owner, Henrik Borgström, was an enthusiastic advocate of, and investor in, the Kaivopuisto Park Company.[27] This is a good reminder of the fact that the adoption of previous models and ideas was never carried out in a vacuum, but were connected to a myriad of practical conditions that either favoured or ruled out various alternatives. Whether purposely more or less innovative, the exhibition was in practice an appropriation of various models situated in the prevailing circumstances and, thus, turned out to be a novel hybridized embodiment of the medium

From the Centre to the Margins and Back Again

One of the most acclaimed sections at the Helsinki exhibition was the ethnographic display. It consisted of a collection of peasant utensils and garments, collected by various student unions in their respective home provinces. The truly eye-catching display was the fictitious cottage interiors and human-size mannequins created for the display. The ethnographic section was not a large entity when compared to the industrial products or the fine arts on display. Indeed, in the catalogue the display appeared quite unassuming, as it had been classified as a mere single entry among what appeared like an endlessly long list of handicraft exhibits: 'Finnish Students' Ethnographic Exhibition. Six cottages with folk groups and costumes.'[28] Yet it seems to have attracted plenty of attention and inspired numerous excited descriptions in the printed media all over the country: commentators alternatively eulogized the displays of their respective home provinces, focused on the national perspective or declared their astonishment and pleasure at having witnessed the amazingly life-like appearance of the display.[29]

Ethnographical elements were already powerfully manifested in the *Exposition Universelle* in Paris in 1867, as the participants had been urged to present national traditions in the form of costumes and wax dummies. The Swedish and Norwegian figures at this exposition had attracted special attention, as they consisted of narrative groups and the figures were complemented by landscape paintings. The Scandinavian displays also gained praise and attention six years later in Vienna.[30] The ethnographic displays married scientificity, spectacularity and nationalism and hence became a fashionable part of the industrial and art exhibitions. This element was also envisaged for the Helsinki exhibition. In Finland, the task of collecting the items and of arranging them for display was undertaken by student unions, which were imbued with patriotic and ethnographic interests. According to the most active group in the project – the Viipuri student union – the exhibition committee had expressed its wish that the union would arrange an ethnographic section for the exhibition.[31]

In addition to the great exhibitions of Paris and Vienna, knowledge of ethnographic displays was also conveyed in another manner. Namely, in October 1873 Finnish newspapers reported that a Scandinavian ethnographic collection had been opened in central Stockholm a few days earlier.[32] This small piece of news referred to what was at that time still quite a small collection. It had been arranged by Artur Hazelius and was not only based on narrative groups of mannequins, but importantly presented them in cottage interiors, thereby creating visually impressive dioramas.[33] The following year, a group of Finnish students who were participating in a Nordic student meeting in Sweden had the opportunity to visit the new collection and were impressed with what they saw. Among them was Otto Donner, the curator of the Viipuri student union.

The mannequins of the Hazelian collection were produced by the same sculptor, Carl August Söderman, who had produced the Swedish and Norwegian mannequins for Paris. Hazelius also recommended the same artist to the Finnish student unions, which contacted him for advice and permission to copy his idea.[34] Hence, some mannequins in the Helsinki exhibition were modelled by Sjöström, whilst others were produced by Finnish artists in a similar manner.

The Finnish student unions followed in an international trend evident in the great exhibitions in their ethnographical interest and in the manner they displayed their collections. However, in a most detailed and explicit manner they sought advice and permission from Hazelius and his newly established permanent collection. Thus, in this case, inter-exhibition models were evidently complemented by connections to a permanent museum-like collection.

The paths of exhibitions and museums have in fact intersected numerous times. Some exhibitions gave birth to entire new museum institutions such as the South Kensington, later Victoria and Albert, Museum in London, the heir of the Great Exhibition of 1851, and Pennsylvania Museum and School of Industrial Art, the successor of the Philadelphia Centennial Exhibition in 1876. Furthermore, a plethora of single exhibits and groups of exhibits were incorporated in various existing private and public collections, for example in the Smithsonian Institution.[35] The Helsinki ethnographic display not only followed the example set by both temporary exhibitions and permanent collections, but also contributed to both fora. The collections of the student unions and their cottage-like sets were opened to the public as a small permanent museum – in more or less Hazelian style – soon after the closure of the exhibition and some of the very same mannequins and cottage interiors were sent two years later to the *Exposition Universelle* in Paris. In the French capital they were exhibited in the Trocadero Palace as part of a large ethnographic section to which many countries had sent mannequins dressed in 'national style'. The Finnish display received the highest possible prize – and so did the very similar but larger and more renowned Swedish section near to it.[36] The international recognition then further burnished the reputation of the collection in Finland and thus contributed to ensuring its conservation.

The ethnographic displays provide evidence of how connections between exhibitions consisted of multiple intersections. The contacts were not unilateral, as the direction of flows – those of ideas, models, objects – was not only from great to small exhibitions. Indeed, something produced for and exhibited in a smaller exhibition or in a permanent collection could soon be transferred to be part of a larger exposition. The interpretations and meanings given to any specific display could of course vary, when exhibited in a different context. This has recently been demonstrated by Daniel Alan DeGroff in his comparison of the Swedish and French conceptions of the Hazelian displays in Paris in 1878 and in Stockholm thereafter.[37] Yet, the fact remains that great exhibitions, which were usually praised as the newest and grandest spectacles of mankind, could actually consist of glimpses of other, minor exhibitions, something already presented elsewhere, even in the margins.

Continuity from minor to great exhibitions was also evident through the presence and contributions of individuals. The Finnish commissioner to the Paris universal exhibition in 1878 was Robert Runeberg whose career as an exhibition commissioner would later include the all-Russian exhibition in Moscow in 1882 and another Parisian *exposition universelle* in 1900. The head commissioner of Helsinki, Wahren, served as an international juror in 1878.[38] In a similar manner, a number of companies which had exhibited their products in Helsinki also participated in the Paris exhibition. The exact reproductions of the same displays may not have been possible as Finnish products were scattered among Russian exhibits, but similarities are not lacking. Arabia once again placed an eye-catching vase replete with ornamental serpentine handles, for example, on the highest point of its display.[39] Some times Arabia was even publicly criticized for having exhibited the same objects in different occasions.[40] Furthermore, large and small exhibitions were afterwards juxtaposed in prize-winning product labels, advertisements and letter papers produced by exhibitors in subsequent years and decades. These products displayed the medals of universal, international and national exhibitions side by side.

Thus, various exhibitions were not only combined by shared practices, models and ideas, but were also mobile vis-à-vis the transit of people and objects. Instead of clear itineraries, the paths were twisting and could often intersect.

A National Exhibition – An International Act

Immediately after the inauguration of the Helsinki exhibition, an excited commentator compared it to a mirror in which the Finnish people could not only see their own achievements and handicaps, but were also able to keep abreast with other peoples.[41] The commentary was not very original in the sense that its emphasis on peaceful competition and the comparison between nations was the well-established crux of the exhibition narrative and ideology. However, the comment is interesting and revealing in one respect: the international comparison was supposed to take place in an exhibition explicitly entitled and defined as 'national' and in which no other countries participated.

Although the Helsinki exhibition was marginal in the European and global context – both due to its location and size – the exhibition genre itself seems to have guaranteed that commentators perceived it to have overcome any sense of marginality. The internationality of the exhibition medium was essential in this aspect. It was one of the established and accepted conventions of the day how to participate in international interaction. Thus, the hosting of a national exhibition was fundamentally an international act, which temporarily nullified, or at least reduced, differences and distances between centres and margins.

The first Finnish national industrial and art exhibition was embedded in a varied landscape of displays. It was connected to a number of colossal exhibitions – in London, Paris and Vienna – as well as to more limited international and all-Russian expositions. In addition to the realized exhibitions, the expectations, speculations about, and plans for future events were also a significant part of the international exhibition scene. The expositions had not only been followed via newspaper reports, but a number of Finns – at least hundreds of them – had also gained some personal experience as visitors or exhibitors. A few had even served as representatives of the Finnish delegations or as international jurors.

When planning and constructing the Helsinki exhibition it was self-evident that earlier solutions and practices would be adopted. Antecedent exhibitions created the framework in which the organizers – and visitors – envisaged how an exposition should appear and how it should function.

The examination of the Helsinki case shows how ideas and practices not only flowed outwards from the great exhibitions, but the interconnectedness of exhibitions was complex and multilateral. This is not to say that the relationship between large and small exhibitions would have been equal. The fame and renown of the grandest expositions was undisputed, but other exhibitions could also contribute significantly to subsequent events. The Nordic exhibition, in particular, bore a significant influence on the outcome of the Finnish national exhibition.

Explanations may not only be discovered in the exhibition itself. The Stockholm exhibition enjoyed abundant attention and publicity in Finland and Scandinavia at the time, as it was the largest and most international event arranged in the Nordic countries up to that time. However, on many levels it had not been considered as a great success. The Finnish participation had been especially harshly criticized, both in terms of quantity and quality. Yet, despite these hardships the Stockholm exhibition was a valuable source of knowledge and a model for the Finnish exhibition organizers. In fact, it is also essential to take into consideration the deeper cultural and social context.

The Finnish–Swedish relationship continued to be particularly close due to the long common history of the countries. Despite Finland's new position in the Russian realm, manifold commercial and cultural connections remained in place and anything Swedish was keenly followed in Finland.[42] As mentioned earlier, the organizers of the Helsinki exhibition all belonged to the Swedish-speaking elite and had multiple personal connections to Sweden and Scandinavia.

However, the Swedish models were a double-edged sword as they tended to polarize public debate. Tensions grew especially when the Fennoman movement that emphasized the Finnish language – spoken by the majority of the population – as the foundation of Finnishness gradually gained in strength during the latter part of the nineteenth century. Indeed, many Finnish-speaking visitors of the Helsinki exhibition bitterly criticized it for its Swedishness.[43] Thus, the connections and explicit co-operation in preparing the ethnographic collection and display was also somewhat contradictory. Especially in the Finnish-speaking context, the Finnish elements of the collections were emphasized and the linkage to Hazelius and Sweden was not highlighted.

Instead, the role of Russian exhibitions seems to have remained slight, although Finns did participate in a number of exhibitions arranged in Russia. At the time of the Helsinki exhibition there were no major schisms between Finland and Russia – in fact Emperor Alexander II enjoyed notable popularity and his visit to the first national Finnish exhibition was greeted with enthusiasm. Yet the experiences in Russian exhibitions seem not to have been inspiring. For example, the Finnish commissioner to the agricultural and industrial exhibition in Moscow in 1864 had presented many criticisms of the practical realization of the exhibition, as well as of the related bureaucracy of customs.[44] The criticisms levelled by Robert Runeberg were very similar when he reported on another Muscovite exhibition 18 years later.[45] In Finnish exhibition circles there was also a sense of discontentment at how Finnish exhibits were on many occasions located amongst those from Russia and the request to have a separate Finnish section were declined by the Russian authorities. Altogether, from the Finnish perspective Russia was interesting as a promising, immense regional market area and as the centre of the empire. Yet, it was not considered a worthy source of inspiration for boosting the Finnish economy and propelling Finnish society into modernity. For that purpose, the Finnish gaze was turned elsewhere.

Conclusion

No exhibition could only be a replica of a preceding exhibition or merely a pastiche of mixed models. The final display also necessarily included appropriation, adaptation and, thus, modification. Practice is a powerful model for action, but it is also dynamic, always containing some space for creativity and variation, as Michel de Certeau has shown.[46] As the case of the Helsinki pavilion and its rotunda illustrates, there were numerous potential sources and models. The intersecting paths could be even more multifold, as was the case with the ethnographic displays where the flows zigzagged between Stockholm, Paris, Vienna and Helsinki and between temporary and permanent exhibitions.

Although the encounters between large and small exhibitions were not equal, the latter displays could and did contribute to the medium. Indeed, echoes of any exhibition could be found in subsequent expositions, such as in the form of objects or display arrangements. The collective sum of the diverse exhibitions made the medium more visible and able to exist temporarily and geographically at times and in places other than at single great exhibitions. Thus, instead of a clear chronological chain, diverse exhibitions formed more like a bunch of threads. Together the various embodiments of the exhibition medium could create a presence and sense of durability which surpassed national frontiers as well as the ephemerality inherent in any single exhibition.

Notes

1 On the generic outlines of the Helsinki exhibition, see Harry Röneholm, *Markkinat messut ja näyttelyt* (Helsinki 1945), 41–56; Taina Syrjämaa, 'Making Difference, Seeking Sameness. Negotiating Finnishness and Foreignness in an Exhibition', in *Nordic Perspectives on Encountering Foreignness*, ed. Anne Folke Henningsen, Leila Koivunen and Taina Syrjämaa (Turku, 2009), 27–40; Taina Syrjämaa, 'Näyttelypaviljonki uudenlaisena kansainvälisenä toiminnan ja tulkinnan tilana Kaivopuistossa vuonna 1876', *Historiallinen Aikakauskirja* 1 (2010): 29–46.

2 On the history of the great exhibitions, see, for example, Paul Greenhalgh, *Ephemeral Vistas. The Expositions Universelles, Great Exhibitions and World's Fairs, 1851–1939* (Manchester 1988, 2000). On exhibitions as a mass medium, see Anders Ekström, Solveig Jülich and Pelle Snickars, eds, *1897 Mediehistorier kring Stockholmsutställningen* (Stockholm, 2006).

3 Alexander C. T. Geppert, *Fleeting Cities. Imperial Expositions in Fin-de-Siècle Europe* (Basingstoke, 2010), 4.

4 For a recent synthesis of transnationalism as a research concept, see Peter Kivisto and Thomas Faist, *Beyond a Border. The Causes and Consequences of Contemporary Immigration* (Los Angeles, 2010), 127–143.

5 The area of present-day Finland was a part of the Swedish kingdom for centuries. After the Russo-Swedish war in 1808–09 Finland was incorporated into the Russian Empire as an autonomous Grand Duchy preserving its earlier legislation and the Protestant religion. Furthermore, Swedish continued to be the language of its public administration.

6 Kerstin Smeds, *Helsingfors–Paris. Finland på världsutställningarna 1851–1900* (Helsingfors 1996), 63, 66–67, 71–76, 82–95, 117–122.

7 Smeds, 64–67.

8 For example, 'Helsingfors den 24 Juli', *Helsingfors Dagblad*, 24 July 1875. On the financial basis of the exhibition, see Robert Runeberg, *Berättelse med anledning af Finlands deltagande i Pariser verldsutställningen 1878* (Helsingfors, 1879), 6.

9 Smeds, 68, 375; 'Förteckning å de finska exponenter', *Finlands Allmänna Tidning*, 16 November 1864; 'Belögningar vid den allmänt ryska industriexposition', *Finlands Allmänna Tidning*, 4 April 1866. Smeds' pioneering study on the Finnish participation in exhibitions has given valuable aid for this study. She has, for example, identified these articles in *Finlands Allmänna Tidning*.

10 J. A. von Born's collection, Sarvlax collection, The Society of Swedish Literature in Finland (SLS).

11 Timo Herranen, 'Wahren Axel Wilhelm (1814–1885)', in *The National Biography of Finland*, at http://www.kansallisbiografia.fi/kb/artikkeli/4356/.

12 Copy of Wahren's letter to an unknown person, 7 July 1866. Central archives for Finnish business records (ELKA), Oy Forssa Ab, F1.39.

13 Fredrika Runeberg's letter to Walter Runeberg, 18 June 1866, 7 August 1867, 13 May 1873, 2 September 1874. Published in Fredrika Runeberg, *Brev till sonen Walter 1861–1879* (Helsingfors, 1971); Robert Runeberg's correspondence from Vienna to *Helsingfors Dagbladet*, 10–28 August 1873.

14 On the exhibition and its visiting practices, see Syrjämaa, 'Näyttelypaviljonki'.

15 See Charles Riis's photographs of the exhibition. Photograph collection, Helsinki City Museum. See also the maps in the exhibition catalogue *Yleinen Suomen Näyttelö Helsingissä Vuonna 1876. Wirallinen luettelo* (Helsinki, 1876); *Allmänna Finska Utställningen i Helsingfors År 1876. Officiel katalog* (Helsingfors, 1876).

16 'Suomen yleinen näyttelö', *Hämäläinen*, 24 August 1876.

17 Leila Koivunen, 'Exemplary Foreignness. Foreign Material Cultures in the Service of Finnish Taste and Industry in the 1870s', in Anne Folke Henningsen, Leila Koivunen and Taina Syrjämaa, eds, *Nordic Perspectives on Encountering Foreignness* (Turku, 2009), 41–53.

18 Eeva Maija Viljo, *Theodor Höijer. En arkitekt under den moderna storstadsarkitekturens genombrottstid i Finland från 1870 till sekelskiftet* (Helsinki, 1985), 23–33.

19 Ulf Sörenson, *När tiden var ung. Arkitekturen och Stockholmsutställningarna 1851, 1866, 1897, 1909* (Stockholm, 1999), 54–63, 79.

20 *Yleinen Suomen Näyttelö*, front cover.

21 On Molin's fountain, see Sörenson, 77, 102–103.

22 'Expositionbyggnaden i Stockholm', *Helsingfors Dagblad*, 31 March 1866; 'Utställningen i Stockholm', *Folkvännen*, 27 June 1866; 'Bref från Swerige. Stockholm den 25 Sept. 1873', *Åbo Underrättelser*, 29 September 1873.

23 Elise Kovanen, 'Arabia: 135 Years', in *Arabia: Ceramics, Art, Industry*, ed. Marianne Aav (Helsinki, 2009), 16–18; Marjut Kumela, 'The History of Tableware Design at Arabia 1874–1948', in *Arabia: Ceramics, Art, Industry*, 42–50.

24 'Suomen yleinen näyttely', *Uusi Suometar*, 2 September 1876.

25 Geppert.

26 'Kollisionen ur en annan synpunkt', *Helsingfors Dagblad*, 30 January 1875.

27 Victor Hoving, *Henrik Borgström. En storborgare i det gamla Helsingfors* (Helsingfors, 1949), 103–121.

28 *Yleinen Suomen Näyttelö; Allmänna Finska Utställningen.*

29 'Från Utställningen', *Folkvännen*, 23 August 1876; 'Suomen Yleinen Näyttelö', *Hämäläinen*, 14 September 1876; 'Suomen yleinen näyttelö 1876', *Satakunta*, 23 September 1876.

30 Bjarne Stocklund, 'Between Scenography and Science: Early Folk Museums and Their Pioneers', *Ethnologia Europaea. Journal of European Ethnology* 33, no. 1 (2003): 25–26. On Scandinavian ethnographic interests, see also Mattias Bäckström, *Hjärtats härdar. Folkliv, folkmuseer och minnesmärken i Skandinavien, 1808–1907* (Örlinge, 2012).

31 Minutes of the Wiborg student union, 22 January 1876, National Library of Finland, Special collections, WiO:Ca1.6.

32 'Skandinavisk-ethnografiska samlingen', *Hufvudstadsbladet*, 30 October 1873.

33 Magdalena Hillström, 'Contested Boundaries: Nation, People and Cultural History Museums in Sweden and Norway 1862–1909', in *Culture Unbound. Journal of Current Cultural Research* 2 (2010): 583–607.

34 Artur Hazelius' letter to E. G. Palmén, 11 April 1876. National Library of Finland, Special collections, Coll. 165.5.

35 Robert W. Rydell, 'World Fairs and Museums', in *A Companion to Museum Studies*, ed. Sharon Macdonald (Chichester, 2011), 135–151; Bruno Giberti, *Designing the Centennial. A History of the 1876 International Exhibition in Philadelphia* (Lexington, 2002), 179–182, 194–197.

36 Runeberg, 13.

37 Daniel Alan DeGroff, 'Artur Hazelius and the Ethnographic Display of the Scandinavian Peasantry: A Study in Context and Appropriation', *European Review of History/Revue europénne d'histoire* 19, no. 2 (2012): 229–248.

38 Letter from the French Minister of agriculture and commerce to A. W. Wahren, Paris, 8 June 1878. Central archives for Finnish business records (ELKA), Oy Forssa Ab, F1.14; Robert Runeberg, *Berättelse om Finlands deltagande i allmänna ryska konst- och industriutställningen i Moskva 1882* (Helsingfors, 1883).

39 Photographs of the Finnish displays in the Paris world exhibition. Photographic collection, National Board of Antiquities.

40 Kumela, 50.

41 'Första Allmänna finska utställningen i Helsingfors', *Folkvännen*, 5 July 1876.

42 Max Engman, *Pitkät jäähyväiset. Suomi Ruotsin ja Venäjän välissä vuoden 1809 jälkeen* (Helsinki, 2009).

43 See, for example, 'Suomen ensimmäinen yleinen näyttelö Helsingissä', *Tampereen Sanomat*, 22 August 1876.

44 *De finska kommissariernes berättelse on expositionen af jordbruks- och landtmannaindustrialster, i Moskwa om hösten år 1864* (Helsingfors, 1865).

45 Runeberg.

46 Michel de Certeau, *The Practice of Everyday Life* (Berkeley, Los Angeles and London, 1984, 2008).

Bibliography

Manuscript and Photographic Sources

Central archives for Finnish business records (ELKA), Oy Forssa Ab, F1.14, F1.39.

Helsinki City Museum, Photograph collection.

National Board of Antiquities, Photographic collection.

National Library of Finland, Special collections, Coll. 165.5; WiO:Ca1.6.

The Society of Swedish Literature in Finland, Sarvlax collection, J. A. von Born's collection.

Newspapers

Finlands Allmänna Tidning

Folkvännen

Hämäläinen

Helsingfors Dagblad

Hufvudstadsbladet

Satakunta

Tampereen Sanomat

Uusi Suometar

Åbo Underrättelser

Printed Primary Sources

Allmänna Finska Utställningen i Helsingfors År 1876. Officiel katalog. Helsingfors, 1876.

De finska kommissariernes berättelse on expositionen af jordbruks- och landtmannaindustrialster, i Moskwa om hösten år 1864. Helsingfors, 1865.

Runeberg, Robert. *Berättelse med anledning af Finlands deltagande i Pariser verldsutställningen 1878*. Helsingfors: Hufvudstadsbladets tryckeri 1879.

Runeberg, Robert. *Berättelse om Finlands deltagande i allmänna ryska konst- och industriutställningen i Moskva 1882*. Helsingfors: J. Simelii arfvingars tryckeri 1883.

Yleinen Suomen Näyttelö Helsingissä Vuonna 1876. Wirallinen luettelo. Helsinki 1876.

Secondary Sources

Aav, Marianne, ed. *Arabia: Ceramics, Art, Industry*. Helsinki: Designmuseo, 2009.

Bäckström, Mattias. *Hjärtats härdar. Folkliv, folkmuseer och minnesmärken i Skandinavien, 1808–1907*. Örlinge: Gidlunds Förlag, 2012.

de Certeau, Michel. *The Practice of Everyday Life*. Berkeley, Los Angeles and London: University of California Press, 1984, 2008.

DeGroff, Daniel Alan. 'Artur Hazelius and the Ethnographic Display of the Scandinavian Peasantry: A Study in Context and Appropriation'. *European Review of History/Revue européenne d'histoire* 19, no. 2 (2012): 229–248.

Ekström, Anders, Solveig Jülich and Snickars, Pelle, eds. *1897 Mediehistorier kring Stockholmsutställningen*. Stockholm: Statens ljud- och bildarkiv, 2006.

Engman, Max. *Pitkät jäähyväiset. Suomi Ruotsin ja Venäjän välissä vuoden 1809 jälkeen*. Helsinki: WSOY, 2009.

Geppert, Alexander C. T. *Fleeting Cities. Imperial Expositions in Fin-de-Siècle Europe*. Basingstoke: Palgrave Macmillan, 2010.

Giberti, Bruno. *Designing the Centennial. A History of the 1876 International Exhibition in Philadelphia*. Lexington: University Press of Kentucky, 2002.

Greenhalgh, Paul. *Ephemeral Vistas. The Expositions Universelles, Great Exhibitions and World's Fairs, 1851–1939*. Manchester: Manchester University Press, 1988, 2000.

Herranen, Timo. 'Wahren Axel Wilhelm (1814–1885)'. In *The National Biography of Finland*, at http://www.kansallisbiografia.fi/kb/artikkeli/4356.

Hillström, Magdalena. 'Contested Boundaries: Nation, People and Cultural History Museums in Sweden and Norway 1862–1909'. *Culture Unbound. Journal of Current Cultural Research* 2 (2010): 583–607.

Hoving, Victor. *Henrik Borgström. En storborgare i det gamla Helsingfors*. Helsingfors: Söderström & Co., 1949.

Kivisto, Peter, and Thomas Faist. *Beyond a Border. The Causes and Consequences of Contemporary Immigration*. Los Angeles: Pine Forge Press, 2010.

Koivunen, Leila. 'Exemplary Foreignness. Foreign Material Cultures in the Service of Finnish Taste and Industry in the 1870s'. In *Nordic Perspectives on Encountering Foreignness*, edited by Anne Folke Henningsen, Leila Koivunen and Taina Syrjämaa, 41–53. Turku: University of Turku, 2009.

Kumela, Marjut. 'The History of Tableware Design at Arabia 1874–1948'. In *Arabia: Ceramics, Art, Industry*, edited by Marianne Aav, 42–50. Helsinki: Designmuseo, 2009.

Röneholm, Harry. *Markkinat messut ja näyttelyt*. Helsinki: Suomen Messut Osuuskunta, 1945.

Runeberg, Fredrika. *Brev till sonen Walter 1861–1879*, edited by Karin Allardt Ekelund. Helsingfors: Svenska Litteratursällskapet i Finland, 1971.

Rydell, Robert W. 'World Fairs and Museums'. In *A Companion to Museum Studies*, edited by Sharon Macdonald, 135–151. Chichester: Wiley-Blackwell, 2011.

Smeds, Kerstin. *Helsingfors–Paris. Finland på världsutställningarna 1851–1900*. Helsingfors: Svenska Litteratursällskapet i Finland & Finska Historiska Samfundet, 1996.

Stocklund, Bjarne. 'Between Scenography and Science: Early Folk Museums and Their Pioneers'. *Ethnologia Europaea. Journal of European Ethnology* 33, no. 1 (2003): 21–35.

Syrjämaa, Taina. 'Making Difference, Seeking Sameness. Negotiating Finnishness and Foreignness in an Exhibition'. In *Nordic Perspectives on Encountering Foreignness*, edited by Anne Folke Henningsen, Leila Koivunen and Taina Syrjämaa, 27–40. Turku: University of Turku, 2009.

Syrjämaa, Taina. 'Näyttelypaviljonki uudenlaisena kansainvälisenä toiminnan ja tulkinnan tilana Kaivopuistossa vuonna 1876'. *Historiallinen Aikakauskirja* 1 (2010): 29–46.

Sörenson, Ulf. *När tiden var ung. Arkitekturen och Stockholmsutställningarna 1851, 1866, 1897, 1909*. Stockholm: Stockholmia Förlag, 1999.

Viljo, Eeva Maija. *Theodor Höijer. En arkitekt under den moderna storstadsarkitekturens genombrottstid i Finland från 1870 till sekelskiftet*. Helsinki: Suomen muinaismuistoyhdistys, 1985.

14

The Last Exhibition of the Italian Colonial Empire: Naples 1938–40

Giovanni Arena

Introduction: The History of Italian Colonial Exhibitions

The story of the *Triennale d'Oltremare* (1938–40), the last Italian colonial exhibition, is closely linked to the political and the cultural politics of Fascism. The event was conceived as a great colonial exhibition which was to enhance the economic and cultural function of its host, Naples, within the new colonial empire that was being born at the time. The *Triennale d'Oltremare* aimed to document the work for the conquest of the empire, to illustrate the characteristics of the various zones of the Italian lands overseas, and present the resources of the Motherland in respect of the valuation of lands overseas.

The show was meant as a response to the great Colonial Exposition in Paris of 1931, in which the Italian Fascist government participated, while it had still not completely resolved the problem of Libyan anti-colonial resistance and when it did not yet rule Ethiopia. The number of visitors and the resonance of the 1938–40 show were, however, much lower than the organizers hoped for. Just a month after the opening on 9 May 1940, Italy entered the war and the exhibition that was supposed to crown a victorious empire was closed. The celebration of the new empire corresponded with the eve of its collapse. What was supposed to represent the cutting edge of colonial Fascist propaganda came to be representative of the gap between the aspirations and the means of the Italian colonial policy and the gap between Italy and other colonial powers.

This *Triennale* also integrates various aspects of the art world of its time: art and propaganda; strategy of acquiring consensus and aesthetics of the colonial policies. The exhibition of Naples has a very specific format, powerfully symbolic, and shows an image that overwhelms the architecture. Created in a moment in which each creative act had to be above all a political act, the show was based on the supremacy of the observer through the visual impact of the setting. This text therefore examines the political motivations of staging a great exhibition project, likely to be repeated every three years, in the time of a looming world crisis and looks at the ways the exhibition was intended as a tool for colonial propaganda. At the same time I will try to analyse some aspects related to the museographic and artistic project.

Colonial Exhibitions in Italy between the 1920s and the 1940s

During the 1920s, the interaction between colonial discourse and colonial propaganda in Italy had a special significance for two reasons. First, colonial propaganda was crucial because the Italian colonial empire was not large and soon proved to be expensive to maintain. Therefore it became necessary to create a strong consensus about the imperial status in Italy. The conquest of Ethiopia in 1936 increased the need to create an image of a legendary victory. Secondly, the Italian national identity was weak compared to many other identities circulating in the peninsula, both pre-existing and co-present: regional identities, ideological and religious identities were often stronger than the national one. For reasons intrinsic and extrinsic, the space for colonial propaganda in Italy was large, especially in the period between the two world wars.

What made such propaganda especially important overseas for both liberal and Fascist Italy was the aura of prestige that the State emanated toward the public, and outward to the ensemble of great powers. Italian fascism introduced several novelties, most importantly in this context the breadth and character of the centralist institutions and media which were enabled through colonial propaganda. Exhibitions and colonial museums played an important part in this and they needed to achieve a double aim – to empower the acculturation process overseas and to amplify the support to the colonial policy.

Fascism, which had been on the rise in Italy from the 1920s, strongly accentuated the role of exclusively colonial exhibitions in relation to the construction of a national identity and gave it an unrealistic colonial policy. Prior to that, the important colonial exhibition of Genoa in 1904 was organized to support and promote the cause of expansion events that had been held also in Palermo (1891), Turin (1898), Florence (1903) and Milan (1906).[1] Various events were organized even after the end of the First World War, but it was during the Fascist period that the Ministry of Colonies gave a particular impetus to the organization of colonial exhibitions both in Italy and abroad. Between 1920 and 1940, many book fairs, colonial expositions and exhibitions of colonial art were held for specialists and scholars. Very successful were the First Colonial Art Exhibition in Rome in 1931 and the Second Colonial Art Exhibition in Naples in 1934–35. There were also agricultural fairs and exhibitions of colonial crafts as well as more general colonial exhibitions targeted at a non-specialist public that aimed at strengthening the 'colonial consciousness' of the Italians and at reinforcing the myth of Africa. Objects collected by explorers in Africa, 'relics' of soldiers who fell in the field, and special ethnographic collections that began to arrive in provincial towns, universities and specialized colonial institutions were organized in the meantime in permanent exhibitions or museums. Simultaneously, African fauna was also brought to zoos around the peninsula.

The structure and content of the early colonial exhibitions that were organized in Italy before 1920 was subjected to general interest in the colonies which took precedence over economic and commercial goals and, in essence, propaganda. In contrast, in the colonial exhibitions of the 1930s, a historical-ideological significance moved above the exclusively commercial role of the previous exhibitions. In fact, from simple 'fairs' of the exotic, these events became complex narratives with different levels of communication and reading.

The effectiveness of the new Fascist propaganda in Italy was also put on display in the colonies. An example can be seen in the first *Fiera Campionaria di Tripoli* (The National Exhibition of Tripoli, 1927), initiated by Benito Mussolini, which, therefore, became the first major statement of the economic revival of Tripoli after the military reconquest. The Exhibition aimed at informing the visitors about the colony of Libya – the idea was to create an ephemeral city where small presentations of the local life and labour were constructed alongside many stalls and various regional and community halls. 'Orderly' and 'cleaned-up', the Arabs and Jews of Tripoli were brought within the enclosure of the Exhibition.[2]

The village featured a 'typical' Tripolitan market (*sūq*). The space was full of palm trees and populated by native Libyans who were distributed according to the different activities that they carried out in all the areas of the Fair as attractions for visitors. Silversmiths, goldsmiths, carpet weavers, ivory craftsmen and manufacturers of thousands of decorative objects worked there in the small space of their shops for two months. Using traditional instruments, they produced silk fabrics, wool and cotton, articles of gold, silver, leather, carpets, Misurata Tauorga mats, household items and ceramics. Next to the *sūq*, there was a reproduction of an Arab house in all its details, an Arabic mill in animal traction and a space used for spinning wool as well as a space for dancers and musicians. Professor Renato Bartoccini, an expert on Tripolitania classical antiquities, was responsible for the ethnographic exhibition in the Colony – the Arab life was framed here in all its manifestations.

The Fair of Tripoli was repeated in subsequent years and was organized by the *Ente Autonomo Fiera Campionaria di Tripoli* (The Autonomous Institute of the Tripoli Trade Fair). Starting in 1930, the *Fiera Campionaria di Tripoli* was officially approved by the Union of International Fairs (UFI), an association founded in Milan in 1925 to bring together the most important national and international organizations in the exhibition sector. With this acknowledgment the *Fiera Campionaria di Tripoli* was transformed into an international fair and *interafricana*, the first international exhibition in Africa.

The Fair occupied an area of 52,000 square metres and differed in many ways from the earlier trade-oriented exhibitions which, like the Great Exhibition of

1851, used the paradigm of a 'city on show'. Such places participated in the dynamics of the industrial society. Through these colonial exhibitions Fascism gradually appropriated the model used to promote modernist exhibition and its own political ideology. The Fair of Tripoli became a propaganda tool of the privileged regime, which tried to show its colonial conquests and celebrate its colonial history. These early events in Africa continued until the outbreak of World War II and the regime was able to take inspiration from them to achieve similar exposures in Italy. Italian exhibitions were organized using old ideas but with extremely modern forms of communication.

After the 'general rehearsals' of the two events of Tripoli in 1927 and 1928, the first significant colonial exposition organized in Italy was the *Mostra Nazionale Coloniale* (The Colonial National Exhibition) show in Turin in 1928. Many descriptive articles and images published in colonial magazines describe the exhibition, which occupied an extensive area on the left bank of the river Po in the Pilonetto area.[3] There appears to be a large colonnaded portico divided symmetrically by a pavilion from which the streets radiate and connect the exhibition of animal colonies and four colonial villages, each inhabited by natives of the respective colony. Enclosed between the two arms of the arches, the hall of honour dominated the entire complex, the site of which was characterized by marble columns with the fasces. Around the arcade, decorated in tiles, a reconstruction of an African town with a market and craft shops opened to the visitors' eyes.

After the Turin exhibition, the dialectic between the colonial discourse and propaganda became exceptionally stronger. In 1930, Belgium celebrated the centenary of its independence with the organization of two exhibitions: the *Exposition Coloniale d'Anvers* (The Colonial Exposition of Antwerp) and *L'Exposition Internationale de Liège* (The Industrial Exhibition in Liège).[4] The participation of Italy in both shows, together with 22 other states, can be seen as a part of the large project of intensification of colonial propaganda as well as a move against other colonial powers. The relatively small Italian colonial regime tried to compensate for its size by resorting to a greater presence in international colonial exhibition. The exhibition in Antwerp, for instance, which was divided into national pavilions, included activities of valorization of the colonizing nations which were aimed at enhancing their political, social and economic status and at commercial traffic.

The project of the Italian Pavilion of Antwerp, which, among other things, contained objects from the Colonial Museum of Rome, was designed by architect Giovanni Chevalley. Located near the main entrance to the exhibition, it consisted of a series of buildings with large windows and large skylights. Comparing the aesthetics of the Italian pavilion with the most significant ones of, for example, the Netherlands, France, Persia and the Belgian Congo, the Italian attempt to impose a reminder of the Roman era was evident. The regime chose to highlight the past and imperial Rome as historical legitimacy of its own colonial ambitions by showing symbols of the ancient empire next to images and objects from the colonies.

One of the most significant aspects of the Italian participation in fairs and exhibitions was to demonstrate its connection to the process of modernization of overseas territories, the development that was viewed as an extension of the economic system of the metropolis. The Italian colonies were represented through displays that illustrated the accomplishments of private and government-sponsored companies that were active in the exploitation of natural resources and the organization of local industries. The viability of these industries and the colonial economy in general, however, was questionable. These exhibitions promoted the image of prosperous territories that were systematically undergoing development according to metropolitan standards.[5]

Alongside these exhibits, the results of the substantial financial investment by the Italian government into infrastructures were presented and included the construction of new road networks, the improvement of water supply and health systems and the establishment of public institutions. While neither the political dimension of these exhibitions nor their role as propaganda could have been dismissed, they were more than just visual presentations aimed to encourage further economic development; indeed, they were a marketplace in which the products of Italy's colonial possessions were both put on display and sale.

The second and equally important aspect of the Italian colonial representation in fairs and exhibitions abroad was that they were important vehicles for the representation of indigenous cultures. These cultures were presented in both the content of the exhibits, which included ethnographic studies of the local populations and the display of their indigenous crafts and their means of presentation. From a stylistic point of view, the pavilions were hybrids of vernacular architecture and the architectural conventions of exhibition design, invoking a complex negotiation between indigenous and metropolitan worlds. The identity of the local populations was simultaneously being constructed by the norms of anthropology and ethnography, just as their indigenous craft production was subjected to a substantial redefinition by the Fascist authorities. These exhibitions can thus be understood as an extension of the direct manipulation and control of the native body under the aegis of colonialism.[6] The complex interaction between the political, economic and representational discourses in these colonial displays creates a powerful trajectory that builds up towards the *Triennale d'Oltremare*.

The City and the Colonial Space in 1938–40: From the Project to the Opening of the Triennale

The *Mostra Triennale delle terre italiane d'Oltremare* (The First Triennial Exhibition of Italian Overseas Territories, 1940) is one of the most spectacular examples of the exhibition of public and private colonial collections in Europe. The Bylaws of the *Ente Autonomo 'Triennale d'Oltremare* were approved by the

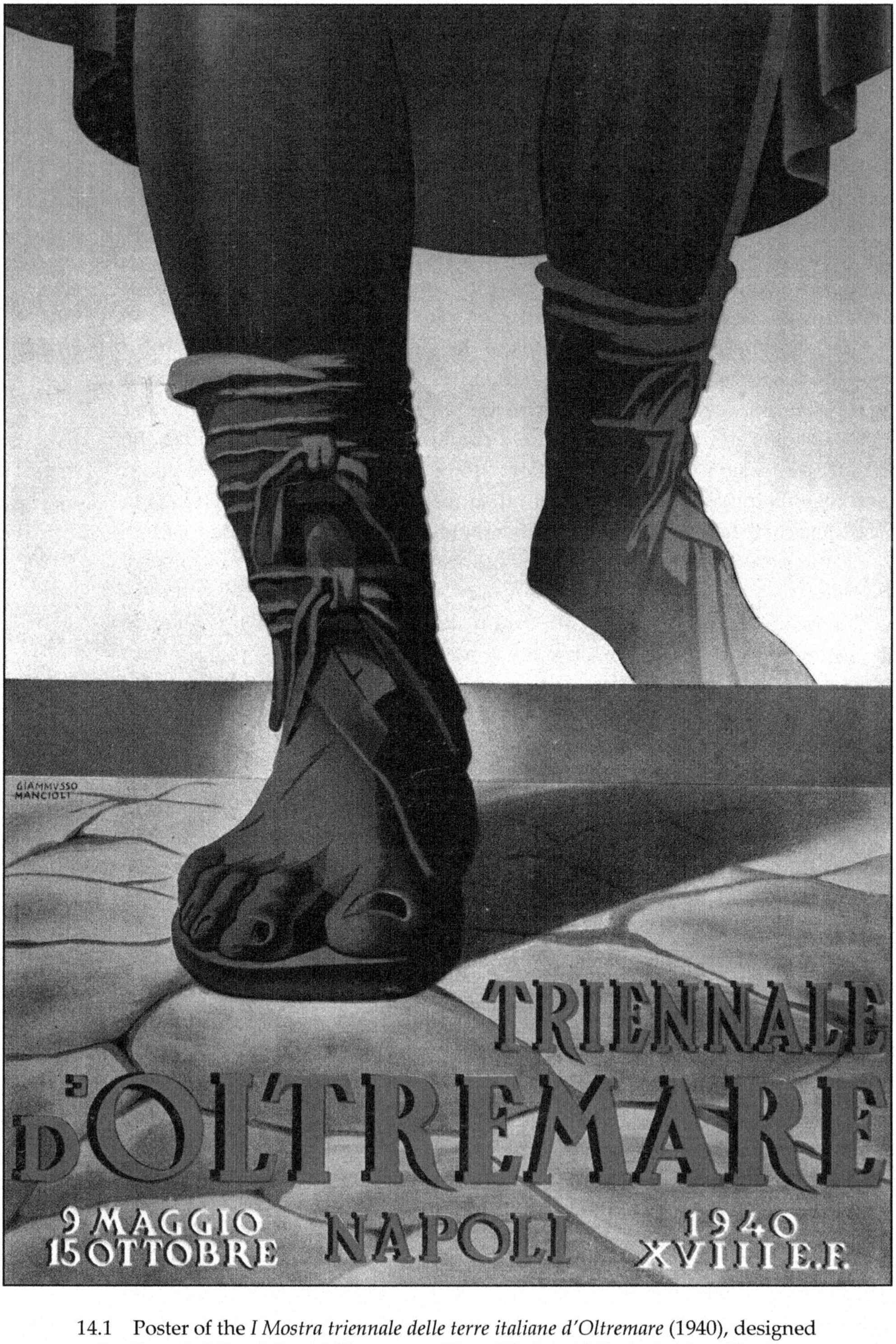

14.1 Poster of the *I Mostra triennale delle terre italiane d'Oltremare* (1940), designed by Corrado Mancioli and Giammusso Ugo, Naples, 1940. Author's collection

Royal Decree of 4 April 1938, no. 2215, and set the exhibition plans, officially opened in 1940, into motion. For its establishment the project received financial contributions from many different institutions, such as the General Government of Africa Orientale (East Africa), the governments of Libya and Somalia, the City of Naples, Naples Province, the Provincial Council of Corporations of Naples, and the *Banco di Napoli* (The Bank of Naples).

The *Triennale d'Oltremare* and its related imperial politics provide the final, and perhaps most telling, example of the evolution that took place in Italian colonial representations in exhibitions and fairs at home and abroad. It was organized to celebrate the expansion of the colonized territory by means of a multifaceted communication approach, abundance of financial resources, modern display techniques and the deployment of well-known propaganda resources concerning the opportunities of Italy's overseas empire.[7]

The *Triennale d'Oltremare* was presented as the largest and most complete survey of the Italian expansion (Fig. 14.1). Its aim was to contextualize Italy's colonial empire in Africa in relation to the historical legacy of conquest and dominance in the region, linking again the accomplishments of Ancient Rome to the Fascist present. It was also the culmination of a process of the development of colonial exhibitions in Italy and abroad that began around the time of Italy's initial colonial conquest of Eritrea in 1890.[8]

As mentioned above, these representations were part of the Italian government's propagandistic effort to create a greater knowledge of these possessions and assert its image as a major colonial power in the international arena. These activities, often undertaken with the participation of groups of colonial supporters like the *Istituto Coloniale Italiano* (The Italian Colonial Institute), included scientific, statistical and historical expeditions. In particular, the *Istituto Coloniale Italiano* was the promoter of the organization of academic conferences and of the publication of books and periodicals reporting on the Italian colonies.[9] Seen in this context, these exhibitions were a carefully measured political programme that responded to the lack of knowledge and interest of Italian society in its colonial empire in Africa.

The intention of the Italian authorities was also to raise support for their colonial activities by promoting awareness of the realities of the colonization process. This would effectively disarm the idea that Italy's colonies were barren lands with no natural or cultural value. However, these exhibitions were not only intended to create popular support within Italy, they were also aimed at the international public and particularly at other colonizing nations such as Britain and France. In this broader European context, these colonial displays can be understood as a metaphorical compensation for the complex of inferiority, which Italy, as the most recent colonizing nation, carried into the colonial enterprise, and a rhetorical expression of Italy's colonial destiny in Africa.[10]

One of the strongest aspects of the idea behind the *Triennale d'Oltremare* was its integration with the surrounding topography in the significant historical and cultural landscape of the Fuorigrotta district in Naples. However, in

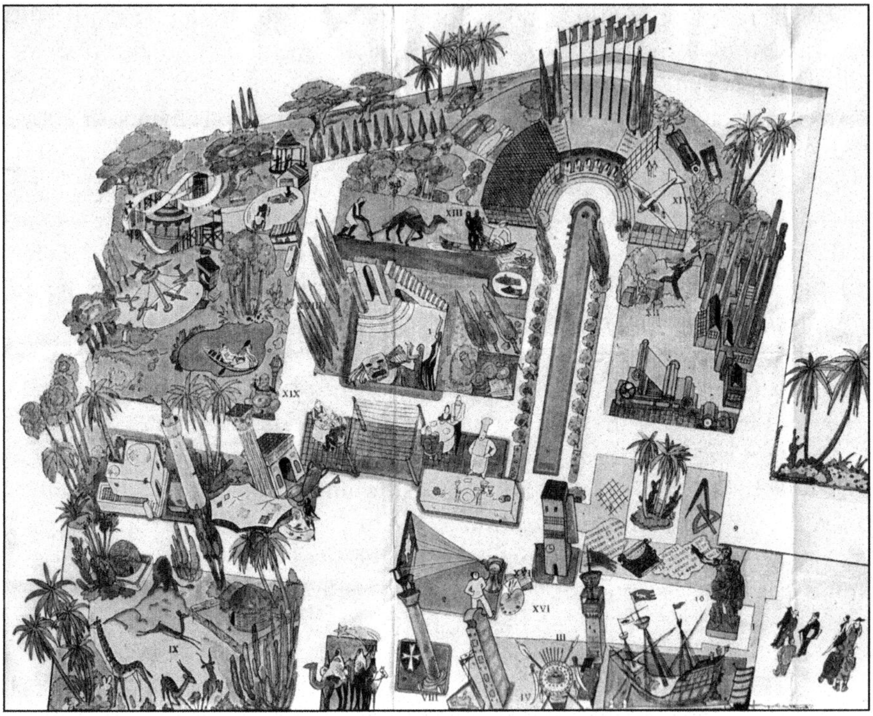

14.2 Planimetry with drawings of *I Mostra triennale delle terre italiane d'Oltremare* (1940), from *The Guide to Individual Exhibitions*, Editions Triennale d'Oltremare, 1940. Author's collection

doing so, this urban proposal radically transformed the existing landscape through a systematic programme of transplanting species from Italy's colonial possession. The result of this effort was a hybrid between metropolitan and colonial landscapes.

The fusion of these two contexts, metropolitan and colonial, is evident in the design and layout of the exhibition. These qualities can be seen in the arrangement of the various pavilions, which were linked through a complex structure of streets, fountains and landscape that lent an excellent quality to the site that was evocative of a colonial environment (Fig. 14.2). On the other hand, the generous use of landscape and relaxed spatial planning of the *piano regolatore* (the development plan) of Carlo Cocchia and Luigi Piccinato, the architects who designed the general project of this exhibition, reflected contemporary planning principles, which sought a well-balanced interaction between the building and the environment.

The Fascist regime put a great effort into the organization of the Exhibition. Vincenzo Tecchio was appointed National Director at the head of the teamwork, and the organization was also followed by the Ministero dell'Africa Italiana (the Ministry of Italian Africa). Each sector benefited from the work of counsellors, for instance Carlo Zaghi, Italian journalist and historian, a scholar of colonial Africa and Italy, who was responsible for the historical sector. Pietro Badoglio, Italian military general and politician responsible for the war of conquest in Ethiopia, who organized the military sector, while Giovanni Guerrini and Giorgio Quaroni, the representatives of the *Sindacato fascista di Belle Arti* (The Fascist Union of Artists of Fine Arts) were asked to run the artistic project. Many initiatives were planned around the Exposition in a way typical for colonial and other large exhibitions. For instance, international conferences on the theme of the colonies, colonial art exhibitions and retrospectives of colonial art were held and a zoo with African fauna and a tropical aquarium was built here.

Content and Ideology of the Exhibition

The influences of the official propaganda on the exhibition design and content were visible around the entire exhibition ground. The exhibition was organized in three sectors – historical, geographical and manufacture. Thirty-six permanent pavilions were built here as showrooms alongside three theatres, two restaurants, an Olympic-size swimming pool and botanical greenhouses as well as an entertainment park (Fig. 14.3). The Exhibition was attended by representatives from the Italian colonies not only in Africa but also from Albania and the Greek islands of the Dodecanese, and the most powerful European countries were also invited to visit. The international outreach was evident from the range of exhibition guides which were available in English, French, German and Spanish.

Three specific typologies of the exhibition sections can be detected in the *Triennale*.[11] The first section was an exhibition dedicated to the cult of Roman civilization, characterized by a dual propagandistic objective: the glorification of the Roman Empire, in close relationship with the foundation of the new

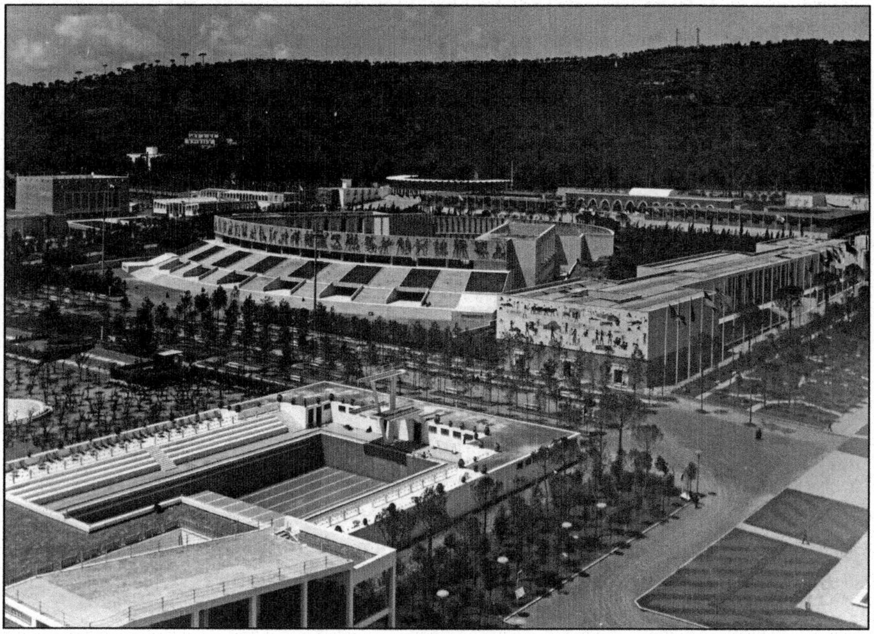

14.3 Naples, *I Mostra triennale delle terre italiane d'Oltremare* (1940), restaurant with an outdoor pool, the Arena Flegrea, next to the Pavilion of Sanità, razza e cultura (Health, Race and Culture). Archives Plinio Marconi, Rome

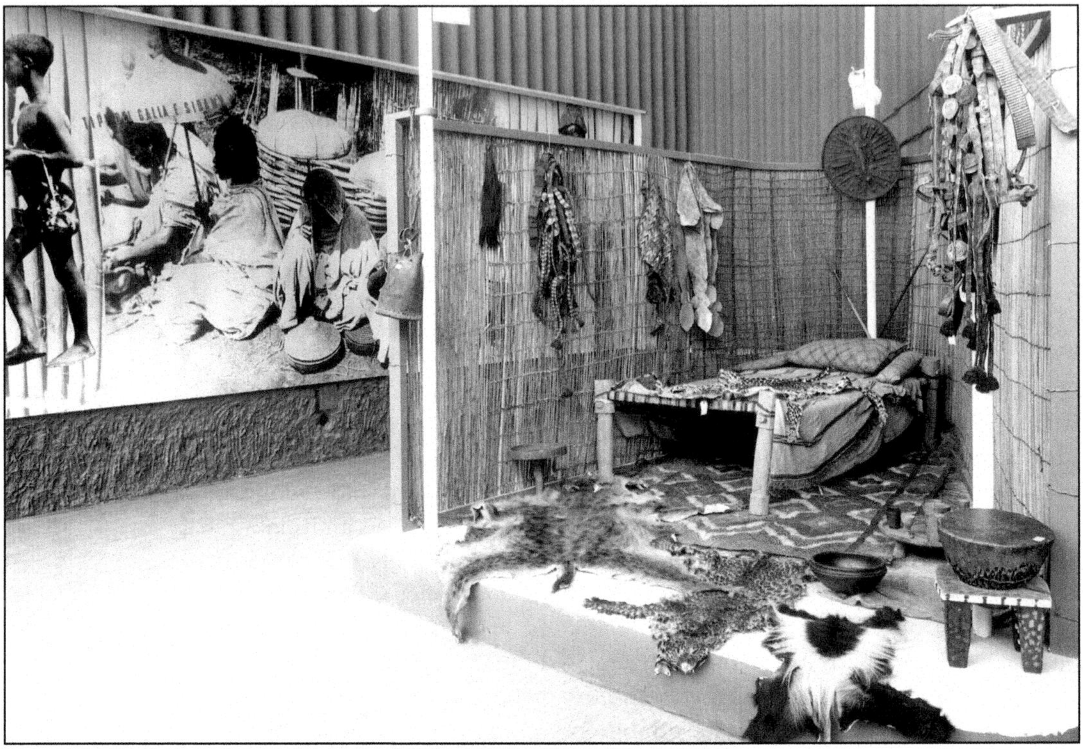

14.4 Pavilion Africa Orientale Italiana (Italian East Africa), installation by the Italian architect Mario Zanetti, from *Emporium* XVLI, no. 8, August 1940, XVIII, vol. XCII, no. 548, Edition Instituto italiano d'arti grafiche, Bergamo, p. 90

Italian Empire, and the epic Fascist conquest of overseas territories.[12] The archaeological findings from the ancient Roman conquests in Africa displayed here were used as a link between the colonial past and present.

The second type is related to the fact that colonial exhibitions were characterized by the overall eclecticism of the collections on display: the memories and photographs of explorers, weapons and local products or artefacts were all selected and exposed with the purpose of demonstrating the wealth of the conquered territories, as well as the existence of new territories, ready to be conquered by Mussolini's new Italy. The core part of these exhibits were pictures and photo-mosaics, which not only covered entire walls of the pavilions, but were incorporated into the architecture, thus becoming an integral part of the exhibition design. The geographical section contained anthropological and ethnographic documentation: each exposition was intended to show the life of the different regions of the Empire, from Africa to Albania and the Dodecanese Islands. Within each territorial partition, the variety of the social and material life was deployed.[13] As in earlier colonial exhibitions, natives, craft shops, reproductions of houses and public places evoking the atmosphere of Africa were an essential element (Fig. 14.4). Such a simulation of scenes of African life had the intent of seducing visitors' emotions.[14] A number of live shows were staged here, not to document the actual complexity of relationships in daily life, but to depict African countries as exotic lands and destinations for heroic deeds and adventures.

The third type contained merchandise exhibits focused mainly on the production and work sectors and offered large wall decorations and unexpected space articulations, for example the spectacular layout of the Mostra dell'Elettrotecnica (the Electrotechnics Show) realized by the Futurist artist Enrico Prampolini: he exhibited a lit model of the Italian peninsula, showing the potential of electricity production in Italy by a large illuminated spiral that started in the centre of the peninsula and led up to the ceiling. These exhibits, homed in permanent buildings divided into five sub-sectors, were meant to show both the economic situation of the Italian colonies and examine their reinforcing role in the Italian national plan for self-sufficiency.

During the interwar period, Naples was historically and geographically disconnected from the current debates in the field of art and architecture. The *Triennale d'Oltremare* offered the city the chance to develop a unique concept of territorial planning, of the role of the city as well as of the ephemeral architecture and artistic features.[15] The Exhibition was not only conceived as the finishing intervention into the urban district of Fuorigrotta but, mainly, as a future vision for the renewal of the whole city.[16] The *Triennale* became the amplifying core of the function of Naples in relation to activities closely linked to the life of the empire: it was thought of as a real 'idea of the city' whose trade and touristic sectors where going to play a key role.[17]

Architecture and art played a fundamental role in the *Triennale*, not only in the general project but also in the specific exposition choices, which were aimed at making the visit simple and enjoyable. The wide range of displays undertaken by the designers of the exhibitions, documented by a few fragments that have survived, became an exemplary model for future Italian exposition design.[18] The recurring subjects, such as imperial eagles, fasces, Roman archaeological fragments combined with the standard display apparatus, including large photomontages and written comments on the exhibited works, did not restrain some artists and architects from achieving high-quality exposition stands.

Propaganda campaigns of the thirties and forties in Italy were mainly based on several aesthetic tendencies: the first one was the model of the Exhibition of the Fascist Revolution (1932), inspired by the experimental futuristic art and implemented through the extensive use of photomontages and typographic techniques.[19] The other tendency, introduced a little later, was inspired by rationalism and was promoted by those artists and architects who worked at the Milan Triennial Expositions including Edoardo Persico, Marcello Nizzoli, Giuseppe Terragni and Giuseppe Pagano, a group of modernist architects who embraced the international architectural language.[20]

The *Triennale d'Oltremare*, too, was designed and built using the model of the 1932 Exhibition of the Fascist Revolution, which was held in Rome to commemorate the first ten years of the Fascist government. Many young architects and artists took part in the preparation and devised new models of construction in Naples. A thematic exhibition was subdivided into several sections, each devoted to a particular aspect of the Exposition theme, covering the main stages in the history of the Fascist government from its birth to the regime's rise to power. From the outset, the exhibition was conceived not as an objective representation of the facts, based solely exposure of historical documents, but as a celebratory work and propaganda that was to influence and emotionally engage visitors. For this reason, in addition to historical figures such as Dino Alfieri and Luigi Freddi were called to cooperate exponents of various artistic movements of the time including Mario Sironi, Enrico Prampolini, Gerardo Doctors, Adalberto Libera and Giuseppe Terragni.

In an attempt to validate Italy's position in Europe and in the world, the *Triennale d'Oltremare* also made a recurrent comparison between the history before the colonial conquest and the colonial situation. Factual data, photomontages and decorative panels were used to show simulations of production and the effects of 'modernizers' on the traditional technologies. In comparison, approach to colonial objects at previous expositions was simpler and similar to the fairs in Tripoli, where exhibits consisted only of commercial products from the colonies without any graphic and photographic support.

The variety of exhibits on show at the Naples *Triennale,* both of a colonialist formulation and produced according to the logic of 'spectacular' modernist settings and organized following the experimental techniques of the Futuristic movement, were not far away from the typical 'propaganda aesthetics'. Such aesthetics was characterized by the use of all possible visual instruments, photographic and artistic, and was intended for the dissemination of Fascist myths as well as for education of the masses. In these particular cases exhibits were aimed at the exaltation of the myth of the Empire and of the colonial conquest. In such a context the exhibits were converted into purely symbolic elements capable of producing emotional reactions; the visitor was flooded by successive waves of environmental reconstructions, images, documents, photographs of the Italian colonies, stylized silhouettes, models, sculptures, murals and giant and mosaic photographs recalling places, events and protagonists of the colonial achievements. Significant in this respect was the Tower of Partito Nazionale Fascista (The Fascist Party), a 46-metre-high 'symbol' of the regime which dominated the entire complex. A decorative frieze with a bas-relief, designed by Pasquale Meconio and Vincenzo Monaco, completely encircled the base of the tower and depicted the exploits of Mussolini's Legionaries in a reference to those of the legionaries of Rome. The theme of 'colonial conquests' of the regime was thus emphasized by the frieze and completed with a giant sculpture of the Victorious Fascist on the west elevation, covered with slabs of travertine by the same designers.[21]

The *Triennale d'Oltremare* developed its own brand of strong symbolic and evocative imagery, which perhaps exceeded the value of the architecture itself. Conceived in a historical context in which every creative act has to be a political gesture, its primary aim was to impress the observer through its visual impact and seductive impression. The arts were organically integrated into the architectural constructions as an essential contribution to most of the features. The exhibition's project involved both the *architetti integrali* (full architects) coming from the most prestigious Italian schools and members of the 'Sarfatti movement', an artistic movement of the so-called 'return to order' which embraced classical art. Roman and Neapolitan artists, including the Futurists Franco Girosi, Edgardo Mannucci, Paolo Ricci, Nicola Fabbricatore, Pietro Barillà and Tommaso Cascella, were also represented.

For this reason the main contribution of the *Mostra d'Oltremare* was not only the intrinsic architectural quality of the buildings. As one of the largest exhibitions in Europe, it also had a vital role in contemporary art and culture. To support the commemorative programme, historical documentation on a variety of topics together with works of art, situated outside and inside the buildings and pavilions, provided symbolic or illustrative visual atmosphere. The arts were organically integrated into architectural constructions, defining their peculiar characteristics.[22] In the *Triennale d'Oltremare* in 1940, the original quest for balance between architectural structures and works of art emerged and gave birth in some cases to a style characterized by a creative spontaneity. The decorative aspect of the pavilions demonstrated that much careful thought had been put into preparations, with artists and architects working together.

About 120 artists, including Enrico Prampolini, Melkiorre Melis, Giovanni Brancaccio, Giulio Rosso and Alberto Chiancone, were invited to collaborate in the production of the artistic decoration of the exhibition. An impressive repertoire of sculptures, wall paintings, large mosaics, interior decorations, ceramic flooring and panels was created and although almost of it was completely destroyed shortly after the exhibition's closure, it made a strong artistic imprint of a great open air art exhibition.[23] Only a few works of this big project have survived as most of it fell victim to bombing by American and retreating German troops during the Second World War.

The simultaneous use of the 'sisters arts' complemented the role of architecture. The whole artistic project was the core of the event and could be considered as an emblematic episode of that period, in which the government experimented with an ambitious path towards a social and political art through the use of Mario Sironi's 'great decorations'.[24] In the period between 1920 and 1940 the official Italian cultural orientation attempted to overcome individualistic aesthetics of the avant-garde and gave art a social function of communication and collective education. In particular, the use of mural painting, the communicative power of which was considered crucial for the popular masses, was theorized in 1933 by Sironi. He drafted the Manifesto of Mural Painting, with contributions from the artists Massimo Campigli, Carlo Carrà and Achille Funi, in which he asserted that mural painting was a quintessentially social phenomenon that operated on the popular imagination more than any other form of painting.

In the *Triennale d'Oltremare*, the governmental patronage of public art allowed artists who both were and were not members of the *Sindacato Fascista di Belle Arti* to participate in the development of the visual image of the Exhibition.[25] The architect and painter Charles Cocchia, for instance, designed a restaurant with a swimming pool as a successful integration between architecture and decoration. The ceramic decorations produced at the factory Posillipo in Naples played a crucial role in giving character to the interior of the restaurant. Interior and exterior ceramic decorations of other buildings and pavilions also used traditional craftsmanship and confirmed its compatibility with the modern setting. And because at previous exhibitions the regime officials had never given space to rationalist architecture, the close relationship between art and architecture comes across as one of the most significant and original contribution of the *Mostra*.[26]

The Pavilion Libya

Another example of such interaction is the Pavilion Libya, designed by Florestano di Fausto. Its *Caffè Arabo,* however, also represents an interesting case study for a different reason, as it demonstrates the Italian relation to its colony in an exhibition setting. The pavilion was divided into three areas, historical-geographical, ethnographic-documentary, and activities related to the metropolis and the regime in Libya. The decorative design of the *Caffè Arabo,* one of the most evocative of the Pavilion Libya, was created by the artist Melkiorre Melis, who was responsible for the Italian School of Arts and Crafts of Tripoli. Melis was appointed to the role of painter, builder and decorator and held the position of artistic advisor at the *Triennale.*

The café therefore demonstrated the special role that the arts, ranging from painting and sculpture to interior decoration, held in the show.[27] Melis worked in collaboration with the architect Florestano di Fausto and the Pavilion Libya was presented to visitors as a composition of architectural volumes with a strong visionary character, in which local African models and African decorative motifs are processed in an original way. The architecture of the pavilion was reminiscent of Libyan cupolas and minarets, although their white colour turned them more into an idealized vision of such buildings.

Melis created numerous decorative projects for the *Triennale,* among which were three large designs for majolica tiles for the exhibition of ceramics in the Pavilion Libya and a design of a tondo with gazelles for a large ceramic panel of 2.6 metres for the *Caffè Arabo* of the Pavilion Libya.[28] In the Pavilion, Melis continued his experimenting with fine art and applied art and created decorative panels for the entrance to the pavilion. It was reminiscent of his previous work for the School of Arts and Crafts of Tripoli and for a Moorish café in the Pavilion Libya, which was reconstructed using a model of a café located in the ancient district of Souk el Muscir in Tripoli.[29]

In the design of the Moorish café, Melis was inspired by the colourful and dynamic paintings and photographs he made in Libya. He selected recurring patterns of gazelles on the run, Arab-Egyptian dances, cypress decorations with animals and other subjects from nature, as well as numerous ornaments such as purely decorative arabesques which were used by local artists.[30] Melis designed not just the ceramics but also decorative furniture, with the aim of turning the café into a universal work of art. It is possible to reconstruct, at least in part, the physiognomy of the Moorish café from the photographic documentation of the inauguration of the *Triennale* and from the sketches of some of the ceramic panels. The walls were decorated with round ceramic panels that were freely inspired by the geometric patterns designed by Melis for the realization of brass plates and embossed copper. The entrance doors were also decorated with ceramic tiles in geometric patterns. A ceramic floor (7.20 × 3.20 metres), using the decorative motif of twisted cypress twigs, surrounded an area for a small music band and Arabic dancers. The whole room was furnished with tables and chairs with simple designs reminiscent of an unspecified oriental style by which Melis created a real brand of Moorish café. Its most characteristic feature was the motif of a six-pointed star and a crescent moon placed on the tables, chairs and walls.

With his artistic production inspired by Libyan local craft, Melis directly fitted the eclectic decorative and colonial project of the *Triennale*. After five years spent in Libya's Muslim School of Arts and Crafts and the School of Artisan Ceramics, Melis was able to design ceramics, decorative furniture, pitchers and plates in a uniform style in order to create a global artistic product.

Conclusion

Within the confines of the *Triennale*, the original and innovative development of the dialectical relationship between architecture and ceramic decoration satisfied the communicative and functional aspect and influenced artistic expression and the promotion of material design. Because of Italy's entrance into the Second World War, the *Triennale* closed less than a month after its opening, in the subsequent years suffering much damage due to the incessant bombings over the city of Naples. The two bombardments in August 1943 seriously affected the area of the Exhibition. In the ambiguous position that ranged from war booty to occupied zone, the fate of the *Triennale d'Oltremare* may be counted among the inevitable tolls of war.

The function and focus of the exhibition complex in Naples had been defined since its creation: its duration, the public, the spectacular combination of educational and recreational purposes, the scope and logic of a route that would make the content readable and understandable; all was a matter of pride for the citizens of Naples to express in that ephemeral event the dream of the colonialist drive, which was expected to be recreated every three years, becoming progressively an open-air museum. The decision to overcome the limits and the frontier of a single building in the exhibition, which distinguished it from the two previous international colonial art exhibitions at Rome in 1931 and Naples in 1934–35, was a clear and spectacular statement of intent. The exhibition was conceived from the beginning as a big park close to the city centre, in strict relationship with the landscape and was defined as the original experience of the Fascist period in Italy. The policy of Fascism in the first half of the twentieth century aimed to connote territories and cities with a 'special presence'. This included streets, squares and buildings, as well as exhibition spaces, and the *Triennale d'Oltremare* can be considered one of these special presences. The *Triennale* today documents an important era in Italian history characterized by an intense cultural activity related to its colonies and territorial and cultural expansion. The works of art, architecture and sculpture from the Italian colonies, but also those created specially for the event, represent an important integration between art and propaganda together with the strategy of involving the masses on the grounds of an exhibition.

Notes

1 Martine Astier Loutfi, *Littérature et colonialisme. L'expansion coloniale vue dans la littérature romanesque française, 1871–1914* (Paris, 1971); Vinigi Lorenzo Grottanelli, *Gerarchie etniche e conflitti culturali. Saggi di etnologia Nord-Est africana* (Milan, 1976); Giovanna Tomasello, *La letteratura coloniale italiana dalle avanguardie al fascismo* (Palermo, 1984); Mario Isneghi, *Il sogno africano*, in *Le guerre coloniali del fascismo* (Rome and Bari, 1991).

2 Angelo Piccioli, 'L'imminente inaugurazione della fiera di Tripoli. La sagra del lavoro nazionale sulla quarta sponda', *L'Italia coloniale* 4, no. 2 (1927): 23–25.

3 'Le colonie all'Esposizione di Torino', *L'Italia coloniale* 5, no. 2 (1928): 26–27; 'La Mostra coloniale di Torino', *L'Italia coloniale* 5, no. 7 (1928): 141–143; Giuseppe Preda, 'Rodi all'Esposizione di Torino', *L'Italia coloniale* 5, no. 8 (1928): 161–163.

4 Guido Abbattista and Nicola Labanca, 'Living Ethnological and Colonial Exhibitions in Liberal and Fascist Italy', in *Human Zoos. Science and Spectacle in the Age of Colonial Empires*, ed. Pascal Blanchard et al. (Liverpool, 2008), 341–352.

5 For a deeper view of the evolution of Italian colonial national and international exhibitions, see Giovanni Arena, *Visioni d'Oltremare, Allestimenti e politica dell'immagine nelle esposizioni coloniali del XX secolo* (Naples, 2011).

6 Zeynep Çelik and Leila Kinney, 'Ethnography and Exhibitionism at the Expositions Universelles', *Assemblage* 13 (1990): 34–59.

7 Francesco Stocchetti, 'La Triennale d'Oltremare', *Le Vie d'Italia* 7 (1938): 832–836; Luigi De Lillo, 'Potenza imperiale dell'Italia fascista nei riflessi della Prima Mostra delle Terre d'Oltremare', *L'illustrazione Italiana* 28 (1940): 103–106; Arrigo Pozzi, 'Orme di legionari sulle Terre d'Oltremare', *Le Vie d'Italia* 6 (1940): 601–614.

8 All of these exhibitions were organized by the *Ministero degli Affari Esteri*. Although Somalia became a colony in 1905, they only presented the colony of Eritrea. For a fairly complete listing of these exhibitions, see Nicola Labanca, *L'Africa in vetrina, storie di musei e di esposizioni coloniali in Italia* (Treviso, 1992), 35.

9 It was not until a year after the initial conquest of the North African colonies of Tripolitania and Cyrenaica in 1911 that the Italian government created

the Ministero delle Colonie (1912) and that Italy's participation in colonial exhibitions and its propaganda efforts related to its colonies was coordinated by a centralized authority. See Nicola Labanca, *Oltremare, storia dell'espansione coloniale italiana* (Bologna, 2002).

10 Angelo Del Boca, *Gli italiani in Africa Orientale, la conquista dell'Impero* (Rome and Bari, 1979).

11 Plinio Marconi, 'Prima Mostra delle Terre Italiane d'Oltremare', *Architettura* 1–2 (1941): 1–7.

12 Amedeo Maiuri, 'Il segno di Roma alla Mostra delle Terre d'Oltremare', *L'illustrazione Italiana* 22 (1940): 34–38.

13 Bruno Ernesto Lapadula Archive (Rome), *Padiglione Libia*, designed for the architect Lapadula, 10 tables, 19 drawings, documents, 36 tables, architect Florestano di Fausto; Archive Central State (Rome), Presidenza del Consiglio dei Ministri, years 1937–39, dossier *Mostra Triennale delle Terre Italiane d'Oltremare*, case 2483.

14 For a broader discussion of colonial exhibitions, see for example Anne Maxwell, *Colonial Photography and Exhibitions, Representations of the 'Native' and the Making of European Identities* (London, 1999); Peter H. Hoffenberg, *An Empire on Display: English, Indian, and Australian Exhibitions from the Crystal Palace to the Great War* (London, 2001); Jeffrey A. Ayerbach and Peter H. Hoffenberg, *Britain, the Empire, and the World at the Great Exhibition of 1851* (Ashgate, 2008); Pramod K. Nayar, *Colonial Voices, The Discourses of Empire* (Chichester, 2012).

15 Giovanni Arena, *Napoli 1940–1952, dalla prima mostra triennale delle terre italiane d'oltremare alla prima mostra triennale del lavoro italiano nel mondo* (Naples, 2012); Antonio Dal Pozzo Gaggiotti, 'Cenni sulla Mostra', *Emporium* 548 (1940): 57–58.

16 Arturo Assante, *Funzione mediterranea del porto di Napoli* (Naples, 1941); Anna Maria Puleo, 'Piano regolatore e progetti architettonici nella Mostra Triennale delle Terre Italiane d'Oltremare', *ArQ* 3 (1990): 74–84.

17 *Giornale Luce B1308, Italia: Napoli, I lavori di preparazione per la Mostra d'Oltremare*, 25 May 1938; *Giornale Luce B1644, Italia: Napoli, I lavori della Triennale d'Oltremare visti in occasione del Congresso Demografia e Statistica indetto a Napoli*, 27 December 1939; *Giornale Luce B1684, Italia: Napoli, 'Scoperte archeologiche nella zona della Mostra Triennale delle Terre d'Oltremare'*, 8 March 1940; *Giornale Luce B1685, Italia: Napoli, 'S. E. Ciano visita la Mostra Triennale delle Terre d'Oltremare'*, 8 March 1940; *Giornale Luce C0031, Triennale d'Oltremare, 'S.M. il Re imperatore inaugura a Napoli la Mostra delle Terre Italiane d'Oltremare'*, 17 May 1940; *Giornale Luce C0032, Nove Maggio a Napoli, 'Le cerimonie e le feste per la Triennale d'Oltremare. S.M. il Re imperatore inaugura un nuovo collegio della Gil', l'Istituto nazionale dei motori e la nuova sede del Banco di Napoli. Rievocazione celebrativa del centenario delle Ferrovie Italiane*, 17 May 1940.

18 *I Mostra Triennale delle Terre Italiane d'Oltremare. Napoli 9 maggio–15 ottobre 1940–XVIII. Guida* (Naples, 1940).

19 Dino Alfieri and Luigi Freddi, *Mostra della Rivoluzione fascista. Guida storica* (Bergamo, 1933).

20 Anty Pansera, *Storia e cronaca della Triennale* (Milan, 1979).

21 Arena, *Visioni d'Otremare*, 117–145.

22 Giovanni Arena, 'Allestimenti e apparati decorativi alla Mostra d'Oltremare: Melkiorre Melis e il Padiglione Libia', *Napoli Nobilissima* 2 (Naples, 2001): 137–152.

23 Michele Biancale, 'La prima Mostra Triennale delle terre italiane d'Oltremare', *Le Arti* 1 (1940): 54–57.

24 Mario Sironi, 'Manifesto of Mural Painting' (1933), reprinted in Charles Harrison and Paul Wood, eds, *Art in Theory: 1900–2000. An Anthology of Changing Ideas* (Malden, MA and London, 2003), 424–425.

25 Laura Malvano, *Fascismo e politica dell'immagine* (Turin, 1988); Emilio Raffaele Papa, *Bottai e l'arte: un fascismo diverso? La politica culturale di Giuseppe Bottai e il Premio Bergamo (1939–1942)* (Milan, 1994).

26 Gaia Salvatori, *Nelle maglie della storia. Produzione artistico-industriale, illustrazione e fotografia a Napoli nel XX secolo* (Naples, 2003), 51–82.

27 Antonello Cuccu, *Melkiorre Melis* (Nuoro, 2004), 92–98.

28 The three paintings are now on display in the Hall of Libya Melkiorre Melis Collection at the Art Gallery of Bosa (OR) Italy, Casa Deriu.

29 For more on the topic, see Gaia Salvatori, 'Forme dell'utile e del superfluo: episodi di storia delle arti applicate in Campania dal 1920 al 1945', in *Gli anni difficili. Arte a Napoli dal 1920 al 1945*, edited by Mariantonietta Picone Petrusa (Naples, 2000), 77–83.

30 The *Caffè Arabo* and *Hall Monopoli* were located in the fourth section of the sector folk documentation. See Archival and architectural Mostra d'Oltremare, Inventory projects 1939–52 (edited by the author), envelope 11, *Triennale d'Oltremare, Padiglione Libia, Monopolio*, plan and elevation, rendering 1/50, 46 × 35 cm.

There are two photographs, bearing the stamp on the back of the Photographic Service of the Triennale, which reproduce the final project of 1940, showing the tondo with gazelles and the ceramic flooring. Three more photographs, bearing the same stamp on the back and taken on the opening day (9 May 1940), which depict some details of the interior are preserved in the Archives Melkiorre Melis, Casa Deriu, Bosa (OR), Italy.

Bibliography

Printed Primary Sources

Giornale Luce B1308, Italia: Napoli, I lavori di preparazione per la Mostra d'Oltremare. Naples: Istituto Luce, Rome, 1938.

Giornale Luce B1644, Italia: Napoli, I lavori della Triennale d'Oltremare visti in occasione del Congresso Demografia e Statistica indetto a Napoli. Naples: Istituto Luce, Rome, 1939.

Giornale Luce B1684, Italia: Napoli, 'Scoperte archeologiche nella zona della Mostra Triennale delle Terre d'Oltremare'. Naples: Istituto Luce, Rome, 1940.

Giornale Luce B1685, Italia: Napoli, 'S.E. Ciano visita la Mostra Triennale delle Terre d'Oltremare'. Naples: Istituto Luce, Rome, 1940.

Giornale Luce C0031, Triennale d'Oltremare, 'S.M. il Re imperatore inaugura a Napoli la Mostra delle Terre Italiane d'Oltremare'. Naples: Istituto Luce, Rome, 1940.

Giornale Luce C0032, Nove Maggio a Napoli, 'Le cerimonie e le feste per la Triennale d'Oltremare. S.M. il Re imperatore inaugura un nuovo collegio della Gil, l'Istituto nazionale dei motori e la nuova sede del Banco di Napoli. Rievocazione celebrativa del centenario delle Ferrovie Italiane. Naples: Istituto Luce, Rome, 1940.

I Mostra Triennale delle Terre Italiane d'Oltremare, catalogo. Naples: Istituto Luce, Rome 1940.

I Mostra Triennale delle Terre Italiane d'Oltremare. Napoli 9 maggio–15 ottobre 1940–XVIII. Guida. Naples: Triennale d'Oltremare, 1940.

Incartamento Fiera Campionaria di Tripoli, Archivio Centrale di Stato. Rome: Central State Archive of Rome, unpublished, 1937–39.

Mostre d'arte, fotografiche, della moda, Archivio di Stato di Napoli. Naples: State Archives of Naples, unpublished, 1931–38.

Padiglione Libia, progetto di allestimento dell'architetto Bruno Ernesto Lapadula. Rome: Private archive of architect Bruno Ernesto Lapadula, unpublished, 1938–40.

Prima Mostra Triennale delle Terre Italiane d'Oltremare, Wolfsonian. Genoa: Archive of the Foundation Wolfsonian, 1940.

Secondary Sources

Alfieri, Dino, and Luigi Freddi. *Mostra della Rivoluzione fascista. Guida storica*. Bergamo: Istituto Italiano Arti grafiche, 1933.

Arena, Giovanni. 'Allestimenti e apparati decorativi alla Mostra d'Oltremare: Melkiorre Melis e il Padiglione Libia'. *Napoli Nobilissima* 2 (2011): 137–152.

Arena, Giovanni. *Napoli 1940–1952, dalla prima mostra triennale delle terre italiane d'oltremare alla prima mostra triennale del lavoro italiano nel mondo*. Naples: Fioranna, 2012.

Arena, Giovanni. *Visioni d'Oltremare, Allestimenti e politica dell'immagine nelle esposizioni coloniali del XX secolo*. Naples: Fioranna, 2011.

Assante, Arturo. *Funzione mediterranea del porto di Napoli*. Naples: Il Porto, 1941.

Biancale, Michele. 'La prima Mostra Triennale delle terre italiane d'Oltremare'. *Le Arti* 1 (1940): 54–57.

Cefariello Grosso, Gilda. 'La produzione vietrese nel rinnovamento della ceramica italiana tra le due guerre'. In *Il Museo della Ceramica. Raito di Vietri sul Mare*, edited by Matilde Romito, 85–89. Salerno: Edizioni 10, no. 17 (1994).

Cuccu, Antonello. *Melkiorre Melis*. Nuoro: Illisso, 2004.

De Lillo, Luigi. 'Potenza imperiale dell'Italia fascista nei riflessi della Prima Mostra delle Terre d'Oltremare'. *L'illustrazione Italiana* 28 (1940): 103–106.

De Micheli, Mario. *Arte sotto le dittature*. Milan: Feltrinelli, 2000.

Del Boca, Angelo. *Gli italiani in Africa Orientale, la conquista dell'Impero*. Rome and Bari: Laterza, 1979.

Labanca, Nicola. *L'Africa in vetrina, storie di musei e di esposizioni coloniali in Italia*. Treviso: Pagus, 1992.

Labanca, Nicola. *Oltremare, storia dell'espansione coloniale italiana*. Bologna: Il Mulino, 2002.

Maiuri, Amedeo. 'Il segno di Roma alla Mostra delle Terre d'Oltremare'. *L'illustrazione Italiana* 22 (1940): 34–38.

Malvano, Laura. *Fascismo e politica dell'immagine*. Turin: Bollati Boringhieri, 1988.

Marconi, Plinio. 'Prima Mostra delle Terre Italiane d'Oltremare'. *Architettura* 1–2 (1941): 1–7.

Mignemi, Adolfo. *Lo sguardo e le immagini, la fotografia come documento storico*. Turin: Bollati Boringhieri, 2003.

Pansera, Anty. *Storia e cronaca della Triennale*. Milan: Longanesi, 1979.

Papa, Emilio Raffaele. *Bottai e l'arte: un fascismo diverso? La politica culturale di Giuseppe Bottai e il Premio Bergamo (1939–1942)*. Milan: Electa, 1994.

Pozzi, Ariggo. 'Orme di legionari sulle Terre d'Oltremare'. *Le Vie d'Italia* 6 (1940): 601–614.

Puleo, Anna Maria. 'Piano regolatore e progetti architettonici nella Mostra Triennale delle Terre Italiane d'Oltremare'. *ArQ* 3 (1990): 74–84.

Salvatori, Gaia. *Nelle maglie della storia. Produzione artistico-industriale, illustrazione e fotografia a Napoli nel XX secolo*. Naples: Luciano Editore, 2003.

Salvatori, Gaia. 'Forme dell'utile e del superfluo: episodi di storia delle arti applicate in Campania dal 1920 al 1945'. In *Gli anni difficili. Arte a Napoli dal 1920 al 1945*, edited by Mariantonietta Picone Petrusa, 77–83. Naples: Electa Napoli, 2000.

Stocchetti, Francesco. 'La Triennale d'Oltremare'. *Le Vie d'Italia* 7 (1938): 832–836.

15

International Ambitions of an Exhibition at the Margin: Japan's 1903 Osaka Exposition

Jeffer Daykin

As its formal title suggests, the Fifth National Industrial Exhibition of 1903 held in the city of Osaka was not Japan's first endeavour in hosting an exposition. Following the 1866 publication of the enormously popular *Seiyo Jijo* (*Conditions in The West*) in which the famed interpreter of all things Western, Fukuzawa Yukichi, described his experiences in America and Europe (including attendance at the 1862 London Exposition), an exposition craze emerged in Japan.[1] The nation became enthusiastic about these new *hakurankai* (expositions) and participated in all the major fairs held abroad – from the 1867 Paris *Exposition Universelle* on – and began experimenting with this new modernizing technology at home.

Where the events held in Japan in the following years were termed domestic expositions (*naikoku hakurankai*) as opposed to international expositions (*bankoku hakurankai*), the international focus of Japan's domestic fairs slowly emerged and it is through this process that one can see how international expositions could serve as a marker of the host country's industrial and technological capabilities. More than this, international expositions provided a technology for asserting a nation's standing in the fiercely competitive late nineteenth- and early twentieth-century world where the consequences of being deemed a 'civilized' and 'modern' power by the West could mean the difference between being accepted as a legitimate member of the international community or a potential target of imperialism.

Only five decades removed from its position as a nearly isolated country forcibly opened to trade with the West in 1854 by an American expeditionary fleet, Japan had undertaken a dramatic transformation in which virtually every aspect of its society was refashioned along the lines of the successful Western powers in order to guarantee its safety as an independent nation. One might argue that Japan essentially achieved its place as a world power by 1903 through its participation in a coalition sent to suppress China's Boxer Rebellion in 1900 and through the abrogation of virtually all unequal treaties it had been forced to sign with the US and European nations. Yet Japan's unique position as a modern Asian power in a world dominated by Western notions of 'The White Man's Burden' and 'Manifest Destiny' left it potentially vulnerable. The 1903 Osaka Exposition can therefore be seen as an event in which Japan existed in the margin between the industrialized Western powers on one hand and, perhaps more importantly, other non-Western peoples who could be subjected to imperial domination. The story of the Osaka Exposition and its formulation as an event in the model of the West's world fairs can help illuminate the ways in which international expositions served to define a nation's legitimacy as an actor on the global stage.

Japan's Domestic Expositions

While exposition-like events existed in Japan prior to the opening of the country in 1854, Japan's first Western-style exposition was held in 1872 under the auspices of the Ministry of Education.[2] The event was used to launch the Tokyo National Museum and had the additional purpose of collecting items for display that could then be exhibited at the world's fair in Vienna the following year. After 1872 the Japanese government undertook a series of national expositions starting with the First National Industrial Exposition (*Daiichi Naikoku Kangyô Hakurankai*) of 1877, held at Ueno Park in Tokyo. Noriko Aso notes that this occurred as planned despite the fact that Japan was embroiled in the Satsuma Rebellion which, beyond the significant loss of lives, cost the Japanese government approximately 80 per cent of its annual budget to suppress. The Japanese government clearly placed significant importance on hosting its own exposition.[3]

Japan valued the international potential of expositions, yet the aim of its early domestic expositions was to realize the goals of the government's official slogan *shokusan kôgyô* (encouraging domestic production and enterprise).[4] A manifestation of this approach can be found in the fair's regulations regarding exhibits. In an effort to promote Japan's own burgeoning industries and sense of national pride, foreign products could only be exhibited by exposition officials whereas Japanese domestic goods, even those that copied international products, could be exhibited by their manufacturers.[5]

Following the success of the 1877 Exposition which received 500,000 visitors in only three months, the state pronounced a domestic exposition was to be held every four years.[6] Where the Second National Industrial Exposition for the Promotion of Industry took place as planned in 1881, the government postponed the following exposition an additional four years. It was then argued that the Third National Industrial Exposition should be further postponed until 1890 to coincide with the 2,550th anniversary of the first Yamato Emperor's enthronement.[7]

The Fourth National Industrial Exposition (*Dai Yonkai Naikoku Kangyô Hakurankai*) was held in Okazaki district in Kyoto in 1895 and was timed to coincide with the 1,100th anniversary of Kyoto's founding. The Main Exhibition Hall – later to become the Kyoto National Museum – was built in a neoclassical interpretation of French Baroque style and publications such as the *Official Guide-Book to Kyoto and the Allied Prefectures* were printed in English to facilitate the foreign tourist experience at the Exposition and in the region.[8] Despite these efforts and that it was the most attended Japanese exposition to date with gate receipts of 1,137,000, official international participation was not a feature; there were no buildings allocated for the exhibition of foreign manufacture.

A Modern International Exposition in Japan

The Fifth National Industrial Exhibition held in Osaka's Tennoji Park between 1 March 1903 and 31 July 1903 was vastly different from its predecessors. With an area of 105,000 *tsubo* (roughly 85 acres, or 34 hectares), the Osaka Exposition was more than twice as large as any of the previous domestic *hakurankai* and within this massive area the products of the nation's post-Sino-Japanese War (1894–95) boom in industrialization were showcased to its almost 4.5 million visitors.[9] Beyond its sheer size and attendance, the 1903 Osaka Exposition was strikingly different from Japan's earlier domestic expositions in another important respect: international participation.

The Japanese government sent out invitations to participate in the Osaka Exposition through its consulates abroad, but the Foreign Ministry and exposition planners had not anticipated much interest and only planned for 300 *tsubo* (roughly 10,600 square feet, 985 square metres) of display space for foreign exhibits. When the applications from foreign exhibitors arrived, however, it was clear that more than 2,000 *tsubo* (71,000 square feet, or 6,600 square metres) of exhibit space was needed.[10] Following the organization of fairs in the West, Osaka's exposition featured categorized exhibition pavilions; the Machinery, Manufacture, Education and Fine Arts Buildings, among others, all hosted Japan's domestic exhibits (Fig. 15.1). For non-Japanese exhibitors, a Foreign Samples Building was constructed and, when the space within was fully booked, those who still desired to exhibit were left to construct their own pavilions, which the nation of Canada as well as five Western firms were obliged to do. Ultimately, 14 foreign countries and many Western firms participated in the Osaka Exposition, transforming a domestic exposition into an international event.

Because of its industry, Osaka was commonly referred to as the Manchester of Japan, making it a fitting setting for a National Industrial Exposition along the lines of those held in the West.[11] Japan's impressive industrial progress was showcased in the Exposition's Machinery Building. A piece in the *Japan Weekly Chronicle*, part of a regular feature by a special foreign correspondent to the fair, does well to introduce this perhaps surprising aspect of the Osaka Exposition:

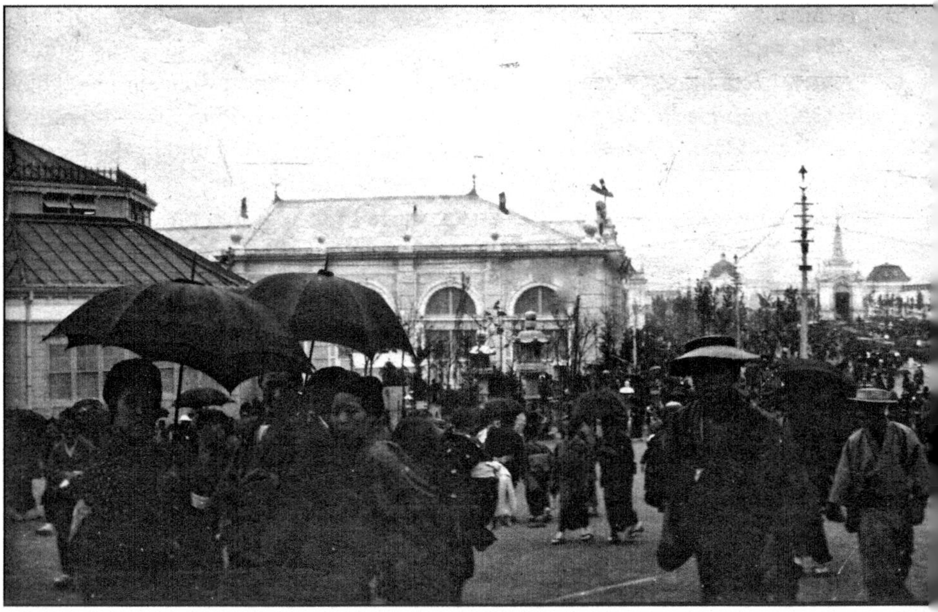

15.1 Osaka Exposition grounds. Courtesy of John Wilson Special Collections, Multnomah County Library, Portland, Oregon

> When one considers that but some thirty years ago not only was there no such institution as a factory in Japan, but that iron foundries and mechanics' workshops as now understood were unknown, while engineering was an alien art, the display beneath the roof of the Machinery Building at the Exhibition is little short of marvelous. The machinery section of any Exhibition has always [been] a fascination for the public, but that at Osaka seems to have a specially attractive effect. A very large number of visitors is almost always to be found there, and they do no merely pass through it and glance at the exhibits; they cluster around them, especially those in motion, and the section is certainly very popular.[12]

Though fraught with hyperbole regarding the limitations of Japan's industry prior to the 1854 opening of the country and subsequent 1868 Meiji Restoration, the article illuminates Japan's demonstrable and remarkable gains in industrial capability.

The article further details the great variety of such modern marvels like the 'massive hoisting engine' of the Kobukuro Iron Works of Fukuoka-ken, the electrical motors and generators of the Mitsui Shibaura Engineering Works of Tokyo, the Osaka Iron Works' 'huge steam navvy [excavator] which towers above the other exhibits' and other products that were described as being 'well worth attention as indicating what can be done in the engineering line in Japan'.[13] Singular inventions such as Osaka resident Mr. S. Tada's compact rice-cleaning machine also demonstrated what the small entrepreneur could accomplish; machines for the production of matches, socks, lemonade and soap, as well as tea refining, spirit distilling and printing – all of Japanese manufacture – were proudly displayed and examined by Japanese and foreigners alike.

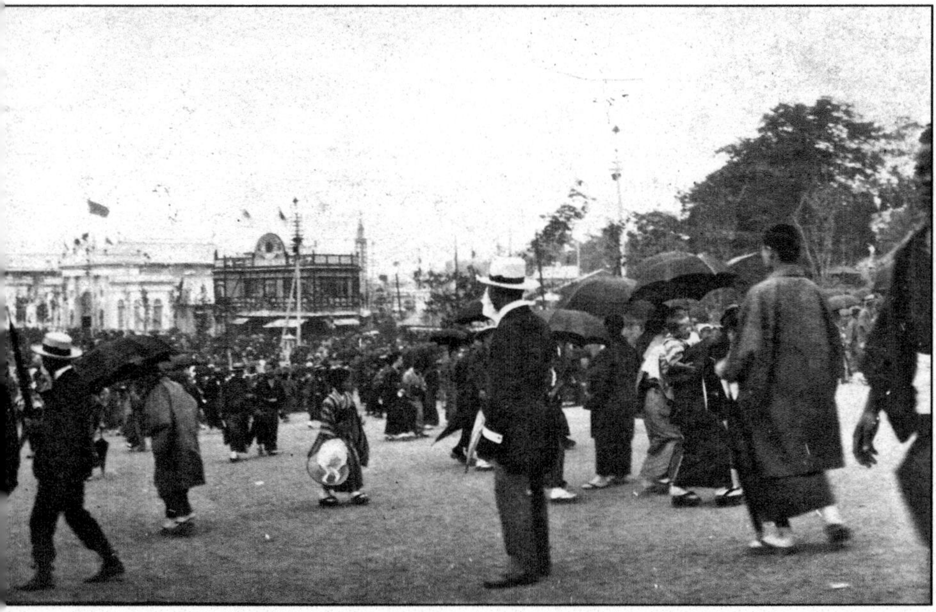

International Challenges for an Asian Exposition

Foreign enjoyment was hampered during the early days of the Osaka Exposition because no English- or other European-language guidebooks were published and foreign language signage and explanations to accompany Japanese exhibits were a rarity. This fact was lamented in the *Japan Weekly Chronicle* though they acknowledged it might be an unfair criticism given that American world's fairs, for example, did not bother to accommodate foreign visitors with multi-lingual materials. The problem, as they saw it, was something unique to Japan's Exposition because of the inherent challenges posed by written Japanese to unfamiliar Westerners.[14] This point, buttressed by a hope foreign visitors would purchase some of the Japanese products exhibited, was well-taken and foreign-language guides and English signage were quickly offered within the first month of the Exposition. One such guide, *The Photographic Reproduction of the Fifth National Industrial Exhibition, 1903*, was published in monthly instalments in both English and Japanese, providing a souvenir that showed the scope of the exposition to a range of foreign and domestic visitors.[15]

Only a few of the Western exhibitors came prepared to broadly reach out to Japanese fairgoers beyond the standard visual exhibits of products. The Netherlands-India Section had the foresight – undoubtedly a result of the Netherland's long-standing relationship even through Japan's period of *sakoku* (seclusion) from the seventeenth to the nineteenth centuries – to publish its catalogue in both Dutch and Japanese. Similarly, the American firm of Messrs Andrews and George distributed free souvenir mirrors with bilingual advertising to those who entered their company's pavilion.[16] Gradually other

Western exhibitors began to follow suit and *A Guide to the Foreign Samples Building and Its Annexe* was issued in early June, published in both English and Japanese, to facilitate the full and unhampered exploration of the foreign exhibits.[17]

The issuance of awards solely to domestic exhibitors also reveals a general failure to prepare for the international nature of the Osaka Exposition. This all-important function of period world's fairs gave additional incentives for exhibitor participation because the quality of their products would be concretely validated and subsequently used in marketing. Japan's decision to exclude foreign products from awards was a major oversight. Though it is possible the Japanese felt it inappropriate to be the arbiter of products of foreign manufacture as their own industrialized status had yet to be confirmed by the world community, the stated reason was that the Exposition was 'not considered an International one'.[18] Foreign exhibitors voiced their protest at not being placed on an equal footing with Japan's domestic exhibitors and so the exposition officials, presented with the unavoidable fact that the event had indeed become international, arrived at a compromise: award juries carefully examined the foreign exhibits, wrote reports and distributed Letters of Thanks.

A World-Class Exposition

Where perhaps it is fair to say that both the Japanese and the Western exhibitors had stumbled into an international exposition, it was clearly Japan's intention to hold a modern exposition along the lines of those being held in the West. The Osaka Exposition followed the general organization and architecture of Western world's fairs which typically included a landmark structure on the fairgrounds. For Osaka, the 92-feet-high (28-metre) Obayashi Tower served this purpose. As the tower provided convenient elevator transport to an observation deck, fairgoers were treated to the kinds of spectacles and amenities available at Western expositions. The Osaka Exposition also hosted the same kinds of amusements that could be found at any of the fairs abroad. The Water Chute concession was a popular attraction at Osaka and its variants had become a regular feature of European and American expositions such as France's *Exposition internationale de Lille* (1902), America's Louisiana Purchase Exposition in St Louis (1904) and Belgium's *Exposition universelle et internationale de Liège* (1905).[19]

Another class of attraction found in the Western expositions was also a feature at Osaka: the live exhibition of foreign peoples. Perhaps beyond the Western trappings of the exposition buildings, landmark Obayashi Tower and even exhibits of domestically produced machinery, the fact that it too could place other nations on display can best indicate how 'modern' Japan's Osaka Exposition in fact was.

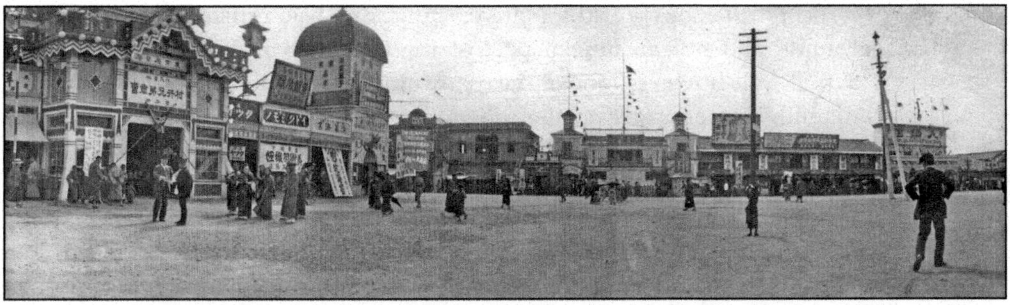

15.2 Entrance to the Osaka Exposition – *Jinrui Kan* to far right. Courtesy of John Wilson Special Collections, Multnomah County Library, Portland, Oregon

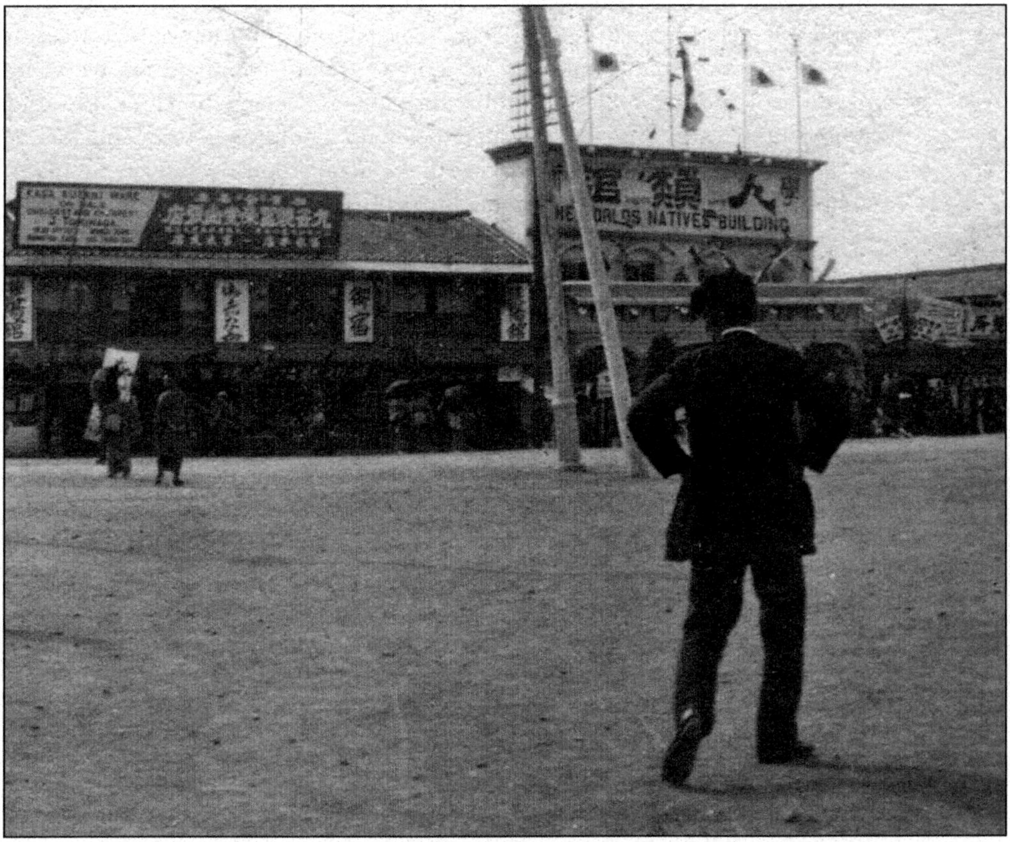

15.3 Close-up of *Jinrui Kan* (The World's Natives Building). Courtesy of John Wilson Special Collections, Multnomah County Library, Portland, Oregon

The *Jinrui Kan* (World's Natives Building) was outside the actual fair grounds and not an official part of the Osaka Exposition (Figs 15.2 and 15.3). It was, however, one of the many amusements that stood as a phalanx adjoining the entrance to the fair and so could be regarded as a *de facto* exposition attraction and it was certainly experienced as such at the time.[20] As in similar concessions at Western expositions, the *Jinrui Kan* intended to display live exhibits of various native peoples from Africa and Asia – including those from newly acquired territorial possessions – as curiosities to be observed and as exemplars of their race for purposes of comparison.

Controversy, however, swamped the exhibit even before the Osaka Exposition opened. Despite its attempt to cast itself as an edifying anthropological concession by changing its name to the *Gakujutsu Jinrui Kan* (Science of Anthropology Pavilion), the choice to include particular 'exhibits' triggered a storm of protest.[21] Though the collection of native peoples from India, Malaya, Java, Turkey and Zanzibar exhibited in their native dress in virtual zoo-like conditions were accepted without comment (as were the exhibits of indigenous Ainu from the Japanese island of Hokkaido and representatives of three Formosan tribes who would have been considered as falling within Japan's Empire with its acquisition of Taiwan in 1895), a storm of protest began to swell regarding the *Jinrui Kan*'s inclusion of exhibits related to Chinese, Koreans and Ryūkūans (Okinawans).[22]

The planned Chinese exhibits which included mannequins depicting China's 'uncivilized' culture as evidenced by life-size dioramas of practices such as opium smoking and foot binding were cancelled before the Exposition opened. Chinese students in Japan became aware of the inclusion of their nation as *yabanjin* (uncivilized savages) and protested.[23] Official pressure from China was then immediately relayed by the Japanese Foreign Ministry to the President of the Exposition who then secured the removal of these features.[24] Soon after the start of the fair, Korean students protested the live exhibition of fellow nationals and their calls were followed by official pressure from Seoul. This, coupled with the reports of the cancellation of Osaka Exposition trips by would-be fairgoers from China and Korea, managed to convince Komura Jutarō, Japanese Minister of Foreign Affairs, to present a paper to the Osaka government on 19 March 1903 calling for the removal of the remaining objects related to China (such as an opium pipe and the picture of a compressed foot of a Chinese woman) and ending the exhibition of Koreans.[25] By the end of March the *Jinrui Kan* complied.

In April criticism emerged regarding the *Jinrui Kan*'s exhibition of two Ryūkyūan (Okinawan) women by that Japanese prefecture's press. The *Ryūkyū Shimpo* complained that the show portrayed two young women who were prostitutes as being representative of Okinawan women as a whole. Also, more fundamentally, they protested that by including any Okinawans in the *Jinrui Kan*, the message was sent that their culture was as primitive as that of the Ainu or the Formosan tribes that were also on display. Though it is unclear whether the Okinawan women decided to leave the *Jinrui Kan* on their own or that their 'exhibit' was cut by the concession's proprietor, Okinawans were no longer included in the live exhibits of indigenous peoples after 17 May 1903.[26]

What these controversies related to the *Jinrui Kan* reveal is that the contested issue was not whether such displays were degrading in general, but which nations were privileged to be the spectators and which were left as the spectacle. Protests over such designations were not unheard of at Western expositions, but the case of the *Jinrui Kan* is illuminating because Japan was pressed to fully accommodate such complaints. In contrast to its own participation in Western expositions, the Japanese could remain safely on the voyeur's side of the exhibit. This was accomplished in part because of the nation's impressive feats of modernization but more basically because the exposition host was always afforded this privileged position.

Where the outrages of the *Jinrui Kan* lay outside the official space of the Osaka Exposition, the Formosa Section within the fairgrounds clearly demonstrates how Japan used the Exposition to solidify its position among the privileged and protected class of nations (Fig. 15.4). A further instalment by the special correspondent to the Osaka Exposition introduced the pavilion as 'not merely a building [but a] veritable Formosan village'.[27] Without a trace of irony to acknowledge the stunning similarities with how the period's 'Japanese Village' concessions at Western world's fairs were typically depicted, the special correspondent served as a narrating guide:

> Entering by the main gateway, beneath the little square veranda, the visitor finds himself in another world, so totally different is everything here to that to which the eye had become accustomed in looking at the many interesting things elsewhere… Things are so strange even to the Japanese, so that if the Westerner is left to draw his own conclusions more in this section even than elsewhere, he cannot complain; after all, we are in Formosa now and English would 'spoil the romance' somewhat.[28]

One cannot help but wonder to what degree this difference would have been understood by a visiting Westerner who had not yet developed a strong enough sense of Japanese culture with which to discern the distinctive Formosan features of this section, but regardless, the message to be derived was plain even to foreign reporters. Japan was thus representative of the modern world through its contrast with the 'totally different' world of its colonial possession Formosa. The precise nature of that difference comes into greater focus as the guide continued the tour past the 'sturdy Formosan [who] was sleeping peacefully on a wooden chair outside the "house"' and beyond the bazaar where straw hats and native-produced walking sticks were sold for souvenirs, to the Formosan section's Tea House (another standard feature of the Japanese Village concessions at expositions held abroad):

> The visitors swarm round the doors and windows in order to get a glimpse of the interior of this interesting room with its 'willow pattern' ceiling; but if one

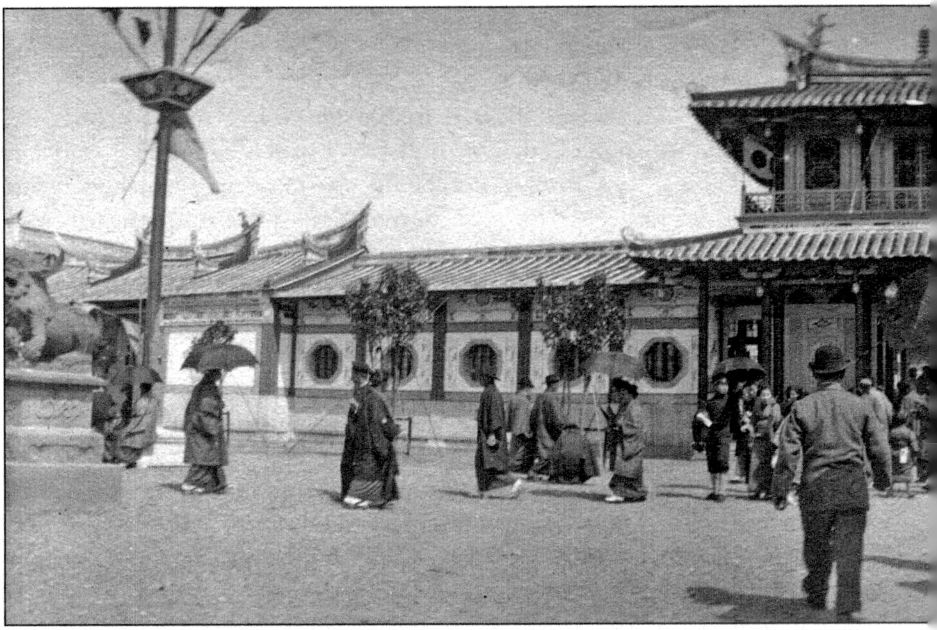

15.4 Formosa Building at the Osaka Exposition. Courtesy of John Wilson Special Collections, Multnomah County Library, Portland, Oregon

is possessed of a normal amount of perseverance and patience, access may be ultimately gained to it, and most delicious tea will be rewarded therefore. Tea and *kwashi* [cakes] are brought to you by pretty little Formosan girls, not much higher than the tables, and they stand on tiptoe and fill your cup from the big teapot, and replenish it *ad libitum*. These tiny waitresses are beautifully dressed in the Chinese style and are under the superintendence of a Formosan lady who is evidently the manageress of this cozy tea room. Europeans will be somewhat surprised to find that one of the Formosan men waiters converses in excellent English and is most courteous and anxious to do everything for one's convenience.[29]

What is fascinating about the guide's descriptions is, though it is made clear that they were written by a European, there would be no inherent conflict were these words written by a Japanese reporter (of which there were several on the staff of the *Japanese Weekly Chronicle*).[30] The Chinese-style of the clothes worn by the 'tiny waitresses' and the 'surprising' excellent English of the male Formosan waiter were likely just as readily remarked upon by Japanese visitors to the Exposition and would doubtless reinforce, if not create, a sense of China and the rest of Asia as being backward relative to modernizing Japan.

Serving as an exact mirror image of the Japanese Village concessions in the Western expositions, the Formosan Section can be seen as Japan's effort to apportion an exotic Otherness to Formosans that could then help differentiate itself – as an Asian nation – from 'backward' Asian peoples. This process of asserting a nation's 'civilization' and justifying its imperial projects through the exoticization and/or dehumanization of perceived Others is something that has been well-studied as a Western phenomenon and there is a great deal of research on how Western expositions served as, at best, an arena where these perspectives could become reified and, at worst, a mechanism to

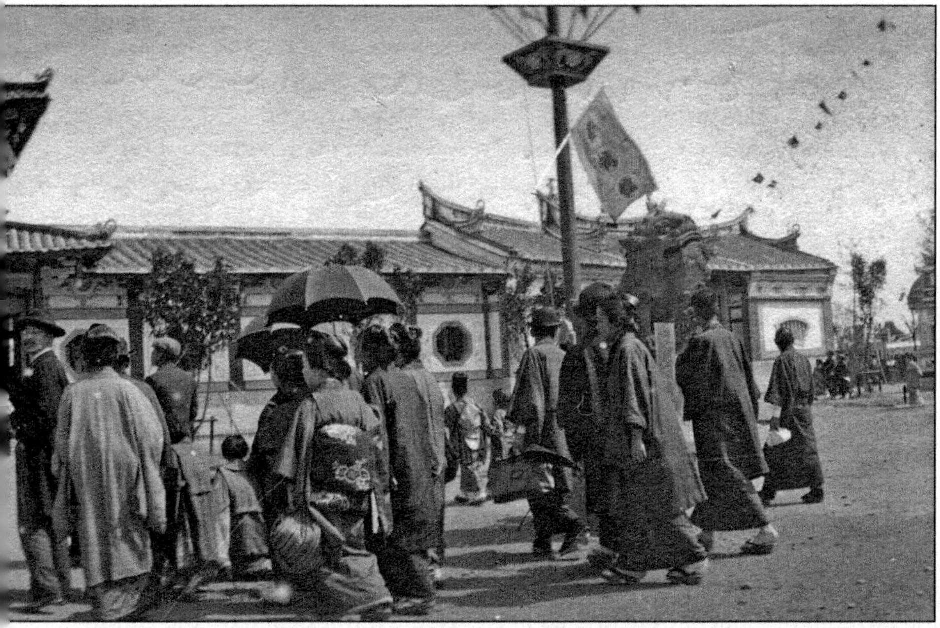

inculcate such views.[31] What makes the Osaka Exposition unique, however, is its position as an event constructed by a non-Western people. In this regard, perhaps Japan's need to define itself in sharp contrast to Asians deemed 'uncivilized' – and therefore potential targets of imperialism – was perhaps even more pressing than it was for those holding expositions in the West. As the host of an internationally attended exposition, Japan could present itself as a civilized peer of the modern Western powers and therefore immune from the indignity of being on the wrong side of the glass in the broader world's *Jinrui Kan*.

An International Exposition at the Margin

The Osaka Exposition was a brilliant success for Japan, boasting of 4,350,693 attendees including 14,443 'Europeans and Americans' and 8,677 'Chinese and Koreans'.[32] In many ways, the event surpassed Japan's greatest ambitions for international interest, participation and attendance. A poignant example can be seen in the events of Osaka Day where over 100,000 visitors hoped to enter but could not be accommodated despite the construction of two additional ticketing booths. After one of the booths was overturned in the rush of would-be fairgoers, the Exposition decided to hold a raffle for the remaining tickets. Ultimately, the gates were closed to further entry at 2:20 p.m. with the tally of paid admissions at 93,464.[33] In just one day, the Osaka Exposition had served what amounted to about one fifth of the total attendance at Japan's First National Industrial Exposition held in 1877.

The Exposition can also be said to have successfully highlighted Japan's advancing industrial abilities which was no doubt a source of pride for the fair's Japanese visitors. The *Japan Weekly Chronicle* observed at the close of the fair that 'the various Japanese sections have revealed to foreigners many products in the way of machinery and manufactures that, but for the Exhibition, might have remained in comparative obscurity for many years to come'.[34] There was of course commercial potential in this, but the Exposition's inclusion of international participation also made a statement about Japan's emerging right to claim itself as a world power. In a way that was perhaps impossible except through military victories, international events such as these could provide tangible evidence of a nation's standing in the world.

The question remains, however, whether the Osaka Exposition can rightfully be considered an international exposition in the fullest sense. Though it clearly had not effectively been planned as one, the Osaka Exposition ultimately managed to adapt to this role. The guide to the Foreign Samples Building, which was published in both English and Japanese some months into the exposition, can serve as an example of this kind of effort, but its contents can also illuminate the perceived status of the event. Several foreign commissioners were invited to contribute introductions to the volume and their brief overtures can provide insight into the range of opinions on what the Osaka Exposition signified.

Henry E. Dosch, the commissioner representing the American State of Oregon and its exhibits in the Foreign Samples Building, only tentatively supported the notion that Japan's exposition was more than a marginal one. Though he praised the fair's planners and the value of the event itself, he depicted it as a mere stepping stone to some future grand-scale exposition. He observed:

> We venture the prediction, that the next Exhibition to be held in Japan, and which will be at no distant day, and to the furtherance of which this publication will be a great promoter, will be Inter-national in the fullest sense of the word, and owing to the progressiveness of the Japanese people, all the nations on earth will rival each other in their displays, and this Fifth National Industrial Exhibition of Osaka, will go down into history as having laid the foundation and made possible the greatest Exhibition ever held in the Far East.[35]

In contrast, the Canadian commissioner, who represented the only foreign country to erect its own national pavilion, proclaimed that the fair was definitively an 'International Exposition' due to its size and that many foreign nations, colonies and firms exhibited.[36]

Perhaps the best synthesis of these ideas was the view offered by M. Yikami, the Special Commissioner representing Korea (a country to be made officially a Japanese protectorate only two years later in 1905 and then annexed in 1910). After briefly outlining the history of Japan's first four National Industrial Exhibitions, he asserts that the incorporation of the Foreign Samples Building represented a revolutionary change for Japan's *hakurankai*:

> No observing mind will ever fail to see the significance of this. It means a modest but first step toward the International Exhibition. If anyone in the future should write a history of the Exhibitions in Japan he must look upon this as the First International Exhibition held within the boundaries of Japan. It also shows that the people of Japan are no longer satisfied with the position that their country has previously held among the nations of the world. Since her formal admittance into the family of nations, Japan did not let chance slip by to disclose the fact that she is not only theoretically but really a member of the International Family. She has shown it in war at the time of the Boxers' Uprising. Now she wants to show it in peace.[37]

From a historical vantage point, the Osaka Exposition was held just two years prior to Japan's 1905 victory over a European power in the Russo-Japanese War. The host of the event was on the verge of attaining a concrete measure of parity with the West so it is unsurprising Japan's exposition – like the nation – in 1903 should exist on the margin between domestic and international status.

Conclusion

The Fifth National Industrial Exhibition ultimately transformed into an international event and the hosting of international expositions materially signified a nation's capacity to belong to the community of modern world powers. Japan demonstrated the industrial attributes necessary for such status through the exhibition of its industrial products, but the privilege of holding an international exposition entailed the ability to construct official pavilions and amusements to showcase the nation's imperial successes and assert a distinction from peoples who could be deemed subordinate. Although Japan was obliged to respond to diplomatic pressure by modifying its presentation of foreign nations and subject colonial peoples in the *Jinrui Kan*, the Exposition was able to establish Japan's position of privilege quite clearly at home and to the participating international community. With these outcomes in mind, perhaps it is justified, as the Japanese commissioner for Korea M. Yikami suggested, understanding the event as Japan's First International Exposition and a marker that signified the end of the nation's marginal status.

Notes

1 Angus Lockyer, 'Japan at the Exhibition, 1867–1970' (PhD diss., Stanford University, 2000), 82. Lockyer notes that the term *hakurankai* was coined by Kurimoto Jôun (a Bakufu official) in 1865 who, upon hearing what an international exposition was, imagined it as something between the indigenous *yakuhinkai* (fairs of medical products) and the imported *hakubutsukan* (Western-style museum). However, Lockyer also points out that Fukuzawa Yukichi can be credited with popularizing the term.

2 For information on the native Japanese antecedents of expositions, see P. F. Kornicki, 'Public Display and Changing Values: Early Meiji Exhibitions and Their Precursors', *Monumenta Nipponica* 49, no. 2 (1994): 167–196.

3 Noriko Aso, 'New Illusions: The Emergence of a Discourse on Traditional Japanese Arts and Crafts, 1868–1945' (PhD diss., University of Chicago, 1997), 68.

4 Aso, 69.

5 Lockyer, 100.

6 Lockyer, 111.

7 Lockyer, 112–113.

8 Alice Y. Tseng, 'Styling Japan: The Case of Josiah Conder and the Museum at Ueno, Tokyo', *Journal of the Society of Architectural Historians* 63, no. 4 (2004): 472–497, at 491; *The Official Guide-Book to Kyoto and the Allied Prefectures: Prepared Specially for the Eleven-Hundredth Anniversary of the Founding of Kyoto and the Fourth National Industrial Exhibition by the City Council of Kyoto With Three Maps and Sixty-Nine Engravings. Meishinsha (1895).* JWSC, Henry Dosch Collection.

9 Aso, 67.

10 Undated/unsigned 1902 Foreign Ministry Memorandum, NAJ-AMFA, 3.15.1.10.

11 'An Exhibition Criticism', *Japan Weekly Chronicle*, 11 February 1903, 114.

12 'At the Osaka Exhibition', *Japan Weekly Chronicle*, 25 March 1903, 267–268.

13 'At the Osaka Exhibition', 267–268.

14 'The Osaka Exhibition', *Japan Weekly Chronicle*, 11 March 1903, 231.

15 Untitled note, *Japan Weekly Chronicle*, 1 April 1903, 298.

16 'Foreign Exhibits at the Osaka Exhibition', *Japan Weekly Chronicle*, 3 June 1903, 505.

17 'Foreign Exhibits at Osaka Exhibition: Observation of Netherlands-India Section Catalogue', *Japan Weekly Chronicle*, 2 May 1903, 447.

18 'Distribution of Awards at the Exhibition', *Japan Weekly Chronicle*, 8 July 1903, 37.

19 Other expositions mentioned in this volume with a Water Chutes concession are the Wolverhampton Arts and Industrial Exhibition (1902), the German Universal Exhibition, Ústí nad Labem, Bohemia (1903) and the National General Exhibition in Bucharest (1906).

20 World's fairs in the West had initially excluded the amusement sections that sprang up along their exteriors but exposition planners soon came to see the value of incorporating concessions into the exposition grounds where they could both control and profit from them. These 'Midways' became integral to the exposition experience. For the evolution of this incorporation of amusement zones, see Robert Rydell, *All the World's a Fair: Visions of Empire at American International Expositions* (Chicago, 1987), 236.

21 Kyoko Matsuda, *Teikoku no Shisen: Hakurankai to Ibunka Hyosho* (Tokyo, 2003), 123.

22 Steve Rabson, 'Memories of Okinawa: Life and Times in the Greater Osaka Diaspora', in *Islands of Discontent: Okinawan Responses to Japanese and American Power*, ed. Laura Hein and Mark Selden (New York, 2003), 113.

23 Ayako Hotta-Lister, *The Japan-British Exhibition of 1910: Gateway to the Island Empire of the East* (Richmond, UK, 1999), 142.

24 Matsuda, 121. She relates in her text that 'it is said' that the private owner of the concession was an American, but this important detail is unverified except through oral history.

25 Matsuda, 122. Though the *Japan Weekly Chronicle* reported that the exhibition of Chinese people at the *Jinrui Kan* was an 'unfounded' rumour (apparently being unaware of the pre-Exposition modifications), they did observe that the opium pipe and foot-binding photograph had been removed in 'Chinese as Exhibits at Osaka', *Japan Weekly Chronicle*, 25 March 1903, 277.

26 Matsuda, 123–124; Matsuda also notes that the Ainu were being compensated for their role as 'exhibits' with money to help in the building of a modern school on Hokkaido.

27 'At the Exhibition: The Formosa Exhibit', *Japan Weekly Chronicle*, 1 April 1903, 291.

28 'At the Exhibition', 291.

29 'At the Exhibition', 291.

30 Alexis Dudden, *Japan's Colonization of Korea: Discourse and Power* (Honolulu, 2005), 122.

31 There are a number of studies that specifically address the role of Western expositions and museums in ethnographic displays. A few to note are Raymond Corbey, 'Ethnographic Showcases, 1870–1930', *Cultural Anthropology* 8 (1993): 338–369; Zeynep Çelik, *Displaying the Orient: Architecture of Islam at Nineteenth-Century World's Fairs* (Berkeley, CA, 1992); Ivan Karp and Steven D. Levine, eds, *Exhibiting Cultures: The Poetics and Politics of Museum Display* (Washington, DC, 1990).

32 'The Closing of the Exhibition', *Japan Weekly Chronicle*, 5 August 1903, 132.

33 '"Osaka Day" at the Exposition', *Japan Weekly Chronicle*, 17 June 1903, 556.

34 'The Exhibition and its Results', *Japan Weekly Chronicle*, 5 August 1903, 126.

35 Henry Dosch, Introduction to *A Guide to the Foreign Samples Building and Its Annexes of the Fifth National Industrial Exhibition at Osaka*, JWSC, Henry E. Dosch Collection.

36 W. A. Burns, Introduction to *A Guide to the Foreign Samples Building and Its Annexes of the Fifth National Industrial Exhibition at Osaka*, JWSC, Henry E. Dosch Collection.

37 M. Yikami, Introduction to *A Guide to the Foreign Samples Building and Its Annexes of the Fifth National Industrial Exhibition at Osaka*, JWSC, Henry E. Dosch Collection.

Bibliography

Manuscript Sources

John Wilson Special Collections (JWSC), Multnomah County Public Library, Henry E. Dosch Collection.

National Archives of Japan, Diplomatic Archives of the Ministry of Foreign Affairs, Records of the Foreign Ministry of Affairs (NAJ-AMFA), Series 3: Commercial Relations, Category 15: Exhibitions, Section1: Japan, Subseries: B-3-15-1-1, B-3-15-1-2, B-3-15-1-4.

National Diet Library Website, Japan, Expositions, Part 3: Japanese Industries in the Meiji Period – Seen from Exhibits, http://www.ndl.go.jp/exposition/e/index.html.

Oregon Historical Society Research Library, Lewis and Clark Centennial Exposition Manuscript Records, MSS 1609.

United States National Archives Microfilm Publications, Dispatches from US Consuls in Osaka and Hioho (Kobe), Japan, 1868–1906, M460. Roll number 6.

Printed Primary Sources

A Guide to the Foreign Samples Building and Its Annexe of the Fifth National Industrial Exhibition at Osaka Compiled by Yokichi Kobayash. Osaka, 1903.

The Japan Weekly Chronicle. Kobe, Japan, 1902–05.

The Official Guide-Book to Kyoto and the Allied Prefectures: Prepared Specially for the Eleven-Hundredth Anniversary of the Founding of Kyoto and the Fourth National Industrial Exhibition by the City Council of Kyoto With Three Maps and Sixty-Nine Engravings. Meishinsha (1895). Kyoto, 1895.

The Photographic Reproduction of the Fifth National Industrial Exhibition. Osaka, 1903.

Printed Secondary Sources

Aso, Noriko. 'New Illusions: The Emergence of a Discourse on Traditional Japanese Arts and Crafts, 1868–1945'. PhD diss., University of Chicago, 1997.

Çelik, Zeynep. *Displaying the Orient: Architecture of Islam at Nineteenth-Century World's Fairs*. Berkeley: University of California Press, 1992.

Corby, Raymond. 'Ethnographic Showcases, 1870–1930'. *Cultural Anthropology* 8, no. 3 (1993): 338–369.

Dudden, Alexis. *Japan's Colonization of Korea: Discourse and Power*. Honolulu: University of Hawaii Press, 2005.

Gilbert, James. 'World's Fairs as Historical Events'. In *Fair Representations: World's Fairs and the Modern World*, edited by Robert Rydell and Nancy E. Gwinn, 13–27. Amsterdam: VU University Press, 1994.

Hosley, William. 'Selling Japan in the West: The Centennial and Beyond'. In *The Japan Idea: Art and Life in Victorian America*, edited by William Hosley, 29–46. Hartford, CT: Wadsworth Atheneum, 1990.

Hotta-Lister, Ayako. *The Japan-British Exhibition of 1910: Gateway to the Island Empire of the East*. Richmond, UK: Japan Library, 1999.

Karp, Ivan, and Steven D. Lavine, eds. *Exhibiting Cultures: The Poetics and Politics of Museum Display*. Washington, DC: Smithsonian Institution Press, 1990.

Kornicki, P. F. 'Public Display and Changing Values: Early Meiji Exhibitions and Their Precursors'. *Monumenta Nipponica* 49, no. 2 (1994): 167–196.

Lockyer, Angus. 'Japan at the Exhibition, 1867–1970'. PhD diss., Stanford University, 2000.

Matsuda, Kyoko. *Teikoku no Shisen: Hakurankai to Ibunka Hyosho*. Tokyo: Yoshikawa Kobunkan, 2003.

Rabson, Steve. 'Memories of Okinawa: Life and Times in the Greater Osaka Diaspora'. In *Islands of Discontent: Okinawan Responses to Japanese and American Power*, edited by Laura Hein and Mark Selden, 99–134. New York: Rowman and Littlefield, 2003.

Rydell, Robert. *All the World's a Fair: Visions of Empire at American International Expositions*. Chicago: University of Chicago Press, 1987.

Tseng, Alice Y. 'Styling Japan: The Case of Josiah Conder and the Museum at Ueno, Tokyo'. *Journal of the Society of Architectural Historians* 63, no. 4 (2004): 472–497.

Index